Pathwise Estimation and Inference for Diffusion Market Models

Pathwise Estimation and Inference for Diffusion Market Models

Nikolai Dokuchaev
Lin Yee Hin

CRC Press
Taylor & Francis Group
Boca Raton London New York

CRC Press is an imprint of the
Taylor & Francis Group, an **informa** business

A CHAPMAN & HALL BOOK

CRC Press
Taylor & Francis Group
6000 Broken Sound Parkway NW, Suite 300
Boca Raton, FL 33487-2742

First issued in paperback 2020

© 2019 by Taylor & Francis Group, LLC
CRC Press is an imprint of Taylor & Francis Group, an Informa business

No claim to original U.S. Government works

ISBN-13: 978-1-138-59164-6 (hbk)
ISBN-13: 978-0-367-73121-2 (pbk)

Visit the Taylor & Francis Web site at
http://www.taylorandfrancis.com

and the CRC Press Web site at
http://www.crcpress.com

Contents

Legend of Notations and Abbreviations   **ix**

Preface   **xi**

1 Some background on stochastic analysis   **1**
 1.1 Basics of probability theory 1
  1.1.1 Probability space 1
  1.1.2 Random variables 2
  1.1.3 Expectations 3
  1.1.4 Conditional probability and expectation 5
  1.1.5 The σ-algebra generated by a random vector 6
 1.2 Basics of stochastic processes 7
  1.2.1 Special classes of processes 8
  1.2.2 Wiener process (Brownian motion) 9
 1.3 Basics of the stochastic calculus (Ito calculus) 10
  1.3.1 Ito formula 13
  1.3.2 Stochastic differential equations (Ito equations) 15
  1.3.3 Some explicit solutions for Ito equations 16
  1.3.4 Diffusion Markov processes and related parabolic equations 17
  1.3.5 Martingale representation theorem 19
  1.3.6 Change of measure and Girsanov theorem 20

2 Some background on diffusion market models   **23**
 2.1 Continuous time model for stock price 23
 2.2 Continuous time bond-stock market model 24
 2.3 Discounted wealth and stock prices 25
 2.4 Risk-neutral measure 27
 2.5 Replicating strategies 29
 2.6 Arbitrage possibilities and the arbitrage-free market 30
 2.7 The case of a complete market 31
 2.8 Completeness of the Black–Scholes model 31
 2.9 Option pricing 33
  2.9.1 Options and their prices 33
  2.9.2 Option pricing for a complete market 35
  2.9.3 Black–Scholes formula 36
 2.10 Pricing for an incomplete market 37

2.11 A multi-stock market model   38

3 Some special market models   **41**
3.1   Mean-reverting market model   41
      3.1.1   Basic properties of a mean-reverting model   41
      3.1.2   Absence of arbitrage and the Novikov condition   42
      3.1.3   Proofs   43
3.2   A market model with delay in coefficients   48
      3.2.1   Existence, regularity, and non-arbitrage properties . . .   48
      3.2.2   Time discretization and restrictions on growth   50
3.3   A market model with stochastic numéraire   51
      3.3.1   Model setting   51
      3.3.2   Replication of claims: Strategies and hedging errors . . . 55
      3.3.3   On selection of θ and the equivalent martingale measure . .   58
      3.3.4   Markov case   61
      3.3.5   Proofs   63
3.4   Bibliographic notes and literature review   70

4 Pathwise inference for the parameters of market models   **73**
4.1   Estimation of volatility   73
      4.1.1   Representation theorems for the volatility   73
      4.1.2   Estimation of discrete time samples   75
      4.1.3   Reducing the impact of the appreciation rate   78
      4.1.4   The algorithm   80
      4.1.5   Some experiments   81
4.2   Modeling the impact of the sampling frequency   84
      4.2.1   Analysis of the model's parameters   85
      4.2.2   Monte Carlo simulation of the process with delay   86
      4.2.3   Examples for dependence of volatility on sampling fre-
              quency for historical data   89
      4.2.4   Matching delay parameters for historical data   93
4.3   Inference for diffusion parameters for CIR-type models   96
      4.3.1   The underlying continuous time model   96
      4.3.2   A representation theorem for the diffusion coefficient . . . 97
      4.3.3   Estimation based on the representation theorem   98
      4.3.4   Numerical experiments   102
      4.3.5   On the consistency of the method   106
      4.3.6   Some properties of the estimates 107
4.4   Estimation of the appreciation rates 108
4.5   Bibliographic notes and literature review   111

5 Some background on bond pricing   **113**
5.1   Zero-coupon bonds 113
5.2   One-factor model   115
      5.2.1   Dynamics of discounted bond prices 116

	5.2.2	Dynamics of the bond prices under the original measure	117
	5.2.3	An example: The Cox–Ross–Ingresoll model	119
5.3	Vasicek model		119
5.4	An example of a multi-bond market model		122

6 Implied volatility and other implied market parameters **125**
6.1	Risk-neutral pricing in a Black–Scoles setting		125
6.2	Implied volatility: The case of constant r		129
6.3	Correction of the volatility smile for constant r		130
	6.3.1	Imperfection of the volatility smile for constant r	130
	6.3.2	A pricing rule correcting the volatility smile	131
	6.3.3	A class of volatilities in a Markovian setting	132
6.4	Unconditionally implied volatility and risk-free rate		140
	6.4.1	Two calls with different strike prices	141
6.5	Bond price inferred from option prices		141
	6.5.1	Definitions	142
	6.5.2	Inferred ρ from put and call prices	142
	6.5.3	Application to a special model	143
6.6	A dynamically purified option price process		144
6.7	The implied market price of risk with random numéraire		146
	6.7.1	The risk-free bonds for the market with random numéraire	146
	6.7.2	The case of a complete market	147
	6.7.3	The case of an incomplete market	149
6.8	Bibliographic notes		151

7 Inference of implied parameters from option prices **153**
7.1	Sensitivity analysis of implied volatility estimation		154
	7.1.1	An under-defined system of nonlinear equations	154
	7.1.2	Numerical analysis using cross-sectional S&P 500 call options data	156
	7.1.3	Numerical analysis using longitudinal S&P500 call options data	160
7.2	A brief review of evolutionary optimization		169
	7.2.1	The original differential evolution algorithm	171
	7.2.2	The Zhang–Sanderson adaptive differential evolution algorithms	171
7.3	Inference of implied parameters from over-defined systems		175
	7.3.1	An over-defined system of nonlinear equations	175
	7.3.2	Computational implementation	176
	7.3.3	Construction of the estimation uncertainty bounds for the estimated implied discount rates and implied volatilities	177
	7.3.4	Numerical experiment with synthetic test data	178
	7.3.5	Numerical analysis using historical S&P500 call options data	181
7.4	Bibliographic notes and literature review		184

8   Forecast of short rate based on the CIR model   **191**

   8.1   The model framework 192

        8.1.1   General setting 192

        8.1.2   The CIR model 194

   8.2   Inference of the implied CIR model parameters based on cross-
sectional zero coupon bond prices 196

   8.3   Numerical framework for the inference 196

   8.4   Computational implementation 198

   8.5   Forecast of short rate using the implied CIR model parameters . . . 198

        8.5.1   Forecast within the multi-curve framework 198

        8.5.2   Forecast within the single-curve framework 199

   8.6   Numerical analysis using historical data 200

        8.6.1   Short rate prediction in the multi-curve framework 201

        8.6.2   Short rate prediction in the single-curve framework 205

   8.7   Bibliographic notes and literature review 208

Bibliography   **213**

Index   **223**

Legend of Notations and Abbreviations

- a.e. - almost everywhere, or for almost every

- a.s. - almost surely

- $\forall$ - for all

- $\mathbf{C}$ - the set of all complex numbers

- $\mathbf{E}X$ - expectation of X

- $\mathbf{E}X^2 = \mathbf{E}(X^2), \mathbf{E}XY = \mathbf{E}(XY)$

- iff - if and only if

- $\mathbb{I}_A$ is the indicator function of an event A, $\mathbb{I}_D(x)$ is the indicator function of a set D

- $L_p(\Omega, \mathcal{F}, \mathbf{P})$, $p \in [1, +\infty)$ - the set of classes of $\mathbf{P}$-equivalent random variables on a probability space $(\Omega, \mathcal{F}, \mathbf{P})$ such that $\mathbf{E}|\xi|^p < +\infty$ for $p \in [1, +\infty)$, or that there exists a (non-random) constant $c = c(\xi) > 0$ such that $|\xi| \leq c$ a.s.. for $p = +\infty$

- $N(a, \sigma^2)$ - the normal distribution with the expectation a and the variance σ^2

- $\emptyset$ - empty set

- $\mathbf{P}(A)$ - probability of an event A

- $\mathbf{R}$ - the set of all real numbers

- $\mathbf{R}^n$ - the set of all real vectors (vector columns) with n components

- $\mathrm{Var}\,X$ - variance of X

- $x \stackrel{\triangle}{=} X$ - means that x is defined as X

- $x^+ \stackrel{\triangle}{=} \max(x, 0), x^- \stackrel{\triangle}{=} \max(-x, 0).$

- $|x|$ - the Euclidean norm $\left(\sum_{i=1}^m x_i^2\right)^{1/2}$ for $x \in \mathbf{R}^m$ or the Frobenius matrix norm $\left(\sum_{i,j=1}^m x_{ij}^2\right)^{1/2}$ for matrices $x \in \mathbf{R}^{m \times m}$.

Preface

This monograph discusses contemporary techniques for inferring, from options and bond prices, the market participants' aggregate view on important financial parameters such as implied volatility, discount rate, future interest rate, and the uncertainty thereof. The inference is considered for the parameters of the so-called diffusion market models, where the prices of the underlying risky assets are represented by pathwise continuous stochastic processes described by stochastic differential equations. The focus is on the pathwise inference methods that are applicable to a sole path of the observed prices and do not require the observation of an ensemble of such paths. In particular, one of these methods explores the fact that the volatility process for the price is uniquely defined by any path of the price process in the diffusion model. Another method requires a solution to the problems that are the inverse of the pricing problems; the solution allows the recovery of unknown parameters of the pricing model using observation of the market prices of the underlying derivatives.

The book provides a self-contained introduction to stochastic processes relevant to the study of classic continuous time market model that represent modifications of models with log-normal price processes, including mean-reverting processes. In addition, this book offers a self-contained introduction to pricing theory for options and bonds for the diffusion models, and a review of important special market models. Furthermore, this book presents a collection of methods for pathwise inference of the model parameters from observed historical prices.

Putting into perspective the industrial practice of choosing a beskope discount rate to reflect the funding cost in pricing contingent claims, this book considers numerical techniques for sensitivity analysis of implied volatility estimation in the presence of discount rate uncertainty. This book describes implementation of a strategy to infer the implied discount rate and the implied volatility simultaneously using the Black–Scholes option pricing formula as a mapping tool. This can be achieved via solution to a set of over-defined nonlinear equations, the approximate solution of which is sought using differential evolution, a computationally intensive stochastic-based multi-point direct-search optimization procedure. A separate chapter demonstrates how to apply this strategy to map the market participants' aggregate view of future short rate at different horizons to sets of model parameters for the Cox–Ingersoll–Ross process in a non-classical multiple yield curve framework.

The suggested methods are illustrated by description of numerical experiments for historical financial market data.

From a pedagogical perspective, this book is pitched at the level of senior undergraduate students pursuing research at honors year, and postgraduate candidates pursuing Masters or PhD degree by research. From a research perspective, this book reaches out to academic researchers from backgrounds as diverse as mathematics and probability, econometrics and statistics, and computational mathematics and optimization whose interests lie in analysis and modelling of financial market data from a multi-disciplinary approach. Additionally, this book is also aimed at financial market practitioners participating in capital market facing businesses who seek to keep abreast with and draw inspiration from novel approaches in market data analysis.

The book is organized as follows. Chapter 1 provides a background in stochastic analysis. Chapter 2 provides a background in mathematical finance and presents classical diffusion market models. Chapter 3 introduces several unconventional diffusion market models. Chapter 4 describes some pathwise inference methods for market parameter. Chapter 5 provides a background on pricing methods for zero-coupon bonds. Chapter 6 describes some unconventional types of implied market parameters. Chapter 7 describes numerical methods for inferring market parameters from large sets of market stock prices. Chapter 8 describes numerical methods for inferring market parameters from large sets of market bond prices and applications for forecasting short rates.

1

Some background on stochastic analysis

In probability theory based on Kolmogorov's probability axioms, the model of randomness is the following. It assumes that there exists a set Ω, and it is assumed that subsets $A \subseteq \Omega$ are random events. Some value $\mathbf{P}(A) \in [0, 1]$ is attached to any event as the probability of an event, and $\mathbf{P}(\Omega) = 1$. To make this model valid, some axioms about possible classes of events are accepted such that the expectation can be interpreted as an integral.

1.1   Basics of probability theory

1.1.1   Probability space

We denote by $|x|$ the Euclidean norm of a vector $x \in \mathbf{R}^k$.

σ-algebra of events

Let Ω be a non-empty set. We denote by 2^Ω the set of all subsets of Ω.

Example 1.1 *Let $\Omega = \{a, b\}$, then $2^\Omega = \{\emptyset, \{a\}, \{b\}, \Omega\}$.*

Definition 1.2 *A system of sub-sets $\mathcal{F} \subset 2^\Omega$ is called an algebra of subsets of Ω if*

(i) $\Omega \in \mathcal{F}$;

(ii) If $A \in \mathcal{F}$ then $\Omega \backslash A \in \mathcal{F}$;

(iii) If $A_1, A_2, \ldots, A_n \in \mathcal{F}$, then $\cup_{i=1}^{n} A_i \in \mathcal{F}$.

Note that (i) and (ii) imply that the empty set $\emptyset$ always belongs to an algebra.

Definition 1.3 *A system of subsets $\mathcal{F} \subset 2^\Omega$ is called a σ-algebra of subsets of Ω if*

(i) It is an algebra of subsets;

(ii) If $A_1, A_2, \ldots \in \mathcal{F}$ (i.e., $\{A_i\}_{i=1}^{+\infty} \subset \mathcal{F}$), then $\cup_{i=1}^{+\infty} A_i \in \mathcal{F}$.

Definition 1.4 *Let Ω be a set, let $\mathcal{F}$ be a σ-algebra of subsets, and let $\mu : \mathcal{F} \rightarrow [0, +\infty]$ be a mapping.*

(i) We said that μ is a σ-additive measure if $\mu(\cup_{i=1}^{+\infty} A_i) = \sum_{i=1}^{+\infty} \mu(A_i)$ for any $A_1, A_2, \ldots \in \mathcal{F}$ such that $A_i \cap A_j = \emptyset$ if $i \neq j$. In that case, the triplet $(\Omega, \mathcal{F}, \mu)$ is said to be a measure space.

(ii) If $\mu(\Omega) < +\infty$, then the measure μ is said to be finite.

(iii) If $\mu(\Omega) = 1$, then the measure μ is said to be a probability measure.

To make notations more visible, we use the symbol $\mathbf{P}$ for the probability measures.

Definition 1.5 *Consider a measure space* $(\Omega, \mathcal{F}, \mu)$. *Assume that some property holds for all* $\omega \in \Omega_1$, *where* $\Omega_1 \in \mathcal{F}$ *is such that* $\mu(\Omega \backslash \Omega_1) = 0$. *We say that this property holds a.e. (almost everywhere). In the case of a probability measure, we say that this property holds with probability 1, or a.s. (almost surely).*

In probability theory based on Kolmogorov's probability axioms, the following definition is accepted.

Definition 1.6 *A measure space* $(\Omega, \mathcal{F}, \mathbf{P})$ *is said to be a probability space if* $\mathbf{P}$ *is a probability measure, i.e.,* $\mathbf{P}(\Omega) = 1$. *Elements* $\omega \in \Omega$ *are said to be elementary events, and sets* $A \in \mathcal{F}$ *are said to be events (or random events). Correspondingly,* $\mathcal{F}$ *is the* σ-*algebra of events.*

Under these axioms, $A \cap B$ means the event "A and B" (or $A \cdot B$), and $A \cup B$ means the event "A or B" (or $A + B$), where A and B are events.

A random event $A = \{\omega\}$ is a set of elementary events.

Completeness

Definition 1.7 *A* σ-*algebra* $\mathcal{F}$ *is said to be complete (with respect to a measure* μ : $\mathcal{F} \to \mathbf{R}$) *if the following is satisfied: if* $A \in \mathcal{F}$, $B \subset A$, *and* $\mu(A) = 0$, *then* $B \in \mathcal{F}$.

Definition 1.8 *Let* $(\Omega, \mathcal{F}, \mu)$ *be a measure space. Let* $\overline{\mathcal{F}}$ *be a minimal* σ-*algebra such that* $\mathcal{F} \subseteq \overline{\mathcal{F}}$ *and* $\overline{\mathcal{F}}$ *is complete with respect to* μ. *Then* $\overline{\mathcal{F}}$ *is called the completion with respect to* μ *(in the literature, sometimes, it is called sometimes the* μ-*augmentation of* $\mathcal{F}$).

1.1.2   Random variables

Let X and Y be two sets, let $f : X \to Y$ be a mapping, and let $B \subset Y$. We denote $f^{-1}(B) \triangleq \{x \in X : f(x) \in B\}$. *(Note that we do not exclude the case when the inverse function* $f^{-1} : Y \to X$ *does not exist.)*

Definition 1.9 *Let* $(\Omega, \mathcal{F}, \mathbf{P})$ *be a measure space. A mapping* $\xi : \Omega \to \mathbf{R}$ *is said to be measurable if* $\xi^{-1}(D) \in \mathcal{F}$ *for any open set* D. *If* $(\Omega, \mathcal{F}, \mathbf{P})$ *is a probability space, then a measurable mapping* $\xi : \Omega \to \mathbf{R}$ *is said to be a random variable (on this probability space).*

As can be seen from the definitions, a mapping may be a random variable for some $\mathcal{F}$ and be not a random variable for some different $\mathcal{F}$.

Below, *iff* means if and only if.

Definition 1.10 *(i) Let $(\Omega, \mathcal{F}, \mu)$ be a complete measure space. A mapping $\xi : \Omega \to \mathbf{R}^n$ is said to be measurable (with respect to $\mathcal{F}$), if $\xi^{-1}(D) \subset \mathcal{F}$ for any open set $D \subseteq \mathbf{R}^n$. (ii) Let $(\Omega, \mathcal{F}, \mathbf{P})$ be a probability space. A measurable mapping $\xi : \Omega \to \mathbf{R}^n$ is said to be a random vector.*

Note that $\xi = (\xi_1, \ldots, \xi_n)$ is a random vector iff all components ξ_i are random variables.

1.1.3 Expectations

Let $(\Omega, \mathcal{F}, \mathbf{P})$ be a probability space.

Definition 1.11 *(i) A random variable $\xi : \Omega \to \mathbf{R}$ is said to be finitely valued if there exist an integer $n > 0$, $c_1, \ldots, c_n \in \mathbf{R}$, and $A_1, A_2, \ldots, A_n \in \mathcal{F}$ such that*

$$\xi(\omega)|_{\omega \in A_i} \equiv c_i \qquad \xi(\omega)|_{\Omega \setminus (\cup_i A_i)} = 0.$$

(ii) The value

$$\sum_{i=n} c_i \mathbf{P}(A_i) \, ,$$

is said to be the integral $\int_\Omega \xi(\omega) \mathbf{P}(d\omega)$ (i.e., the integral of ξ over Ω with respect to the measure $\mathbf{P}$). It is also said to be the expectation $\mathbf{E}\xi$ of ξ (or the expected value, or the mathematical expectation, or the mean).

Let $\mathbb{I}_A(\omega)$ denotes the indicator function of a set A: $\mathbb{I}_A(\omega) = 1$ if $\omega \in A$ and $\mathbb{I}_A(\omega) = 1$ if $\omega \notin A$.

Definition 1.12 *A non-negative random variable $\xi : \Omega \to \mathbf{R}$ is said to be integrable if there exists a non-decreasing sequence of non-negative finitely valued random variables $\xi_i(\cdot)$ such that $\xi_i(\omega) \to \xi(\omega)$ as $i \to +\infty$ a.s. (almost surely) (i.e., with probability 1), and such that*

$$\sup_n \int_\Omega \xi_n(\omega) \mathbf{P}(d\omega)$$

is finite.

Lemma 1.13 *Under the assumptions of the previous definition, there exists the limit of $\mathbf{E}\xi_n = \int_\Omega \xi_n(\omega) \mathbf{P}(d\omega)$ as $n \to +\infty$. This limit is uniquely defined (i.e., it does not depend on the choice of $\{\xi_k\}$), and this limit is said to be the integral $\int_\Omega \xi(\omega) \mathbf{P}(d\omega)$, or the expectation $\mathbf{E}\xi$.*

We denote $x^+ \overset{\Delta}{=} \max(x, 0)$ and $x^- = \max(-x, 0)$.

Definition 1.14 *A random variable* $\xi : \Omega \to \mathbf{R}$ *is said to be integrable if* ξ^+ *and* ξ^- *are integrable (note that* $\xi = \xi^+ - \xi^-$ *). In that case, the value*

$$\mathbf{E}\xi^+ - \mathbf{E}\xi^- = \int_\Omega \xi^+(\omega)\mathbf{P}(d\omega) - \int_\Omega \xi^-(\omega)\mathbf{P}(d\omega)$$

is said to be $\mathbf{E}\xi = \int_\Omega \xi(\omega)\mathbf{P}(d\omega)$.

Resuming, we may say that, for a probability space $(\Omega, \mathcal{F}, \mathbf{P})$, a measurable (with respect to $\mathcal{F}$) function $\xi : \Omega \to \mathbf{R}$ is a random variable, and the integral $\mathbf{E}\xi = \int_\Omega \xi(\omega)\mathbf{P}(d\omega)$ is the expectation (or mathematical expectation, or *mean*).

Let $p \in [1, +\infty)$. We denote by $\mathcal{L}_p(\Omega, \mathcal{F}, \mathbf{P})$ the set of all random variables ξ on a probability space $(\Omega, \mathcal{F}, \mathbf{P})$ such that $\mathbf{E}|\xi|^p < +\infty$. In addition, we denote by $\mathcal{L}_\infty(\Omega, \mathcal{F}, \mathbf{P})$ the set of all random variables ξ on a probability space $(\Omega, \mathcal{F}, \mathbf{P})$ such that there exists a (non-random) constant $c = c(\xi) > 0$ such that $|\xi| \leq c$ a.s.

With these notations, the set of all integrable random variables is $\mathcal{L}_1(\Omega, \mathcal{F}, \mathbf{P})$.

If $\xi \in \mathcal{L}_2(\Omega, \mathcal{F}, \mathbf{P})$, then $\mathbf{E}\xi^2 < +\infty$. In that case, the variance of ξ is defined: $\mathrm{Var}\, \xi \stackrel{\Delta}{=} \mathbf{E}\xi^2 - (\mathbf{E}\xi)^2 = \mathbf{E}(\xi - \mathbf{E}\xi)^2$.

Note that it can happen that $\xi, \eta \in \mathcal{L}_p(\Omega, \mathcal{F}, \mathbf{P})$, $\xi \neq \eta$, and $\mathbf{P}(\xi \neq \eta) = 0$ (in other words, they are $\mathbf{P}$-indistinguishable). Formally, ξ and η are different elements of $\mathcal{L}_p(\Omega, \mathcal{F}, \mathbf{P})$. This can be inconvenient, so we introduce the following notation.

For $p \in [1, +\infty]$, we denote by $L_p(\Omega, \mathcal{F}, \mathbf{P})$ the set of classes of random variables from $\mathcal{L}_p(\Omega, \mathcal{F}, \mathbf{P})$ that are $\mathbf{P}$-equivalent. In other words, if $\mathbf{P}(\xi \neq \eta) = 0$, then $\xi = \eta$, meaning that they represents the same element of $L_p(\Omega, \mathcal{F}, \mathbf{P})$, i.e., they are in the same class of equivalency.

It can be noted that $L_2(\Omega, \mathcal{F}, \mathbf{P})$ is a Hilbert space with natural liner operations and with the inner product (scalar product) $\langle \xi, \eta \rangle \stackrel{\Delta}{=} \mathbf{E}\xi\eta$.

We say that a random variable ξ has the probability distribution $N(a, \sigma^2)$ and write $\xi \sim N(a, \sigma^2)$ if ξ is a Gaussian random variable such that $\mathbf{E}\xi = a$ and $\mathrm{Var}\, \xi = \sigma^2$ (see Section 1.1.5 below).

In fact, the definitions above give a brief description of the measure theory and integration theory that cover the theory of Lebesgue's integral.

Equivalent probability measures

Definition 1.15 *Let* $(\Omega, \mathcal{F}, \mathbf{P}_i)$ *be two probability spaces with the same* $(\Omega, \mathcal{F})$ *and with different* $\mathbf{P}_i$, $i = 1, 2$. *The measures* $\mathbf{P}_i$ *are said to be equivalent if they have the same sets of zero sets, i.e.,*

$$\mathbf{P}_1(A) = 0 \Leftrightarrow \mathbf{P}_2(A) = 0, \quad A \in \mathcal{F}.$$

Theorem 1.16 *(Radon–Nikodim Theorem). The measures* $\mathbf{P}_1$ *and* $\mathbf{P}_2$ *are equivalent iff there exist* $Z \in \mathcal{L}_1(\Omega, \mathcal{F}, \mathbf{P}_2)$ *such that* $Z(\omega) > 0$ *a.s., and*

$$\mathbf{P}_1(A) = \int_A Z(\omega)\mathbf{P}_2(d\omega) \quad \forall A \in \mathcal{F}.$$

We say that Z is the Radon–Nikodim derivative and denote it as $Z = \frac{d\mathbf{P}_1}{d\mathbf{P}_2}$.

In that case, $\mathbf{E}_1\xi = \mathbf{E}_2 Z\xi$ for any $\mathbf{P}_1$-integrable random variable ξ, where $\mathbf{E}_i$ is the expectation under the measure $\mathbf{P}_i$.

1.1.4 Conditional probability and expectation

Definition 1.17 *Let* A, B *be random events. The conditional probability* $\mathbf{P}(A \mid B)$ *is defined as* $\mathbf{P}(A \mid B) \triangleq \mathbf{P}(A \cdot B)/\mathbf{P}(B)$. *(In fact, it is the probability of* A *under the condition that the event* B *occurs.)*

Definition 1.18 *Let* $\xi \in \mathcal{L}_2(\Omega, \mathcal{F}, \mathbf{P})$. *Let* $\mathcal{G}$ *be a* σ-*algebra such that* $\mathcal{G} \subseteq \mathcal{F}$. *A random variable* $\mathbf{E}\{\xi \mid \mathcal{G}\}$ *from* $\mathcal{L}_2(\Omega, \mathcal{G}, \mathbf{P})$ *such that*

$$\mathbf{E}|\xi - \mathbf{E}\{\xi \mid \mathcal{G}\}|^2 \leq \mathbf{E}|\xi - \eta|^2 \quad \forall \eta \in \mathcal{L}_2(\Omega, \mathcal{G}, \mathbf{P})$$

is called the conditional expectation.

(Note that $(\Omega, \mathcal{G}, \mathbf{P})$ is also a probability space.)

In addition to the formal definition, it may be useful to keep in mind the following intuitive description: the conditional expectation $\mathbf{E}\{\xi \mid \mathcal{G}\}$ is the expectation of a random variable ξ in an imaginary universe, where an observer knows about all events from $\mathcal{G}$, if they occur or not.

Theorem 1.19 *Let a* σ-*algebra* $\mathcal{G}$ *and a random vector* ξ *be given. Then:*

    (i) The conditional expectation $\mathbf{E}\{\xi \mid \mathcal{G}\}$ *is uniquely defined (up to* $\mathbf{P}$-*equivalency; i.e., all versions of* $\mathbf{E}\{\xi \mid \mathcal{G}\}$ *are* $\mathbf{P}$-*indistinguishable);*

    (ii) $\mathbf{E}(\xi - \mathbf{E}\{\xi \mid \mathcal{G}\})\eta = 0$ *for all* $\eta \in \mathcal{L}_2(\Omega, \mathcal{G}, \mathbf{P})$;

    (iii) $\mathbf{E}\xi = \mathbf{E}\,\mathbf{E}\{\xi \mid \mathcal{G}\}$;

    (iv) Let $\mathcal{G}_0$ *be a* σ-*algebra such that* $\mathcal{G}_0 \subseteq \mathcal{G}$. *Then* $\mathbf{E}\{\xi \mid \mathcal{G}_0\} = \mathbf{E}\{\mathbf{E}\{\xi \mid \mathcal{G}\}|\mathcal{G}_0\}$.

By this theorem, the conditional expectation may be interpreted as a projection. The space $L_2(\Omega, \mathcal{G}, \mathbf{P})$ is a linear subspace of $L_2(\Omega, \mathcal{F}, \mathbf{P})$, and $\mathbf{E}\{\xi \mid \mathcal{G}\}$ is the projection of ξ on this subspace. As usual, statement (ii) of Theorem 1.19 above means that $\xi - \mathbf{E}\{\xi \mid \mathcal{G}\} \perp \eta$ for all $\eta \in L_2(\Omega, \mathcal{G}, \mathbf{P})$; here $\perp$ means the orthogonality, i.e., $\xi \perp \eta$ means that $\langle \xi, \eta \rangle \triangleq \mathbf{E}\xi\eta = 0$.

Example 1.20 *If* $\mathcal{G} = \{\emptyset, \Omega\}$ *(the trivial* σ-*algebra), then* $\mathbf{E}\{\xi \mid \mathcal{G}\} = \mathbf{E}\xi$.

Definition 1.21 *The conditional probability measure* $\mathbf{P}\{\cdot \mid \mathcal{G}\} : \mathcal{F} \to [0, 1]$ *is defined as*

$$\mathbf{P}(A \mid \mathcal{G}) \triangleq \mathbf{E}\{\mathbb{I}_A \mid \mathcal{G}\}.$$

It can be shown that it is a probability measure, and that $(\Omega, \mathcal{F}, \mathbf{P}(\cdot \mid \mathcal{G}))$ is a probability space.

In fact, $(\Omega, \mathcal{F}, \mathbf{P}(\cdot \mid \mathcal{G}))$ is a probability space, where the probabilities are calculated by an observer who knows if any event from $\mathcal{G}$ occurs or not. We illustrate this statement below via σ-algebras generated by a random vector.

1.1.5   The σ-algebra generated by a random vector

Definition 1.22 *Borel σ-algebra of subsets in $\mathbf{R}^n$ is the minimal σ-algebra that contains all sets $\{x \in \mathbf{R}^n : a_i < x_i \leq b_i\}$. (In fact, it is also the minimal σ-algebra that contains all open sets.) A function $f : \mathbf{R}^n \to \mathbf{R}$ is said to be measurable (or Borel measurable) if it is measurable with respect to this σ-algebra.*

Definition 1.23 *Let $\xi : \Omega \to \mathbf{R}^n$ be a random vector (i.e., all its components are random variables). Let $\mathcal{F}_\xi$ be the minimal σ-algebra such that $\mathcal{F}_\xi$ includes all events $\{\xi \in B\}$ for all Borel sets B in $\mathbf{R}^n$ (or for all open sets). We say that $\mathcal{F}_\xi$ is the σ-algebra generated by ξ.*

We denote by $\sigma(\xi)$ the σ-algebra generated by ξ; (ii) $\overline{\sigma}(\xi)$ denotes the completion of the σ-algebra generated by ξ. (Sometime we use other notations, for instance, $\mathcal{F}_\xi$.)

Definition 1.24 *Let ξ and η be random variables. Then $\mathbf{E}\{\eta \mid \xi\} \triangleq \mathbf{E}\{\eta \mid \mathcal{F}_\xi\}$.*

In fact, the σ-algebra represents the set of all random events generated by ξ, and $\mathbf{P}(\cdot \mid \mathcal{F}_\xi)$ is the modification of the original probability $\mathbf{P}$ for an observer for whom ξ is known.

Remember that $\mathbf{E}\{\eta \mid \mathcal{F}_\xi\}$ is the best (in mean variance sense) estimate of η obtained via observations of ξ (see Definition 1.18).

Theorem 1.25 *Let $\xi : \Omega \to \mathbf{R}^n$ be a random vector. Let η be an $\mathcal{F}_\xi$-measurable random variable. Then there exists a (non-random) function $f : \mathbf{R}^n \to \mathbf{R}$ such that $\eta = f(\xi)$.*

Note that any η generates its own f in the previous theorem. Also, f is measurable.

Corollary 1.26 *If $\zeta \in \mathcal{L}_2(\Omega, \mathcal{F}, \mathbf{P})$, then there exists a function $f : \mathbf{R}^n \to \mathbf{R}$ such that $\mathbf{E}\{\zeta \mid \xi\} = \mathbf{E}\{\zeta \mid \mathcal{F}_\xi\} = f(\xi)$.*

Independence

Definition 1.27 *Two random events A and B are said to be independent iff $\mathbf{P}(A \cdot B) = \mathbf{P}(A)\mathbf{P}(B)$.*

Note that if A and B are independent events then $\mathbf{P}(A \mid B) = \mathbf{P}(A)$.

Definition 1.28 *Two σ-algebras of events $\mathcal{F}_1$ and $\mathcal{F}_2$ are said to be independent if any events $A \in \mathcal{F}_1$ and $B \in \mathcal{F}_2$ are independent.*

Definition 1.29 *Two random vectors $\xi : \Omega \to \mathbf{R}^n$ and $\eta : \Omega \to \mathbf{R}^m$ are said to be independent if the σ-algebras $\mathcal{F}_\xi$ and $\mathcal{F}_\eta$ (generated by ξ and η, respectively) are independent.*

Probability distributions

A *probability distribution* on $\mathbf{R}^n$ is a probability measure on the σ-algebra of Borel subsets; i.e., it assigns to every Borel set a probability, so that the probability axioms are satisfied. In fact, this measure is uniquely defined by its values for sets $\{x \in \mathbf{R}^n : a_i < x_i \leq b_i\}$ that generate the Borel σ-algebra; if $n = 1$, then the Borel σ-algebra is generated by intervals. Every random vector gives rise to a probability distribution. On the other hand, for any probability distribution, one can find a random vector with that distribution.

In general, the probability distribution of a random vector is not uniquely defined by the set of probability distributions of its components.

Probability distributions on infinite-dimensional spaces are commonly used in the theory of stochastic processes. For example, the *Wiener process* $w(t)_{t\in[0,T]}$ considered below is a random (infinity-dimensional) vector with values at the space $C(0, T)$ of continuous functions $f : [0, T] \to \mathbf{R}$. It generates a probability distribution on this space.

1.2   Basics of stochastic processes

Definitions of stochastic processes

Sometimes it is necessary to consider random variables or vectors that depend on time.

Definition 1.30 *Let $T \in [0, +\infty]$ be given. A mapping $\xi : [0, T] \times \Omega \to \mathbf{R}$ is said to be a continuous time stochastic (random) process, if $\xi(t, \omega)$ is a random variable for a.e. (almost every) t.*

A random process has two independent variables (t and ω). It can be written as $\xi_t(\omega)$, $\xi(t, \omega)$ or just ξ_t, $\xi(t)$.

Definition 1.31 *A continuous time process $\xi(t) = \xi(t, \omega)$ is said to be continuous (or pathwise continuous), if trajectories $\xi(t, \omega)$ are continuous in t a.s. (i.e., with probability 1, or for a.e. ω).*

It can happen that a continuous time process is not continuous (for instance, a process with jumps).

Filtrations, independent processes, and martingales

In this section, we assume that $t \in [0, +\infty)$.

Filtrations

In addition to evolving random variables, we use evolving σ-algebras.

Definition 1.32 *A set of σ-algebras $\{\mathcal{F}_t\}$ is called a filtration if $\mathcal{F}_s \subseteq \mathcal{F}_t$ for $s < t$.*

Definition 1.33 *Let $\xi(t)$ be a random process, and let $\mathcal{F}_t$ be a filtration. We say that the process $\xi(\cdot)$ is adapted to the filtration $\mathcal{F}$., if any random variable $\xi(t)$ is measurable with respect to $\mathcal{F}_t$ (i.e., $\{\xi(t) \in B\} \in \mathcal{F}$, where $B \subset \mathbf{R}$ is any open interval).*

Definition 1.34 *Let $\xi(t)$ be a random process. The filtration $\mathcal{F}_t$ generated by $\xi(t)$ is defined as the minimal filtration such that $\xi(t)$ is adapted to it.*

Independent processes

Definition 1.35 *Random processes $\xi(\cdot)$ and $\eta(\cdot)$ are said to be independent iff the events $\{(\xi(t_1), \ldots, \xi(t_n)) \in A\}$ and $\{(\eta(\tau_1), \ldots, \eta(\tau_m)) \in B\}$ are independent for all m, n, all times $(t_1, \ldots, t_n)$ and $(\tau_1, \ldots, \tau_m)$, and all sets $A \subset \mathbf{R}^n$ and $B \subset \mathbf{R}^m$.*

In fact, processes are independent iff all events from the filtrations generated by them are mutually independent.

Martingales

Definition 1.36 *Let $\xi(t)$ be a process such that $\mathbf{E}|\xi(t)|^2 < +\infty$ for all t, and let $\mathcal{F}_t$ be a filtration. We say that $\xi(t)$ is a martingale with respect to $\mathcal{F}_t$ if*

$$\mathbf{E}\{\xi(t)|\mathcal{F}_s\} = \xi(s) \quad a.s. \quad \forall s, t : s < t.$$

Note that we require that $\mathbf{E}|\xi(t)|^2 < +\infty$ because, for simplicity, we have defined the conditional expectation only for this case. In the literature, the martingales are often defined under the condition $\mathbf{E}|\xi(t)| < +\infty$ which is less restrictive.

Sometime the term "martingale" is used without mentioning the filtration.

Definition 1.37 *Let $\xi(t)$ be a process, and let $\mathcal{F}_t^\xi$ be the filtration generated by this process. We say that $\xi(t)$ is a martingale if $\xi(t)$ is a martingale with respect to the filtration $\mathcal{F}_t^\xi$.*

Example 1.38 *Let ζ be a random variable such that $\mathbf{E}|\zeta|^2 < +\infty$, and let $\mathcal{F}_t$ be a filtration. Prove that $\xi(t) \triangleq \mathbf{E}\{\zeta|\mathcal{F}_t\}$ is a martingale with respect to $\mathcal{F}_t$.*

1.2.1 Special classes of processes

Markov processes

Definition 1.39 *Let $\xi(t)$ be a process, and let $\mathcal{F}_t^\xi$ be the filtration generated by $\xi(t)$. We say that $\xi(t)$ is a Markov (Markovian) process if*

$$\mathbf{P}(\xi(t_1) \in D_1, \ldots, \xi(t_k) \in D_k \,|\, \mathcal{F}_s^\xi) = \mathbf{P}(\xi(t_1) \in D_1, \ldots, \xi(t_k) \in D_k \,|\, \xi(s))$$

for any $k > 0$, for any times s and t_m such that $t_m > s$, for any system of open sets $\{D_m\}$, $m = 1, \ldots, k$.

This property is said to be the Markov property.

The Markov property means that if we want to estimate the distribution of $\xi(t)|_{t>s}$ using information of the past values $\xi(r)|_{r \le s}$, it suffices to use the last observable value $\xi(s)$ only. Using the values for $r \in [0, s)$ does not give any additional benefits. This property (if it holds) helps to solve many problems.

The following proposition will be useful.

Proposition 1.40 *Under the assumptions and notations of Definition 1.39,*
$\mathbf{E}\{F(\xi(t_1), \xi(t_2), \ldots, \xi(t_k)) \mid \mathcal{F}_s^\xi\} = \mathbf{E}\{F(\xi(t_1), \xi(t_2), \ldots, \xi(t_k)) \mid \xi(s)\}$ *for all measurable deterministic functions F such that the corresponding random variables are integrable.*

Vector processes

Let $\xi(t) = (\xi_1(t), \ldots, \xi_n(t))$ be a vector process such that all its components are random processes. Then ξ is said to be an n-dimensional (vector) random process. All definitions given above can be extended to these vector processes.

Sometimes, we can convert a process that is not a Markov process into a Markov process of higher dimension.

1.2.2   Wiener process (Brownian motion)

Let $T > 0$ be given, $t \in [0, T]$.

Definition 1.41 *We say that a continuous random process $w(t)$ is a (one-dimensional) Wiener process (or Brownian motion) if*

(i) $w(0) = 0$;

(ii) $w(t)$ is Gaussian with $\mathbf{E}w(t) = 0$, $\mathbf{E}w(t)^2 = t$, i.e., $w(t)$ is distributed as $N(0, t)$;

(iii) $w(t + \tau) - w(t)$ does not depend on $\{w(s), s \le t\}$ for all $t \ge 0, \tau > 0$.

Theorem 1.42 *(N. Wiener) There exists a probability space $(\Omega, \mathcal{F}, \mathbf{P})$ such that there is a pathwise continuous process with these properties.*

Corollary 1.43 *Let $\Delta t > 0$, $\Delta w(t) \triangleq w(t + \Delta t) - w(t)$, then* $\operatorname{Var} \Delta w = \Delta t$.

It follows that

$$\mathbf{E}\left(\frac{\Delta w(t)}{\Delta t}\right)^2 = \frac{1}{\Delta t}.$$

It can be interpreted as

$$\frac{\Delta w(t)}{\Delta t} \approx \frac{1}{\sqrt{\Delta t}} \quad \text{as} \quad \Delta t \to 0.$$

This means that a Wiener process cannot have pathwise differentiable trajectories.

Definition 1.44 *We say that a continuous time process $w(t) = (w_1(t), \dots, w_n(t))$: $[0, +\infty) \times \Omega \to \mathbf{R}^n$ is a (standard) n-dimensional Wiener process if*

   (i) $w_i(t)$ is a (one-dimensional) Wiener process for any $i = 1, \dots, n$;

   (ii) the processes $\{w_i(t)\}$ are mutually independent.

Proposition 1.45 *Let $\mathcal{F}_t$ be a filtration such that an n-dimensional Wiener process $w(t)$ is adapted to $\mathcal{F}_t$, and $w(t + \tau) - w(t)$ does not depend on $\mathcal{F}_t$. Then $w(t)$ is a martingale with respect to $\mathcal{F}_t$.*

Corollary 1.46 *A Wiener process $w(t)$ is a martingale. (In other words, if $\mathcal{F}_t^w$ is the filtration generated by $w(t)$, then $w(t)$ is a martingale with respect to $\mathcal{F}_t^w$.)*

Let us list some basic properties of a Wiener process $w(t)$:

- Sample paths are continuous.

- Sample paths are non-differentiable and they are not absolutely continuous.

- It is a Markov process.

Up to the end of this chapter, we assume that we are given an n-dimensional Wiener process $w(t)$ and the filtration $\mathcal{F}_t$ such as described in Proposition 1.45. One may assume that this filtration is generated by the process $(w(t), \eta(t))$, where $\eta(s)$ is a process independent from $w(\cdot)$. We assume also that $t \in [0, T]$, where $T > 0$ is given deterministic terminal time.

1.3   Basics of the stochastic calculus (Ito calculus)

Stochastic integral for step functions

Let $w(t)$ be a one-dimensional Wiener process. Repeat $\mathcal{F}_t$ as a filtration such as described in Proposition 1.45.

Let $\mathcal{L}_{22}^0$ be the set of $\mathcal{F}_t$-adapted functions $f(t, \omega)$ such that there exists an integer $N > 0$, a set of times $0 = t_0 < t_1 < \dots < t_N = T$, and a sequence $\{\xi_k\}_{k=1}^N \subset \mathcal{L}_2(\Omega, \mathcal{F}, P)$, such that $f(t) = \xi_k$ for $t \in [t_k, t_{k+1})$, $k = 0, \dots, N - 1$.

Clearly, all these functions are pathwise step functions.

Example 1.47 *Prove that, in the definition above, ξ_k are $\mathcal{F}_{t_k}$-measurable.*

Definition 1.48 *Let $f(\cdot) \in \mathcal{L}_{22}^0$. The value*

$$I(f) \triangleq \sum_{k=0}^{N-1} f(t_k)[w(t_{k+1}) - w(t_k)]$$

is said to be the Ito integral of f, or stochastic integral, and it is denoted as $\int_0^T f(t)dw(t)$, *i.e.,*

$$\int_0^T f(t)dw(t) = I(f).$$

Stochastic integral (Ito integral) for general functions

We denote by $\mathcal{L}_{22}$ the set of all random processes that can be approximated by processes from $\mathcal{L}_{22}^0$ in the following sense: for any $f \in \mathcal{L}_{22}$, there exists a sequence $\{f_k(\cdot)\}_{k=1}^{+\infty} \subset \mathcal{L}_{22}^0$ such that $\mathbf{E} \int_0^T |f(t) - f_k(t)|^2 dt \to 0$ as $k \to +\infty$.

Note that all processes from $\mathcal{L}_{22}$ are adapted to the filtration $\mathcal{F}_t$ (more precisely, if $f \in \mathcal{L}_{22}$, then $f(t)$ is $\mathcal{F}_t$-measurable for a.e. (almost every) t.

In fact, processes $\xi \in \mathcal{L}_{22}$ are measurable as mappings $\xi : [0, T] \times \mathbf{P} \to \mathbf{R}$ with respect to the completion of the σ-algebra generated by all mappings $\xi_0 : [0, T] \times \mathbf{P} \to \mathbf{R}$ such that $\xi_0 \in \mathcal{L}_{22}^0$.

Theorem 1.49 *Let $f \in \mathcal{L}_{22}$, and let $\{f_k(\cdot)\}_{k=1}^{+\infty} \subset \mathcal{L}_{22}^0$ be such that $\mathbf{E} \int_0^T |f(t) - f_k(t)|^2 dt \to 0$ as $k \to +\infty$. Then $\{I(f_k)\}_{k=1}^{\infty}$ is a Cauchy sequence in $L_2(\Omega, \mathcal{F}, \mathbf{P})$, where $I(f_k) = \int_0^T f_k(t)dw(t)$. This sequence converges in $L_2(\Omega, \mathcal{F}, \mathbf{P})$, and its limit depends only on f and does not depend on the choice of the approximating sequence (in the sense that all possible modifications of the limit are $\mathbf{P}$-equivalent).*

Definition 1.50 *The limit of $I(f_k)$ in $\mathcal{L}_2(\Omega, \mathcal{F}, \mathbf{P})$ from the theorem above is said to be the Ito integral (stochastic integral)*

$$\int_0^T f(t)dw(t).$$

Theorem 1.51 *Let $f, g \in \mathcal{L}_{22}$. Then*

(i) $\mathbf{E} \int_0^T f(t)dw(t) = 0$;

(ii) $\mathbf{E} \left(\int_0^T f(t)dw(t) \right)^2 = \mathbf{E} \int_0^T |f(t)|^2 dt$;

(iii) $\mathbf{E} \int_0^T f(t)dw(t) \int_0^T g(t)dw(t) = \mathbf{E} \int_0^T f(t)g(t)dt$.

Proof follows from the properties for approximating functions from $\mathcal{L}_{22}^0$. $\square$

Theorem 1.52 *Let $f, g \in \mathcal{L}_{22}$. Then*

(i) $\mathbf{E} \left\{ \int_0^T f(t)dw(t) \,\middle|\, \mathcal{F}_s \right\} = \int_0^s f(t)dw(t)$;

(ii) $\mathbf{E} \left\{ \int_s^T f(t)dw(t) \,\middle|\, \mathcal{F}_s \right\} = 0$;

(iii)

$$\mathbf{E}\left\{\int_0^T f(t)dw(t)\int_0^T g(t)dw(t)\,\Big|\,\mathcal{F}_s\right\}$$

$$= \int_0^s f(t)dw(t)\int_0^s g(t)dw(t) + \mathbf{E}\left\{\int_s^T f(t)g(t)dt\,\Big|\,\mathcal{F}_s\right\}.$$

Proof follows from the properties for approximating functions from $\mathcal{L}_{22}^0$. $\square$

Definition 1.53 *A modification of a process $\xi(t,\omega)$ is any process $\xi'(t,\omega)$ such that $\xi(t,\omega) = \xi'(t,\omega)$ for a.e. t,ω.*

Theorem 1.54 *Let $T > 0$ be fixed, and let $f \in \mathcal{L}_{22}$. Then the process $\int_0^t f(s)dw(s)$ is pathwise continuous in $t \in [0,T]$ (more precisely, there exists a modification of the process $\int_0^t f(s)dw(s)$ that is continuous in $t \in [0,T]$ a.s. (i.e., with probability 1, or for a.e. (almost every) ω).*

Note that

(i) A stochastic integral is defined up to $\mathbf{P}$-equivalency;

(ii) It is not defined pathwise, i.e., we cannot construct it as a function of T for a fixed ω.

Vector case

Let $w(t)$ be a n-dimensional Wiener process, and let $\mathcal{F}_t$ be a filtration such as described in Proposition 1.45. Let $f = (f_1, \ldots, f_n) : [0,T] \times \Omega \to \mathbf{R}^{1\times n}$ be a (vector row) process such that $f_i \in \mathcal{L}_{22}$ for all i. Then we can define the Ito integral

$$\int_0^T f(t,\omega)dw(t) \triangleq \sum_{i=1}^n \int_0^T f_i(t,\omega)dw_i(t).$$

The right-hand part is well defined by the previous definitions.

Ito processes

Definition 1.55 *Let $w(t)$ be an n-dimensional Wiener process, $\alpha \in \mathcal{L}_{22}$, $a \in \mathcal{L}_2(\Omega, \mathcal{F}_0, \mathbf{P})$. Let a random process $\beta = (\beta_1, \ldots, \beta_n)$ take values in $\mathbf{R}^{1\times n}$, and let $\beta_i \in \mathcal{L}_{22}$ for all i. Let*

$$y(t) = a + \int_0^t \alpha(r)dr + \int_0^t \beta(r)dw(r).$$

Then the process $y(t)$ is said to be an Ito process. The expression

$$dy(t) = \alpha(t)dt + \beta(t)dw(t)$$

is said to be the stochastic differential (or Ito differential) of $y(t)$. The process $\alpha(t)$ is said to be the drift coefficient, and $\beta(t)$ is said to be the diffusion coefficient.

Theorem 1.56 *An Ito process*

$$y(t) = a + \int_0^t \alpha(r)dr + \int_0^t \beta(r)dw(r)$$

is a martingale with respect to $\mathcal{F}_t$ if and only if $\alpha(t) \equiv 0$ up to equivalency.

Proof. By Theorem 1.52, it follows that if $\alpha \equiv 0$ then y is a martingale. Proof of the opposite statement needs some analysis. $\square$

1.3.1 Ito formula

One-dimensional case

Let us assume first that $\alpha, \beta \in \mathcal{L}_{22}$ are one-dimensional processes, and $w(t)$ is one-dimensional process

$$y(t) = y(s) + \int_s^t \alpha(r)dr + \int_s^t \beta(r)dw(r),$$

i.e., $y(t)$ is an Ito process, and

$$dy(t) = \alpha(t)dt + \beta(t)dw(t).$$

Let $V(\cdot, \cdot) : \mathbf{R} \times [0, T] \to \mathbf{R}$ be a continuous function such that its derivatives V_t', V_x', V_{xx}'' are continuous (and such that some additional conditions on their growth are satisfied).

Theorem 1.57 *(Ito formula, or Ito lemma). The process $V(y(t), t)$ is also an Ito process, and its stochastic differential is*

$$d_t V(y(t), t)$$
$$= \frac{\partial V}{\partial t}(y(t), t)dt + \frac{\partial V}{\partial x}(y(t), t)dy(t) + \frac{1}{2}\frac{\partial^2 V}{\partial x^2}(y(t), t)\beta(t)^2 dt. \quad (1.1)$$

Note that the last equation can be rewritten as

$$d_t V(y(t), t) = \left[\frac{\partial V}{\partial t}(y(t), t) + \mathcal{A}(t)V(y(t), t)\right]dt + \frac{\partial V}{\partial x}(y(t), t)\beta(t)dw(t),$$

where $\mathcal{A}(t)$ is the differential operator

$$\mathcal{A}(t)v(x) = \frac{dv}{dx}(x)\alpha(t) + \frac{1}{2}\frac{d^2v}{dx^2}(x)\beta(t)^2.$$

Remark 1.58 In fact, the formula for the drift and diffusion coefficients of the process $V(y(t), t)$ was first obtained by A.N. Kolmogorov as long ago as 1931 in *Mathematische Annalen* [104] (1931), pp. 415–458 *for the special case when $y(t)$ is a Markov (diffusion) process. It gives (1.1) for this case; see [109] p.263.*

Proof of Theorem 1.57 is based on the Taylor series and the estimate

$$\Delta y \triangleq y(t + \Delta t) - y(t) \sim a(t)\Delta t + \beta(t)\Delta w,$$

where $(\Delta w)^2 \sim \Delta t. \ \Box$

Example 1.59 Let $y(t) = w(t)^2$, then $dy(t) = 2w(t)dw(t) + dt$.

Let $\alpha_i, \beta_i \in \mathcal{L}_{22}$,

$$dy_i(t) = \alpha_i(t)dt + \beta_i(t)dw(t), \quad i = 1, 2.$$

Theorem 1.60 Let $y(t) \triangleq y_1(t)y_2(t)$, then

$$dy(t) = y_1(t)dy_2(t) + y_2(t)dy_1(t) + \beta_1(t)\beta_2(t)dt.$$

The vector case

Let us assume first that $w(t)$ is an n-dimensional Wiener process. Let random processes $a = (a_1, \ldots, a_m)$ and $\beta = \{\beta_{ij}\}$ take values in $\mathbf{R}^m$ and $\mathbf{R}^{m \times n}$, respectively, and let $a_i \in \mathcal{L}_{22}$ and $b_{ij} \in \mathcal{L}_{22}$ for all i, j. Let $y(t)$ be an m-dimensional Ito process, and

$$dy(t) = \alpha(t)dt + \beta(t)dw(t),$$

i.e.,

$$dy(t) = \alpha(t)dt + \sum_{i=1}^{n} \beta_i(t)dw_i(t).$$

Here β_i are the columns of the matrix β. (It is an equation for a vector process that formally was not introduced before; we simply require that the corresponding equation holds for any component.) Let $V(\cdot, \cdot) : \mathbf{R}^m \times [0, T] \to \mathbf{R}$ be a continuous function such that the derivatives V_t', V_x', V_{xx}'' are continuous (and such that some additional conditions on their growth are satisfied). Note that V_x' takes values in $\mathbf{R}^{1 \times m}$, and V_{xx}'' takes values in $\mathbf{R}^{m \times m}$.

Theorem 1.61 *(Ito formula for the vector case.) The process* $V(y(t), t)$ *is also an Ito process, and its stochastic differential is*

$$d_t V(y(t), t)$$
$$= \frac{\partial V}{\partial t}(y(t), t)dt + \frac{\partial V}{\partial x}(y(t), t)dy(t) + \frac{1}{2}\sum_{i=1}^{n} \beta_i(t)^\top \frac{\partial^2 V}{\partial x^2}(y(t), t)\beta_i(t)dt.$$

Note that the last equation can be rewritten as

$$d_t V(y(t), t) = \left[\frac{\partial V}{\partial t}(y(t), t) + \mathcal{A}(t)V(y(t), t)\right]dt + \frac{\partial V}{\partial x}(y(t), t)\beta(t)dw(t),$$

where $\mathcal{A}(t)$ is the differential operator

$$\mathcal{A}(t)v(x) = \frac{dv}{dx}(x)a(t) + \frac{1}{2}\sum_{i=1}^{m}\beta_i(t)^\top\frac{d^2v}{dx^2}(x)\beta_i(t).$$

In addition, it can be useful to note that

$$\sum_{i=1}^{m}\beta_i^\top\frac{\partial^2 V}{\partial x^2}\beta_i \equiv \text{Tr}\left[\beta\beta^\top\frac{\partial^2 V}{\partial x^2}\right],$$

where Tr denotes the *trace* of a martix (i.e., the summa of all eigenvalues).

1.3.2   Stochastic differential equations (Ito equations)

Definitions

Let $f(x,t,\omega) : \mathbf{R}^m \times [0,T] \times \Omega \to \mathbf{R}^m$ and $b(x,t,\omega) : \mathbf{R}^m \times [0,T] \times \Omega \to \mathbf{R}^{m\times n}$ be some functions. Let the processes $f(x,t,\omega)$ and $b(x,t,\omega)$ be adapted to the filtration $\mathcal{F}_t$ for all x.

Let $\mathcal{L}_{22}(s,T)$ be the set of functions $f : [s,T] \times \Omega \to \mathbf{R}$ defined similarly to $\mathcal{L}_{22}$.

Definition 1.62 *Let $s \in [0,T]$, and let $a = (a_1,\ldots,a_m)$ be a random vector with values in $\mathbf{R}^m$ such that $a_i \in L_2(\Omega,\mathcal{F}_s,\mathbf{P})$. Let an m-dimensional process $y(t) = (y_1(t),\ldots,y_m(t))$ be such that $y_i \in \mathcal{L}_{22}(s,T)$ and*

$$y(t) = a + \int_s^t f(y(r),r,\omega)dr + \int_s^t b(y(r),r,\omega)dw(r) \quad \forall t \quad a.s. \qquad (1.2)$$

We say that the process $y(t)$, $t \in [s,T]$, is a solution of the stochastic differential equation (Ito equation)

$$dy(t) = f(y(t),t,\omega)dt + b(y(t),t,\omega)dw(t),$$
$$y(s) = a. \qquad (1.3)$$

Example 1.63 *The following result is immediate. Let $\alpha,\beta \in \mathcal{L}_{22}$, $a \in L_2(\Omega,\mathcal{F}_s,\mathbf{P})$. The equation*

$$\begin{cases} dy(t) = \alpha(t)dt + \beta(t)dw(t), \\ y(s) = a. \end{cases} \qquad (1.4)$$

has a solution

$$y(t) = a + \int_s^t \alpha(r)dr + \int_s^t \beta(r)dw(r), \quad t \in [s,T].$$

Remark 1.64 *For the case when $f(x,t) : \mathbf{R}^m \times [0,T] \to \mathbf{R}^m$ and $b(x,t) : \mathbf{R}^m \times [0,T] \to \mathbf{R}^{m\times n}$ are non-random, the solution $y(t)$ of equation (1.3) is a Markov process. In that case, it is called a diffusion process.*

For the general case of random f or b, the process $y(t)$ is not a Markov process; in that case, it is sometimes called a *diffusion type process* (but not a diffusion process).

In particular, if $n = m = 1$, $f(x,t) \equiv ax$, $b(x,t) \equiv \sigma x$, then the equation for $y(t)$ is the equation for the stock price $dS(t) = S(t)[adt + \sigma dw(t)]$ that we will discuss below.

The existence and uniqueness theorem

Theorem 1.65 *(The existence and uniqueness theorem). Let (random) functions $f(x,t,\omega) : \mathbf{R}^m \times [0,T] \times \Omega \to \mathbf{R}^m$, $b(x,t,\omega) : \mathbf{R}^m \times [0,T] \times \Omega \to \mathbf{R}^{m \times n}$ be continuous in (x,t) with probability 1. Further, let the processes $f(x,\cdot)$ and $b(x,\cdot)$ be $\mathcal{F}_t$-adapted for all x, and let there exist a constant $C > 0$ such that*

$$|f(x,t,\omega)| + |b(x,t,\omega)| \leq C(|x| + 1),$$
$$|f(x,t,\omega) - f(x_1,t,\omega)| + |b(x,t,\omega) - b(x_1,t,\omega)| \leq C|x - x_1|$$

for all $x, x_1 \in \mathbf{R}$, $t \in [0,t]$, a.s. Let $s \in [0,T]$, and let $a \in \mathcal{L}_2(\Omega, \mathcal{F}_s, \mathbf{P})$. Then equation (1.3) has a unique solution $y \in \mathcal{L}_{22}(s,T)$ (unique up to equivalence).

Here and below $|x| = \left(\sum_{i=1}^m x_i^2\right)^{1/2}$ denotes the Euclidean norm for $x \in \mathbf{R}^m$, and $|x| = \left(\sum_{i,j=1}^m x_{ij}^2\right)^{1/2}$ denotes the Frobenius matrix norm for $x \in \mathbf{R}^{m \times m}$.

1.3.3   Some explicit solutions for Ito equations

In the examples below, we assume that $n = m = 1$.

Processes with log-normal distributions

Example 1.66 *Let $a, \sigma \in \mathbf{R}$, $y_s \in \mathcal{L}_2(\Omega, \mathcal{F}_s, \mathbf{P})$. The equation*

$$\begin{cases} dy(t) = ay(t)dt + \sigma y(t)dw(t), \\ y(s) = y_s \end{cases} \tag{1.5}$$

has the unique solution

$$y(t) = y_s \exp\left(\left[a - \frac{\sigma^2}{2}\right](t - s) + \sigma(w(t) - w(s))\right), \quad t \geq s$$

The existence and uniqueness follow from Theorem 1.65. The equation can be derived using the Ito formula for $y(t)$. For instance, set $V(x,t) = e^x$, $\xi(t) = \ln y_0 + at - \frac{\sigma^2}{2}t + \sigma w(t)$. Then the process $y(t) = V(\xi(t),t)$ is such that $y(s) = y_s$, and the Ito formula gives that the stochastic differential equation is satisfied for $y(t) = V(\xi(t),t)$.

A generalization

Example 1.67 *Let $w(t)$ be an n-dimensional Wiener process, $\alpha(\cdot) \in \mathcal{L}_{22}$, let $\sigma(t) = (\sigma_1(t), \ldots, \sigma_n(t))$ be a process with values in $\mathbf{R}^{1 \times n}$ such that $\sigma_i(\cdot) \in \mathcal{L}_{22}$, and and let some conditions on the growth for a, σ be satisfied (it suffices to assume that they are bounded). Let $y_s \in \mathcal{L}_2(\Omega, \mathcal{F}_s, \mathbf{P})$. The equation*

$$\begin{cases} dy(t) = a(t)y(t)dt + y(t)\sigma(t)dw(t), \\ y(s) = y_s \end{cases} \tag{1.6}$$

has a unique solution

$$y(t) = y_s \exp\left(\int_s^t a(r)dr - \frac{1}{2}\int_s^t |\sigma(r)|^2 dr + \int_s^t \sigma(r)dw(r) \right), \quad t \geq s.$$

Ornstein–Uhlenbek process

Example 1.68 *Let $\alpha, \lambda, \sigma \in \mathbf{R}$, $y_s \in \mathcal{L}_2(\Omega, \mathcal{F}_s, \mathbf{P})$. The equation*

$$\begin{cases} dy(t) = (\alpha - \lambda y(t))dt + \sigma dw(t), \\ y(s) = y_s \end{cases} \tag{1.7}$$

has a unique solution

$$y(t) = e^{-\lambda(t-s)}y_s + \int_s^t e^{-\lambda(t-r)}\alpha dr + \int_s^t e^{-\lambda(t-r)}\sigma dw(r), \quad t \geq s.$$

To show this, one can use that

$$\int_s^t e^{-\lambda(t-r)}\sigma dw(r) = e^{-\lambda t}\int_s^t e^{\lambda r}\sigma dw(r).$$

If $\lambda > 0$, then the solution $y(t)$ of (1.7) is said to be an Ornstein-Uhlenbek process. This process converges (in a certain sense) to a stationary Gaussian process as $t \to +\infty$.

1.3.4 Diffusion Markov processes and related parabolic equations

One-dimensional case

Let (non-random) functions $f : \mathbf{R} \times [0, T] \to \mathbf{R}$ and $b : \mathbf{R} \times [0, T] \to \mathbf{R}$ be given. Let $y(t)$ be a solution of the stochastic differential equation

$$\begin{cases} dy(t) = f(y(t), t)dt + b(y(t), t)dw(t), \quad t > s, \\ y(s) = a. \end{cases} \tag{1.8}$$

We shall denote this solution as $y^{a,s}(t)$. As was mentioned above, this process is called a *diffusion process*; it is a Markov process.

Let functions $\Psi : \mathbf{R} \to \mathbf{R}$ and $\varphi : \mathbf{R} \times [0, T] \to \mathbf{R}$ be such that certain conditions on their smoothness and growth are satisfied (for instance, it suffices to assume that they are continuous and bounded).

Let $V(x, s)$ be the solution of the Cauchy problem for the parabolic equation

$$\frac{\partial V}{\partial s}(x, s) + \mathcal{A}(s)V(x, s) = -\varphi(x, s),$$
$$V(x, T) = \Psi(x). \tag{1.9}$$

Here $x \in \mathbf{R}$, $s \in [0, T]$,

$$\mathcal{A}(t)v(x) = \frac{dv}{dx}(x)f(x, t) + \frac{1}{2}\frac{d^2v}{dx^2}(x)b(x, t)^2.$$

Note that (1.9) is a so-called backward parabolic equation, since the Cauchy condition is imposed at the end of the time interval.

We assume that the functions f, b, Ψ, φ are such that this boundary value problem (1.9) has a unique solution V such that it has continuous derivatives V'_t, V'_x, and V''_{xx}.

Theorem 1.69

$$\Psi(y^{x,s}(T)) + \int_s^T \varphi(y^{x,s}(t), t)dt = V(x, s) + \int_s^T \frac{\partial V}{\partial y}(y^{x,s}(t), t)b(y(t), t)dw(t).$$

Proof. By Ito formula,

$$\Psi(y^{x,s}(T)) - V(x, s) = V(y^{x,s}(T), T) - V(y^{x,s}(s), s)$$
$$= \int_s^T \left[\frac{\partial V}{\partial t} + \mathcal{A}V\right](y^{x,s}(t), t)dt + \int_s^T \frac{\partial V}{\partial y}(y^{x,s}(t), t)b(y(t), t)dw(t)$$
$$= -\int_s^T \varphi(y^{x,s}(t), t)dt + \int_s^T \frac{\partial V}{\partial y}(y^{x,s}(t), t)b(y(t), t)dw(t).$$

Then the proof follows. $\square$

Vector case

Let $w(t)$ be an n-dimensional Wiener process. Let (non-random) functions $f : \mathbf{R}^m \times [0, T] \to \mathbf{R}^m$ and $b : \mathbf{R}^m \times [0, T] \to \mathbf{R}^{m \times n}$ be given, $a \in \mathbf{R}^m$. Let $y(t)$ be a solution of the stochastic differential equation

$$\begin{cases} dy(t) = f(y(t), t)dt + b(y(t), t)dw(t), & t > s, \\ y(s) = a. \end{cases} \tag{1.10}$$

We shall denote this solution as $y^{a,s}(t)$.

Let functions $\Psi : \mathbf{R}^m \to \mathbf{R}$ and $\varphi : \mathbf{R}^m \times [0, T] \to \mathbf{R}$ be such that certain conditions on their smoothness and growth are satisfied (it suffices again to assume that they are continuous and bounded).

Let $V(x, s)$ be the solution of the Cauchy problem for the parabolic equation

$$\frac{\partial V}{\partial s}(x, s) + \mathcal{A}(s)V(x, s) = -\varphi(x, s),$$

$$V(x, T) = \Psi(x).$$

(1.11)

Here $x \in \mathbf{R}^m$, $s \in [0, T]$,

$$\mathcal{A}(t)v(x) = \frac{dv}{dx}(x)f(x, t) + \frac{1}{2}\sum_{i=1}^{n} b_i(x, t)^\top \frac{d^2 v}{dx^2}(x)b_i(x, t).$$

Here b_i are the columns of the matrix b.

Again, (1.9) is a so-called backward parabolic equation, since the Cauchy condition is imposed at the end of the time interval.

Theorem 1.70

$$\Psi(y^{x,s}(T)) + \int_s^T \varphi(y^{x,s}(t), t)dt = V(x, s) + \int_s^T \frac{\partial V}{\partial y}(y^{x,s}(t), t)b(y(t), t)dw(t).$$

Proof repeats the proof of Theorem 1.69. □

The following corollary gives the probabilistic representation of the solution V of the Cauchy problem for the parabolic equation.

Corollary 1.71

$$V(x, s) = \mathbf{E}\Psi(y^{x,s}(T)) + \mathbf{E}\int_s^T \varphi(y^{x,s}(t), t)dt.$$

The differential operator $\mathcal{A}$ is said to be the differential operator generated by the process $y(t)$.

Equation (1.9) is said to be the backward Kolmogorov (parabolic) equation for the process $y(t)$ (or Kolmogov–Fokker–Planck equation).

For the examples below, we assume that $n = m = 1$.

1.3.5 Martingale representation theorem

In this section, we assume that $w(t)$ is an n-dimensional vector process, $\mathcal{F}_t$ is the filtration generated by $w(t)$, *and a wider filtration is not allowed.*

The following result is known as the Clark Theorem or Clark–Haussmann–Ocone theorem.

Theorem 1.72 *Let $\xi \in L_2(\Omega, \mathcal{F}_T, \mathbf{P})$. Then there exists n-dimensional process $f = (f_1, \ldots, f_n)$ with values in $\mathbf{R}^{1 \times n}$ such that $f_i \in \mathcal{L}_{22}$ for all i and*

$$\xi = \mathbf{E}\xi + \int_0^T f(t)dw(t).$$

Note that Theorem 1.72 allows equivalent formulation as the following martingale representation theorem.

Theorem 1.73 *Let $\xi(t)$ be a martingale with respect to the filtration $\mathcal{F}_t$ generated by a Wiener process $w(t)$ such that $\mathbf{E}\xi(T)^2 < +\infty$. Then there exists a process $f(t)$ with values in $\mathbf{R}^{1 \times n}$ and with components from $\mathcal{L}_{22}$ such that*

$$\xi(t) = \mathbf{E}\xi(t) + \int_0^t f(s)dw(s), \quad t \in [0, T].$$

Proof. Apply Theorem 1.72 to $\xi(T)$. □

Corollary 1.74 *Any martingale described in Theorem 1.73 is pathwise continuous.* □

The process $f(t)$ in Theorem 1.72 and Theorem 1.73 is uniquely defined up to equivalency.

1.3.6   Change of measure and Girsanov theorem

In this section, we assume again that $\mathcal{F}_t$ is a filtration such that a n-dimensional Wiener process $w(t)$ is $\mathcal{F}_t$-adapted and $w(t + \tau) - w(t)$ does not depend on $\mathcal{F}_t$. In particular, we allow that $\mathcal{F}_t$ is the filtration generated by the process $(w(t), \eta(t))$, where $\eta(t)$ is a random process that does not depend on $w(\cdot)$.

Let $\theta(t) = (\theta_1(t), \dots, \theta_n(t))$ be a bounded $\mathcal{F}_t$-adapted random process with values in $\mathbf{R}^n$ and with components from $\mathcal{L}_{22}$.

Let

$$\mathcal{Z} \triangleq \exp\left(-\int_0^T \theta(t)^\top dw(t) - \frac{1}{2}\int_0^T |\theta(t)|^2 dt\right). \tag{1.12}$$

Change of the probability measure and Girsanov's theorem

Proposition 1.75 *Let the process θ be bounded then   (i) $\mathbf{E}\mathcal{Z} = 1$;   (ii) let mapping $\mathbf{P}_* : \mathcal{F} \to \mathbf{R}$ be defined via the equation*

$$\frac{d\mathbf{P}_*}{d\mathbf{P}} = \mathcal{Z}. \tag{1.13}$$

Then $\mathbf{P}_*$ is a probability measure on $\mathcal{F}$ equivalent to the original measure $\mathbf{P}$.

It can be noted that, instead of boundedness of θ, we could assume that a less restrictive condition is satisfied:

$$\mathbf{E}\exp\left(\frac{1}{2}\int_0^T |\theta(t)|^2 dt\right) < +\infty. \tag{1.14}$$

This condition is called the Novikov condition. Clearly, it is satisfied for all bounded processes θ.

Remember that (1.13) means that

$$\mathbf{P}_*(A) = \int_A \mathcal{Z}(\omega)\mathbf{P}(d\omega) \quad \forall A \in \mathcal{F},$$

i.e.,

$$\mathbf{E}_*\mathcal{Z} = \mathbf{E}\mathbb{I}_A\mathcal{Z} \quad \forall A \in \mathcal{F},$$

and

$$\mathbf{E}_*\xi = \mathbf{E}\mathcal{Z}\xi$$

for any integrable random variable ξ. (See Section 1.1.3.)

Proof of Proposition 1.75. Let us prove (i). We have $\mathcal{Z} = y(T)$, where $y(t)$ is the solution of the equation

$$\begin{cases} dy(t) = -y(t)\theta(t)^\top dw(t), \\ y(0) = 1. \end{cases}$$

Then

$$y(T) = 1 + \int_0^T y(t)\theta(t)^\top dw(t).$$

Hence $\mathbf{E}\mathcal{Z} = \mathbf{E}y(T) = 1$. To prove (ii), it suffices to verify that all probability axioms are satisfied. For instance, we have that $\mathbf{P}_*(\Omega) = \mathbf{E}\mathcal{Z}\mathbb{I}_{\{\omega \in \Omega\}} = \mathbf{E}\mathcal{Z} = 1$. $\square$

Example 1.76 *We have* $\mathbf{E}_*\mathcal{Z} = \mathbf{E}\mathcal{Z} \cdot \mathcal{Z} = \mathbf{E}\mathcal{Z}^2$.

Example 1.77 *Let* $n = 1$. *We have that* $\mathbf{E}_*w(T) = \mathbf{E}\mathcal{Z}w(T)$. *Let* θ *be non-random and constant, then*

$$\begin{aligned} \mathbf{E}_*w(T) &= \mathbf{E}w(T)\exp\left(-\theta w(T) - \frac{T}{2}\theta^2\right) \\ &= \frac{1}{\sqrt{2\pi T}}\int_{-\infty}^{+\infty} e^{-\frac{x^2}{2T}} x\exp\left(-\theta x - \frac{T}{2}\theta^2\right)dx. \end{aligned}$$

Girsanov's theorem

Let

$$w_*(t) \overset{\Delta}{=} w(t) + \int_0^t \theta(s)ds.$$

The following is a special case of the celebrated Girsanov's theorem.

Theorem 1.78 *Let the assumptions of Proposition 1.75 be satisfied, and let the measure* $\mathbf{P}_*$ *be defined via the equation*

$$\frac{d\mathbf{P}_*}{d\mathbf{P}} = \mathcal{Z}.$$

Then $w_*(t)$ *is a Wiener process under* $\mathbf{P}_*$.

Example 1.79 Let us reconsider Example 1.77. We have that $w(T) = w_*(T) - \int_0^T \theta(s)ds$. Hence,

$$\mathbf{E}_* w(T) = \mathbf{E}_* \left[w_*(T) - \int_0^T \theta(s)ds \right] = -\mathbf{E}_* \int_0^T \theta(s)ds.$$

Since w_* is a Wiener process with respect to $\mathbf{P}_*$, it follows that $\mathbf{E}_* w_*(T) = 0$. It can be verified directly that the integral in Example 1.77 has the value $-\mathbf{E}_* \int_0^T \theta(s)ds = -T\theta$ for the case of non-random and constant θ.

2

Some background on diffusion market models

In this chapter, the most mainstream models of markets with continuous time are studied. These models are based on the theory of stochastic integrals (stochastic calculus); stock prices are represented via stochastic integrals. Core concepts and results of mathematical finance are given (including self-financing strategies, replicating, arbitrage, risk-neutral measures, market completeness, and option price).

2.1 Continuous time model for stock price

We assume that we are given a standard complete probability space $(\Omega, \mathcal{F}, \mathbf{P})$. Sometime we will address $\mathbf{P}$ as the *original probability measure*. Some other measures will be also used.

Consider a risky asset (stock, bond, foreign currency unit, etc.,) with time series prices $S_1, S_2, S_3, \ldots$, for example, daily prices. The premier model of price evolution is such that $S_k = S(t_k)$, where $S(t)$ is a continuous time Ito process. (Note that Ito processes are pathwise continuous. For a more general model, continuous time process $S(t)$ may have jumps; this case will not be considered in the present course.)

We consider the evolution of the price $S(t)$ for $t \in [0, T]$, where t is time, and T is some terminal time.

The initial price $S_0 > 0$ is a given non-random value, and the evolution of $S(t)$ is described by the following *Ito equation*:

$$dS(t) = S(t)(a(t)dt + \sigma(t)dw(t)). \tag{2.1}$$

Here $w(t)$ is a (one-dimensional) Wiener process, and a and σ are market parameters.

Sometimes in the literature $S(t)$ is called a *geometric Brownian motion* (for the case of non-random and constant a, σ), sometimes $\ln S(t)$ is also said to be a Brownian motion. Mathematicians prefer to use the term "Brownian motion" for $w(t)$ only (i.e., Brownian motion is the same as a Wiener process).

Definition 2.1 *In (2.1), $a(t)$ is said to be the appreciation rate, and $\sigma(t)$ is said to be the volatility.*

Note that, in terms of more general stochastic differential equations, the coefficient for dt (i.e., $a(t)S(t)$) is said to be the drift (or the drift coefficient), and the coefficient for $dw(t)$ (i.e., $\sigma(t)S(t)$) is said to be the diffusion coefficient.

We assume that there exists a random process $\eta(t)$ that does not depend on $w(\cdot)$. This process describes additional random factors presented in the model besides the driving Wiener process $w(t)$.

Let $\mathcal{F}_t$ be the filtration generated by $(w(t), \eta(t))$, and let $\mathcal{F}_t^w$ be the filtration generated by the process $w(t)$ only.

It follows that $\mathcal{F}_t^w \subseteq \mathcal{F}_t$ and that $w(t + \tau) - w(t)$ does not depend on $\mathcal{F}_t$ for all t and $\tau > 0$. (Note that the case when $\mathcal{F}_t \equiv \mathcal{F}_t^w$ is not excluded.)

We assume that the process $(a(t), \sigma(t))$ is bounded and $\mathcal{F}_t$-adapted. In particular, it follows that $(a(t), \sigma(t))$ does not depend on $w(t + \tau) - w(t)$ for all t and $\tau > 0$.

Without a loss of generality, we assume that $\mathcal{F} = \mathcal{F}_T$.

Remark 2.2 *The assumptions imposed imply that the vector $(r(t), a(t), \sigma(t))$ can be presented as a deterministic function of $(w(s), \eta(s))|_{s \in [0,t]}$.*

Let us discuss some basic properties of the Ito equation (2.1).

Lemma 2.3

$$S(t) = S(0) \exp\left(\int_0^t a(s)ds - \frac{1}{2} \int_0^t \sigma(s)^2 ds + \int_0^t \sigma(s)dw(s) \right).$$

Proof follows from the Ito formula (see Problem 1.67). □

Note that the stochastic integral above is well defined.

2.2 Continuous time bond-stock market model

The case of the market with a non-zero interest rate for borrowing can be described via the following bond-stock model.

We introduce a market model consisting of the risk-free bond or bank account with price $B(t)$ and the risky stock with the price $S(t)$, $t > 0$. The initial prices $S(0) > 0$ and $B(0) > 0$ are given non-random variables. We assume that the bond price is

$$B(t) = B(0) \exp\left(\int_0^t r(s)ds \right), \tag{2.2}$$

where $r(t)$ is the process of the risk-free interest rate. We assume that the process $r(t)$ is $\mathcal{F}_t$-adapted (in particular, it follows that $r(t)$ does not depend on $w(t+\tau) - w(t)$ for all $t, \tau > 0$). Typically, it suffices to consider non-negative processes $r(t)$ (however, we do not assume it, because it can be restrictive for some models, especially for the bond market).

Let $X(0) > 0$ be the initial wealth at time $t = 0$, and let $X(t)$ be the wealth at time $t > 0$. We assume that, for $t \geq 0$,

$$X(t) = \beta(t)B(t) + \gamma(t)S(t). \tag{2.3}$$

Here $\beta(t)$ is the quantity of the bond portfolio, $\gamma(t)$ is the quantity of the stock portfolio. The pair $(\beta(\cdot), \gamma(\cdot))$ describes the state of the bond-stocks portfolio at time t. Each of these pairs is called a *strategy* (portfolio strategy).

We consider the problem of trading or choosing a strategy in a class of strategies that does not use future values of $(S(t), r(t))$. Some constraints will be imposed on current operations in the market, or in other words, on strategies.

Definition 2.4 *A pair $(\beta(\cdot), \gamma(\cdot))$ is said to be an admissible strategy if $\beta(t)$ and $\gamma(t)$ are random processes adapted to the filtration $\mathcal{F}_t$ and such that*

$$\mathbf{E} \int_0^T \left(B(t)^2 \beta(t)^2 + S(t)^2 \gamma(t)^2 \right) dt < +\infty. \tag{2.4}$$

Definition 2.5 *A pair $(\beta(\cdot), \gamma(\cdot))$ is said to be a self-financing strategy, if*

$$dX(t) = \beta(t)dB(t) + \gamma(t)dS(t). \tag{2.5}$$

Note that condition (2.4) ensures that the process $X(t)$ is well defined by equation (2.5) as an Ito process.

We allow negative $\beta(t)$ and $\gamma(t)$, meaning borrowing and short positions.

We shall consider admissible self-financing strategies only.

Example 2.6 A risk-free, "keep-only-bonds," strategy is such that the portfolio contains only the bonds, $\gamma(t) \equiv 0$, and the corresponding total wealth is $X(t) \equiv \beta_0 B(t) \equiv \exp\left(\int_0^t r(s)ds\right) X(0)$.

Example 2.7 A *buy-and-hold strategy* is a strategy when $\gamma(t) > 0$ does not depend on time. This strategy ensures a gain in a scenario where the stock price is increasing.

Remark 2.8 *In the literature, a definition of admissible strategies may include requirements that the risk is bounded. An example of this requirement is the following: there exists a constant C such that $X(t) \geq C$ for all t a.s. For simplicity, we do not require this.*

Remark 2.9 *The case of $r(t) \equiv 0$ corresponds to the market model with free borrowing.*

2.3 Discounted wealth and stock prices

For the trivial, risk-free, "keep-only-bonds" strategy, the portfolio contains only the bonds, $\gamma(t) \equiv 0$, and the corresponding total wealth is $X(t) \equiv \beta_0 B(t) \equiv$

$\exp\left(\int_0^t r(s)ds\right)X(0)$. Some loss is possible for a strategy that deals with risky assets. It is natural to estimate the loss and gain by comparing it with the results for the "keep-only-bonds" strategy.

In mathematical economics, a numéraire is a tradeable economic entity in terms of whose price the relative prices of all other tradeables are expressed. In our market model, we consider the bond with the price $B(t)$ as a numéraire.

Definition 2.10 *The process* $\widetilde{X}(t) \triangleq B(t)^{-1}X(t)$, $\widetilde{X}(0) = X(0)$, *is called the discounted wealth.*

Definition 2.11 *The process* $\widetilde{S}(t) \triangleq B(t)^{-1}S(t)$, $\widetilde{S}_0 = S_0$, *is called the discounted stock price with respect to the numéraire* $B(t)$.

Let $\widetilde{a}(t) \triangleq a(t) - r(t)$.

Proposition 2.12 $d\widetilde{S}(t) = \widetilde{S}(t)(\widetilde{a}(t)dt + \sigma(t)dw(t))$.

The proof is straightforward.

Theorem 2.13 *Property (2.5) of self-financing is equivalent to*

$$d\widetilde{X}(t) = \gamma(t)d\widetilde{S}(t), \tag{2.6}$$

i.e.,

$$\widetilde{X}(t) = X(0) + \int_0^t \gamma(s)d\widetilde{S}(s). \tag{2.7}$$

Proof of Theorem 2.13. Let $(\widetilde{X}(t), \gamma(t))$ be a process such that (2.6) holds. Then it suffices to prove that $X(t) \triangleq \exp\left(\int_0^t r(s)ds\right)\widetilde{X}(t)$ is the wealth corresponding to the self-financing strategy $(\beta(\cdot), \gamma(\cdot)$, where $\beta(t) = (X(t) - \gamma(t)S(t))B(t)^{-1}$.
We have

$$
\begin{aligned}
dX(t) &= \exp\left(\int_0^t r(s)ds\right)d\widetilde{X}(t) + r(t)X(t)dt \\
&= \exp\left(\int_0^t r(s)ds\right)\gamma(t)d\widetilde{S}(t) + r(t)X(t)dt \\
&= \exp\left(\int_0^t r(s)ds\right)\gamma(t)\widetilde{S}(t)[\widetilde{a}(t)dt + \sigma(t)dw(t)] + r(t)X(t)dt \\
&= \gamma(t)S(t)[\widetilde{a}(t)dt + \sigma(t)dw(t)] + r(t)[\gamma(t)S(t) + \beta(t)B(t)]dt \\
&= \gamma(t)S(t)[a(t)dt + \sigma(t)dw(t)] + r(t)\beta(t)B(t)dt \\
&= \gamma(t)dS(t) + \beta(t)dB(t).
\end{aligned}
$$

This completes the proof. $\square$

Thanks to Theorem 2.13, we can reduce many problems for the market with non-zero interest for borrowing to the simpler case of the market with zero interest rate (free borrowing). In particular, it makes the calculation of the wealth for a given strategy easier.

2.4 Risk-neutral measure

Remember that $\mathcal{F} = \mathcal{F}_T$, and $\mathcal{F}_t$ is the filtration generated by the process $(w(t), \eta(t))$, where $\eta(\cdot)$ is a process independent from $w(\cdot)$ that describes additional random factors presented in the model besides the driving Wiener process (see also Remark 2.2).

Definition 2.14 *Let* $\mathbf{P}_* : \mathcal{F} \to [0, 1]$ *be a probability measure such that the process* $\widetilde{S}(t)$ *is a martingale with respect to the filtration* $\mathcal{F}_t$ *for* $\mathbf{P}_*$. *Then* $\mathbf{P}_*$ *is said to be a risk-neutral probability measure for the bond-stock market (2.1),(2.2).*

A risk-neutral measure is also called a martingale measure. As usual, $\mathbf{E}_*$ denotes the corresponding expectation. In particular, $\mathbf{E}_*\{\widetilde{S}_\tau \mid \widetilde{S}(\cdot)|_{[0,t]}\} = \widetilde{S}(t)$ for all $\tau > t$.

Definition 2.15 *If a risk-neutral probability measure* $\mathbf{P}_*$ *is equivalent to the original measure* $\mathbf{P}$, *we call it an equivalent risk-neutral measure.*

Market price of risk

Further, remember that $\widetilde{a}(t) \triangleq a(t) - r(t)$. Let a process θ be a solution of the equation

$$\sigma(t)\theta(t) = \widetilde{a}(t). \tag{2.8}$$

This process θ is called the *market price of risk* process; this term comes from optimal portfolio selection theory. If the market is non-degenerate, i.e., $\sigma(t) \neq 0$, then

$$\theta(t) = \sigma(t)^{-1}\widetilde{a}(t) = \sigma(t)^{-1}[a(t) - r(t)] \quad \text{a.e.}$$

Up to the end of this chapter, we assume that the following condition is satisfied.

Condition 2.16 *The market price of risk process exists for a.e.* t, ω, *and*

$$\mathbf{E}\exp\left(\frac{1}{2}\int_0^T |\theta(t)|^2 dt\right) < +\infty.$$

Condition 2.16 is called *Novikov's condition*, similar to (1.14).

Clearly, this condition ensures that if $\sigma(t) = 0$ then $\widetilde{a}(t) = 0$ a.e., i.e. $a(t) = r(t)$ for a.e. t a.s.

It follows that if the market is non-degenerate (i.e., $|\sigma(t)| \geq \text{const} > 0$), and the process $(r(t), \sigma(t), a(t))$ is bounded, then Condition 2.16 is satisfied.

A measure $\mathbf{P}_*$ defined by the market price of risk

Let

$$\mathcal{Z} \triangleq \exp\left(-\int_0^T \theta(t)dw(t) - \frac{1}{2}\int_0^T |\theta(t)|^2 dt\right). \tag{2.9}$$

By Proposition 1.75, it follows that

(i) $\mathbf{E}\mathcal{Z} = 1$;

(ii) If the mapping $\mathbf{P}_* : \mathcal{F} \to \mathbf{R}$ is defined via the equation

$$\frac{d\mathbf{P}_*}{d\mathbf{P}} = \mathcal{Z}, \tag{2.10}$$

then $\mathbf{P}_*$ is a probability measure on $\mathcal{F}$ equivalent to the original measure $\mathbf{P}$.

Remember that (2.10) means that $\mathbf{P}_*(A) = \int_A \mathcal{Z}(\omega)\mathbf{P}(d\omega)$ for any $A \in \mathcal{F}$, and $\mathbf{E}_*\xi = \mathbf{E}\mathcal{Z}\xi$ for any integrable random variable ξ.

Application of the Girsanov theorem

Let

$$w_*(t) \stackrel{\triangle}{=} w(t) + \int_0^t \theta(s)ds.$$

Here θ is defined by (2.8).

Note that

$$\tilde{a}(t)dt + \sigma(t)dw(t) = \sigma(t)dw_*(t).$$

Hence

$$d\tilde{S}(t) = \tilde{S}(t)[\tilde{a}(t)dt + \sigma(t)dw(t)] = \tilde{S}(t)\sigma(t)dw_*(t).$$

$\mathbf{P}_*$ as an equivalent risk-neutral measure

Theorem 2.17 *Let Condition 2.16 be satisfied, and let the measure $\mathbf{P}_*$ be defined by equation (2.10). Then*

(i) $w_*(t)$ is a Wiener process under $\mathbf{P}_*$;

(ii) $\mathbf{P}_*$ is an equivalent risk-neutral measure.

Proof. Statement (i) follows from the Girsanov Theorem 1.78, as well as the statement that $\mathbf{P}_*$ is equivalent to $\mathbf{P}$. Let us prove the rest part of (ii).

We have that

$$d\tilde{S}(t) = \tilde{S}(t)[\tilde{a}(t)dt + \sigma(t)dw(t)] = \tilde{S}(t)\sigma(t)dw_*(t).$$

By Theorem 1.78, $w_*(t)$ is a Wiener process under $\mathbf{P}_*$. Then

$$
\begin{aligned}
\mathbf{E}_*\{\tilde{S}(t) \,|\, \mathcal{F}_s\} &= \mathbf{E}_*\left\{ S_0 + \int_0^t \tilde{S}(\tau)\sigma(\tau)dw_*(\tau) \,\Big|\, \mathcal{F}_s \right\} \\
&= S(0) + \mathbf{E}_*\left\{ \int_0^t \tilde{S}(\tau)\sigma(\tau)dw_*(\tau) \,\Big|\, \mathcal{F}_s \right\} \\
&= S_0 + \int_0^s \tilde{S}(\tau)\sigma(\tau)dw_*(\tau) \\
&= \tilde{S}(s).
\end{aligned}
$$

This completes the proof. $\square$

Theorem 2.18 *Let Condition 2.16 be satisfied, and let* $\mathbf{P}_*$ *be the equivalent risk-neutral measure defined in Theorem 1.78. For any admissible self-financing strategy, the corresponding discounted wealth* $\widetilde{X}(t)$ *is a martingale with respect to* $\mathcal{F}_t$ *under* $\mathbf{P}_*$.

Proof. We have that

$$d\widetilde{S}(t) = \widetilde{S}(t)\sigma(t)dw_*(t),$$
$$d\widetilde{X}(t) = \gamma(t)d\widetilde{S}(t) = \gamma(t)\widetilde{S}(t)\sigma(t)dw_*(t),$$

where $\gamma(t)$ is the number of shares. By **Girsanov's** theorem, $w_*(t)$ is a Wiener process under $\mathbf{P}_*$. Then

$$
\begin{aligned}
\mathbf{E}_*\{\widetilde{X}(t)\,|\,\mathcal{F}_s\} &= \mathbf{E}_*\left\{X_0 + \int_0^t \gamma(\tau)\widetilde{S}(\tau)\sigma(\tau)dw_*(\tau)\,\Big|\,\mathcal{F}_s\right\} \\
&= X_0 + \mathbf{E}_*\left\{\int_0^t \gamma(\tau)\widetilde{S}(\tau)\sigma(\tau)dw_*(\tau)\,\Big|\,\mathcal{F}_s\right\} \\
&= X_0 + \int_0^s \gamma(\tau)\widetilde{S}(\tau)\sigma(\tau)dw_*(\tau) \\
&= \widetilde{X}(s).
\end{aligned}
$$

This completes the proof. $\square$

2.5 Replicating strategies

Remember that $T > 0$ is given.

Let ψ be a random variable.

Definition 2.19 *Let the initial wealth* $X(0)$ *be given, and let a self-financing strategy* $(\beta(\cdot), \gamma(\cdot))$ *be such that* $X(T) = \psi$ *a.s. for the corresponding wealth. Then the claim* ψ *is called replicable (attainable, redundant), and the strategy is said to be a replicating strategy (with respect to this claim).*

Definition 2.20 *Let the initial wealth* $X(0)$ *be given, and let a self-financing strategy* $(\beta(\cdot), \gamma(\cdot))$ *be such that* $X(T) \geq \psi$ *a.s. for the corresponding wealth. Then the strategy is said to be a super-replicating strategy.*

Theorem 2.21 *Let Condition 2.16 be satisfied, and let* $\mathbf{P}_*$ *be the equivalent risk-neutral measure such as defined in Theorem 2.17. Let* ψ *be an* $\mathcal{F}_T$-measurable random variable such that $\mathbf{E}_*\psi^2 < +\infty$. *Let the initial wealth* $X(0)$ *and a self-financing strategy* $(\beta(\cdot), \gamma(\cdot))$ *be such that* $X(T) = \psi$ *a.s. for the corresponding wealth. Then*

$$X(0) = \mathbf{E}_* \exp\left(-\int_0^T r(s)ds\right)\psi.$$

Proof. Clearly, $X(T) = \psi$ iff $\widetilde{X}(T) = \exp\left(-\int_0^T r(s)ds\right)\psi$ a.s. We have then

$$
\begin{aligned}
\mathbf{E}_*\widetilde{X}(T) = \mathbf{E}_* \exp\left(-\int_0^T r(s)ds\right)\psi &= \mathbf{E}_*\left(X(0) + \int_0^T \gamma(t)d\widetilde{S}(t)\right) \\
&= X(0) + \mathbf{E}_* \int_0^T \gamma(t)d\widetilde{S}(t) \\
&= X(0) + \mathbf{E}_* \int_0^T \gamma(t)\widetilde{S}(t)\sigma(t)dw_*(t) \\
&= X(0).
\end{aligned}
$$

We have used here the fact that $w_*(t)$ is a Wiener process under $\mathbf{P}_*$, and $\int \cdot dw_*$ is an Ito integral under $\mathbf{P}_*$, so $\mathbf{E}_* \int \cdot dw_* = 0$. $\square$

The uniqueness of the replicating strategy

Theorem 2.22 *Let Condition 2.16 be satisfied, and let $\mathbf{P}_*$ be the equivalent risk-neutral measure defined in Theorem 2.17. Let ψ be an $\mathcal{F}_T$-measurable random variable, $\mathbf{E}_*\psi^2 < +\infty$. Let the initial wealth $X(0)$ and a self-financing strategy $(\beta(\cdot), \gamma(\cdot))$ be such that $X(T) = \psi$ a.s. for the corresponding wealth $X(t)$. Then the initial wealth $X(0)$ is uniquely defined. Moreover, the processes $X(t)$ and $\sigma(t)\gamma(t)$ are uniquely defined up to equivalency. If $\sigma(t) \neq 0$ for a.e. t, and the replicating strategy and the corresponding wealth process $X(t)$ are uniquely defined up to equivalency.*

2.6 Arbitrage possibilities and the arbitrage-free market

Similar to the case of a discrete time market, we define arbitrage as the possibility of a risk-free positive gain. The formal definition is the following.

Definition 2.23 *Let $T > 0$ be given. Let $(\beta(\cdot), \gamma(\cdot))$ be an admissible self-financing strategy, and let $\widetilde{X}(t)$ be the corresponding discounted wealth. If*

$$
\mathbf{P}(\widetilde{X}(T) \geq X(0)) = 1, \quad \mathbf{P}(\widetilde{X}(T) > X(0)) > 0, \tag{2.11}
$$

then this strategy is said to be an arbitrage strategy. If there exists an arbitrage strategy, then we say that the market model allows an arbitrage.

We are interested in models without arbitrage possibilities. If a model allows arbitrage, then it is usually not useful (despite the fact that arbitrage opportunities could exist occasionally in real-life market situations).

Example 2.24 *Let there exist t_1 and t_2 such that $0 \leq t_1 < t_2 \leq T$ and $\sigma(t) = 0$, $\widetilde{a}(t) \neq 0$ for $t \in (t_1, t_2)$ a.s. Then this market model allows arbitrage.*

Theorem 2.25 *Let a market model be such that Condition 2.16 is satisfied (in particular, this means that there exists an equivalent risk-neutral probability measure that is equivalent to the original measure* **P***). Then the market model does not allow arbitrage.*

Remark 2.26 The opposite statement to the above theorem "*absence of arbitrage implies the existence of an equivalent risk-neutral measure*" is also valid under some additional requirements on the strategies. The equivalence relation between the existence of equivalent risk-neutral measure and the absence of (certain types of) arbitrage is called the "*fundamental theorem of asset pricing*".

2.7 The case of a complete market

Let $\mathcal{F}_t^{S,r}$ be the filtration generated by the process $(S(t), r(t))$. (Note that $\mathcal{F}_t^{S,r} \subseteq \mathcal{F}_t$, and, for the general case, $\mathcal{F}_t$ is large than $\mathcal{F}_t^{S,r}$.)

In fact, any $\mathcal{F}_T^{S,r}$-measurable random variable ψ can be presented as $\psi = F(S(\cdot), B(\cdot))$ for a certain mapping $F(\cdot) : C(0,T) \times C(0,T) \to \mathbf{R}$ (see the related Theorem 1.25).

Definition 2.27 *A market model is said to be complete if any $\mathcal{F}_T^{S,r}$-measurable random claim ψ such that $\mathbf{E}_* |\psi|^2 < +\infty$ for some risk-neutral measure* **P**$_*$ *is replicable with some initial wealth.*

Theorem 2.28 *If a market model is complete and there exists an equivalent risk-neutral measure, then this measure is unique (as a measure on $\mathcal{F}_T^{S,r}$).*

Proof. Let $A \in \mathcal{F}_T^{S,r}$. By the assumption, the claim $\exp\left(\int_0^T r(t)dt\right) \mathbb{I}_A$ is replicable with some initial wealth $X_A(0)$ ($\mathbb{I}_A$ is the indicator function of A). By Theorem 2.22, this $X_A(0)$ is uniquely defined. By Theorem 2.21, $\mathbf{E}_* \mathbb{I}_A = X_A(0)$ for any risk-neutral measure **P**$_*$. Therefore, **P**$_*$ is uniquely defined on $\mathcal{F}_T^{S,r}$. $\square$

2.8 Completeness of the Black–Scholes model

The so-called Black–Scholes model [24] is such that the vector $(r(t), a(t), \sigma(t)) \equiv (r, a, \sigma)$ is non-random and constant, $\sigma \neq 0$. For this model, we assume that the filtration $\mathcal{F}_t$ is generated by $w(t)$ (or by $S(t)$, or by $\widetilde{S}(t)$).

Let w_* and $\mathbf{P}_*$ be defined as above, i.e., $d\mathbf{P}_*/d\mathbf{P} = \mathcal{Z}$. Remember that $w_*(t)$ is a Wiener process with respect to $\mathbf{P}_*$, and

$$d\widetilde{S}(t) = \widetilde{S}(t)\sigma dw_*(t),$$
$$dS(t) = S(t)[rdt + \sigma dw_*(t)]. \tag{2.12}$$

Theorem 2.29 *The Black–Scholes market is complete.*

Proof. Let $\psi \in \mathcal{L}_2(\Omega, \mathcal{F}_T, \mathbf{P}_*)$ be an arbitrary claim. By the martingale representation theorem (or by Theorem 1.72) applied to the probability space $(\Omega, \mathcal{F}_T, \mathbf{P}_*)$, it follows that there exists a process $f \in \mathcal{L}_{22}$ such that

$$e^{-rT}\psi = e^{-rT}\mathbf{E}_*\psi + \int_0^T f(t)dw_*(t.)$$

(We mean the space $\mathcal{L}_{22}$ defined with respect to the measure $\mathbf{P}_*$.) By (2.12), it follows that

$$e^{-rT}\psi = e^{-rT}\mathbf{E}_*\psi + \int_0^T \gamma(t)d\widetilde{S}(t), \quad \gamma(t) \triangleq f(t)\widetilde{S}(t)^{-1}\sigma^{-1}.$$

Hence the process $\widetilde{X}(t) \triangleq e^{-rT}\mathbf{E}_*\psi + \int_0^t \gamma(s)d\widetilde{S}(s)$ is the discounted wealth generated by a self-financing strategy such that the quantity of the stock portfolio is $\gamma(t)$. Since $f \in \mathcal{L}_{22}$, it is easy to see that this strategy is admissible. $\square$

Corollary 2.30 *The measure $\mathbf{P}_*$ is the only equivalent risk-neutral measure on $\mathcal{F}_T$.*

Theorem 2.29 does not explain how to calculate the replicating strategy and the corresponding initial wealth. The following theorem gives a method of calculation for an important special case.

Let $\widetilde{V}(x, s)$ be the solution of the Cauchy problem for the backward parabolic equation

$$\frac{\partial \widetilde{V}}{\partial s}(x, s) + \mathcal{A}\widetilde{V}(x, s) = -\varphi(x, s),$$
$$\widetilde{V}(x, T) = \Psi(x). \tag{2.13}$$

Here $x > 0$, $s \in [0, T]$,

$$\mathcal{A}v(x) \triangleq \frac{1}{2}\sigma^2 x^2 \frac{d^2 v}{dx^2}(x).$$

Note that the assumptions on Ψ and φ were not specified yet. Starting from now, we assume that they are such that problem (2.13) has a unique classical solution in the domain $\{(x, s)\} = (0, +\infty) \times [0, T]$. [1]

---

[1] At this point, note that the change of variable x for $y = \ln x$ makes this equation a non-degenerate parabolic equation in the domain $\{(y, s)\} = \mathbf{R} \times [0, T]$. This can help to see which conditions for φ and Ψ are sufficient.

Theorem 2.31 *Let functions* $\Psi : \mathbf{R} \to \mathbf{R}$ *and* $\varphi : \mathbf{R} \times [0, T] \to \mathbf{R}$ *be such that certain conditions on their smoothness and growth are satisfied (it suffices to require that* Ψ *and* φ *are continuous and bounded). Let*

$$\psi = e^{rT} \left[\Psi(\widetilde{S}(T)) + \int_0^T \varphi(\widetilde{S}(t), t) dt \right].$$

Then this claim is replicable with the initial wealth $\widetilde{V}(S(0), 0)$ *and with the stock quantity* $\gamma(t) = \frac{\partial \widetilde{V}}{\partial x}(\widetilde{S}(t), t)$*. The corresponding discounted wealth is*

$$\widetilde{X}(t) = \widetilde{V}(\widetilde{S}(t), t) + \int_0^t \varphi(\widetilde{S}(s), s) ds,$$

where $\widetilde{V}$ *is the solution of problem (2.13). In addition,*

$$\widetilde{X}(t) = \mathbf{E}_* \left\{ \Psi(\widetilde{S}(T)) + \int_0^T \varphi(\widetilde{S}(t), t) dt \, \Big| \, \mathcal{F}_t \right\} \quad \forall t \leq T.$$

Moreover,

$$\widetilde{V}(\widetilde{S}(t), t) = \mathbf{E}_* \left\{ \widetilde{V}(\widetilde{S}(\tau), \tau) + \int_t^\tau \varphi(\widetilde{S}(s), s) ds \, \Big| \, \mathcal{F}_t \right\} \quad \forall t, \tau : \ 0 \leq t \leq \tau \leq T.$$

Remark 2.32 *If the volatility process* $\sigma(t)$ *is non-random but time dependent and such that* $|\sigma(t)| \geq \text{const} > 0$ *and some regularity conditions are satisfied, then the market is also complete. The proof of Theorem 2.31 can be repeated for this case with* σ *replaced by time-dependent* $\sigma(t)$ *in the definition for* $\mathcal{A}$*.*

Example 2.33 *Let us find an initial wealth and a strategy that replicates the claim* $\psi = e^{rT} \Psi(\widetilde{S}(T)) = \widetilde{S}(T)^{-1}$ *under the assumptions of Theorem 2.31. We need to find* $\widetilde{V}$ *for* $\Psi(x) = e^{-rT} x^{-1}$ *and* $\varphi \equiv 0$*. In that case,* $\widetilde{V}$ *can be found explicitly:* $\widetilde{V}(x, t) = e^{-rT} e^{\sigma^2(T-t)} x^{-1}$ *(verify that this* $\widetilde{V}$ *is the solution of (2.13)). Then* $\frac{\partial \widetilde{V}}{\partial x}(x, t) = -e^{-rT} e^{\sigma(T-t)} x^{-2}$*, and*

$$\gamma(t) = \frac{\partial \widetilde{V}}{\partial x}(\widetilde{S}(t), t) = -e^{-rT} e^{\sigma^2(T-t)} \widetilde{S}(t)^{-2}.$$

The initial wealth is $X(0) = \widetilde{V}(S(0), 0) = e^{-rT} e^{\sigma^2 T} S(0)^{-1}$*.*

2.9 Option pricing

2.9.1 Options and their prices

Let us repeat the definitions of the most generic options: European call options and European put options. Let terminal time $T > 0$ be given.

A European call option contract traded (contracted and paid) in $t = 0$ is such that the buyer of the contract has the right (not the obligation) to buy one unit of the underlying asset (from the issuer of the option) in $T > 0$ at the strike price K. The market price of option payoff (in T) is $\max(0, S(T) - K)$, where $S(T)$ is the asset price, and K is the strike price.

A European put option contract traded in $t = 0$ gives to the buyer of the contract the right to sell one unit of the underlying asset in $T > 0$ at the strike price K. The market price of option payoff (in T) is $\max(0, K - S(T))$, where $S(T)$ is the asset price, and K is the strike price.

In a more general case, for a given function $F(x) \geq 0$, the European option with payoff $F(S(T))$ can be defined as a contract traded in $t = 0$ that such that the buyer of the contract receives an amount of money equal to $F(S(T))$ at time $T > 0$.

In the most general setting, a non-negative function $F : C(0, T) \times C(0, T) \to \mathbf{R}$ is given. Let $\psi = F(S(\cdot), B(\cdot))$. The European option with payoff $F(S(\cdot), B(\cdot))$ is a contract traded in $t = 0$ such that the buyer of the contract receives an amount of money equal to ψ at time $T > 0$.

The following special cases are covered by this setting:

- (Vanilla) Europian call option: $\psi = (S(T) - K)^+$, where $K > 0$ is the strike price;

- (Vanilla) Europian put option: $\psi = (K - S(T))^+$;

- Share-or-nothing European call option: $\psi = S(T)\mathbb{I}_{\{S(T)>K\}}$;

- An Asian option: $\psi = f_1\left(\int_0^T f_2(S(t))dt\right)$, where f_i are given functions.

There are many other examples, including exotic options of a European type.

Another important class of options is the class of so-called American options. For these options, the option holder can exercise the option at any time $\tau \in [0, T]$ by his/her choice. These options will not be considered in this book.

The concept of the "fair price" of options, or derivatives plays a key role in mathematical finance.

The following pricing rule was suggested by Black and Scholes (1973) [24].

Definition 2.34 *The fair price of an option at time $t = 0$ is the minimal initial wealth such that, for any market situation, it can be raised with some acceptable strategies to a wealth such that the option obligation can be fulfilled.*

In fact, we assume that Definition 2.34 is valid for options of all types. We rewrite it now more formally for European options.

Definition 2.35 *The fair price at time $t = 0$ of the European option with payoff ψ is the minimal wealth $X(0)$ such that there exists an admissible self-financing strategy $(\beta(\cdot), \gamma(\cdot))$ such that*

$$X(T) \geq \psi \quad a.s.$$

for the corresponding wealth $X(\cdot)$.

2.9.2 Option pricing for a complete market

For a complete market, Definition 2.35 leads to replication.

Theorem 2.36 *Let the market be complete, and let a claim ψ be such that $\mathbf{E}_* \psi^2 < +\infty$, and $\psi = F(S(\cdot))$, where $F(\cdot) : C(0,T) \to \mathbf{R}$ is a function. Then the fair price (from Definition 2.35) of the European option with payoff ψ at time T is*

$$e^{-rT} \mathbf{E}_* \psi, \qquad (2.14)$$

and it is the initial wealth $X(0)$ such that there exists an admissible self-financing strategy $(\beta(\cdot), \gamma(\cdot))$ such that

$$X(T) = \psi \quad a.s.$$

for the corresponding wealth.

Proof. From the completeness of the market, it follows that the replicating strategy exists and the corresponding initial wealth is equal to $e^{-rT} \mathbf{E}_* \psi$. Let us show that it is the fair price. Let $X'(0) < c_F$ be another initial wealth, then $\mathbf{E}_* \widetilde{X}'(T) = X'(0) < c_F = e^{-rT} \mathbf{E}_* \psi$ for the corresponding discounted wealth $\widetilde{X}'(\cdot)$. Hence it cannot be true that $e^{rT} \widetilde{X}'(T) \geq \psi$ a.s. $\square$

We shall refer to the price (2.14) as the *Black–Scholes price* of a European option for the case of complete market.

Corollary 2.37 *Let the assumptions of Theorem 2.31 be satisfied. Let functions $\Psi : \mathbf{R} \to \mathbf{R}$ and $\varphi : \mathbf{R} \times [0,T] \to \mathbf{R}$ be such that certain conditions on their smoothness and growth are satisfied such that problem (2.13) has a unique classical solution in $\{x, s\} = (0, +\infty) \times [0, T]$. Let an option claim ψ be such that*

$$\psi = e^{rT} \left[\Psi(\widetilde{S}(T)) + \int_0^T \varphi(\widetilde{S}(t), t) dt \right].$$

Then the fair price (Black–Scholes price) of the option at time $t = 0$ is $\widetilde{V}(S(0), 0)$, where $\widetilde{V}$ is the solution of problem (2.13).

Proof. By Theorem 2.31, $X(0) = \widetilde{V}(S(0), 0)$ is the initial wealth for the replicating strategy. $\square$

Corollary 2.38 *The Black–Scholes price does not depend on the appreciation rate $a(\cdot)$.*

Proof. By Theorems 2.36 and 2.31, the fair price is $\widetilde{V}(S(0), 0)$, and the equation for $\widetilde{V}$ does not include a. Then the proof follows. Another way to prove this corollary is to notice that $\widetilde{S}(t) = S(0) \exp(\sigma w_*(t) - \frac{1}{2}\sigma^2 t)$, i.e., the distribution of $\widetilde{S}(T)$ under $\mathbf{P}_*$ does not depend on a. $\square$

Corollary 2.39 *(Put–call parity). Let $K > 0$ be given. Let H_p be the Black–Scholes price of the call option with payoff $F_c(S(T)) = (S(T) - K)^+$, and let H_c be the Black–Scholes price of the put option with payoff $F_p(S(T)) = (K - S(T))^+$. Then $H_c - H_p = S(0) - e^{-rT}K$.*

Proof. It suffices to note that $(S(T) - K)^+ - (K - S(T))^+ = S(T) - K$, and $H_c - H_p = e^{-rT}\mathbf{E}_*(S(T) - K) = S(0) - e^{-rT}K$. □

Theorem 2.40 *Consider the Black–Scholes model given volatility σ and the risk-free rate r. Let $\psi = e^{rT}\Psi(\widetilde{S}(T))$, where $\Psi : \mathbf{R} \to \mathbf{R}$ is a function. Then the fair price of the option at time $t = 0$ is*

$$e^{-rT}\mathbf{E}_* e^{rT}\Psi(\widetilde{S}(T)) = \mathbf{E}_*\Psi(S(0)e^{\eta_0}),$$

where

$$\eta_0 \overset{\Delta}{=} -\frac{\sigma^2 T}{2} + \sigma w_*(T).$$

We have then $\eta_0 \sim N(-\sigma^2 T/2, \sigma^2 T)$ under $\mathbf{P}_*$. In other words, the price is

$$\mathbf{E}_*\Psi(\widetilde{S}(T)) = \frac{1}{\sqrt{2\pi}} \int_{-\infty}^{+\infty} e^{-\frac{x^2}{2}}\overline{\Psi}\left(S_0 \exp\left[-\frac{\sigma^2 T}{2} + \sigma\sqrt{T}x\right]\right) dx. \quad (2.15)$$

Proof. It suffices to apply the previous result keeping in mind that $\widetilde{S}(T) = S(0)e^{\eta_0}$, $S(T) = e^{rT}\widetilde{S}(T)$. □

2.9.3 Black–Scholes formula

We saw already that the fair option price (Black–Scholes price) can be calculated explicitly for some cases. The corresponding explicit formula for the price of European put and call options is called the *Black–Scholes* formula.

Let $K > 0$, $\sigma > 0$, $r \geq 0$, and $T > 0$ be given. We shall consider two types of options: call and put, with payoff function ψ, where $\psi = (S(T) - K)^+$ or $\psi = (K - S(T))^+$, respectively. Here K is the strike price.

Let $H_{BS,c}(x, K, \sigma, T, r)$ and $H_{BS,p}(x, K, \sigma, T, r)$ denote the fair prices at time $t = 0$ for call and put options with the payoff functions $F(S(T))$ described above given (K, σ, T, r) and under the assumption that $S(0) = x$. Then

$$H_{BS,c}(x, K, \sigma, T, r) = x\Phi(d_+) - Ke^{-rT}\Phi(d_-),$$
$$H_{BS,p}(x, K, \sigma, T, r) = H_{BS,c}(x, K, \sigma, T, r) - x + Ke^{-rT}, \quad (2.16)$$

where

$$\Phi(x) \overset{\Delta}{=} \frac{1}{\sqrt{2\pi}} \int_{-\infty}^{x} e^{-\frac{s^2}{2}} ds,$$

and where

$$d_+ \overset{\Delta}{=} \frac{\ln(x/K) + Tr}{\sigma\sqrt{T}} + \frac{\sigma\sqrt{T}}{2},$$
$$d_- \overset{\Delta}{=} d_+ - \sigma\sqrt{T}. \quad (2.17)$$

It is the celebrated Black–Scholes formula. Note that the formula for put follows from the formula for call from the put-call parity (Corollary 2.39).

Black–Scholes parabolic equation

We have $V(t,x) \triangleq e^{-r(T-t)} \mathbf{E}_* \{F(S(T)) \mid S(t) = x\} = e^{rt} \widetilde{V}(e^{-rt}x, t)$, where $\hat{V}$ is the solution of (2.13) with $\varphi \equiv 0$. It follows that $e^{-rt}V(e^{rt}x, t) = \widetilde{V}(x, t)$. With this change of the variables, parabolic equation (2.13) is converted to the equation

$$\frac{\partial V}{\partial t}(t,x) + \frac{\sigma^2 x^2}{2} \frac{\partial^2 V}{\partial x^2}(t,x) = r\left[V(t,x) - x\frac{\partial V}{\partial x}(t,x)\right],$$
$$V(T,x) = F(x). \tag{2.18}$$

This is the so-called Black–Scholes parabolic equation.

2.10 Pricing for an incomplete market

Typically, an equivalent risk-neutral measure in not unique in the case of random volatility (even if it is constant in time). If an equivalent risk-neutral measure in not unique, then, by Theorem 2.28, the market cannot be complete, i.e., there are claims ψ that cannot be replicated.

In this section, we assume that r is non-random and constant. Let $\mathcal{F}_t^S$ be the filtration generated by the process $S(t)$. (For this case of non-random r, we have that $\mathcal{F}_t^S \equiv \mathcal{F}_t^{S,r}$.) For the general case, the filtration $\mathcal{F}_t$ generated by the process $(w(t), \eta(t))$ is larger than $\mathcal{F}_t^S$.

An example of an incomplete market with $a \equiv r$

Let $a(t) \equiv r(t)$, and let $\sigma = \sigma(t, \eta)$, where η is a random process (or a random vector, or a random variable), independent from the driving Wiener process $w(t)$ (for instance, η may represent another Wiener process). Clearly, any original probability measure $\mathbf{P} = \mathbf{P}_\eta$ is a risk-neutral measure (note that $\mathcal{Z} \equiv 1$ for any η). Any probability measure is defined by the pair (w, η), therefore it depends on the choice of η. In other words, different η may generate different risk-neutral measures. Clearly, it can happen that two of these different measures are equivalent (it suffices to take two $\eta = \eta_i$, $i = 1, 2$, such that their probability distributions are equivalent, i.e., have the same sets of zero probability). Therefore, an equivalent risk-neutral measure depends on the choice of η, and it may not be unique. It is different from the case of non-random σ.

Mean-variance hedging

Similar to the case of the discrete time market, Definition 2.35 leads to super-replication for incomplete markets. Clearly, it is not always meaningful. Therefore, there is another popular approach for an incomplete market.

Definition 2.41 *(Mean-variance hedging). The fair price of the option is the initial wealth $X(0)$ such that $\mathbf{E}|X(T) - \psi|^2$ is minimal over all admissible self-financing strategies.*

In many cases, this definition leads to the option price $e^{-rT}\mathbf{E}_*\psi$, where $\mathbf{E}_*$ is the expectation of a risk-neutral equivalent measure which needs to be chosen in some optimal way, since this measure is not unique to an incomplete market. This measure needs to be found via the solution to an optimization problem. In fact, this method is the latest big step in the development of modern pricing theory. It requires some additional non-trivial analysis outside of our course.

Completion of the market

Sometimes it is possible to make an incomplete market model complete by adding new assets. For instance, if $\sigma(t)$ is random and evolves as the solution of an Ito equation driven by a new Wiener process $W(t)$, then the market can be made complete by allowing trading of any option on this stock (say, a European call with given strike price). All other options can be replicated via portfolio strategies that include the stock, the option, and the bond.

   A similar approach can be used for the case of random r. Remember that, in our generic setting, we called the risk free investment a *bond*, and it was considered to be a risk-free investment. In reality, there are many different bonds (or fixed income securities). In fact, they are risky assets, similar to stocks (it is discussed in the next section). If r is random, then the market can be made complete by including additional fixed income securities.

2.11   A multi-stock market model

Similarly, we can consider a multi-stock market model, when there are N stocks with $N > 1$. Let $\{S_i(t)\}$ be the vector of the stock prices. The most common continuous time model for the prices is based again on Ito equations which now can be written as

$$dS_i(t) = S_i(t)[a_i(t)dt + \sum_{j=1}^{n} \sigma_{ij}(t)dw_j(t)], \quad i = 1, \ldots, N.$$

Here $w(t) = (w_1(t), \ldots, w_n(t))$ is a vector Wiener process; i.e., its components are scalar Wiener processes. Further, $a(t) = \{a_i(t)\}$ is the vector of the appreciation rates, and $\sigma(t) = \{\sigma_{ij}(t)\}$ is the volatility matrix.

We assume that the components of $w(t)$ are independent. The equation for the stock prices may be rewritten in the vector form:

$$dS(t) = \mathbf{S}(t)[a(t)dt + \sigma(t)dw(t)],$$

where $S(t) = (S_1(t), \ldots, S_N(t))$ is a vector with values in $\mathbf{R}^N$, $\mathbf{S}(t)$ is a diagonal matrix in $\mathbf{R}^{N \times N}$ with the main diagonal $(S_1(t), \ldots, S_N(t))$.

Similar to the case of a single stock market, we assume there is also the risk-free bond or bank account with price $B(t)$ such as described in Section 2.2. In particular, we assume that (2.2) holds, where $r(t)$ is a process of risk-free interest rates that is adapted with respect to the filtration generated by $(w(t), \eta(t))$, where $\eta(t)$ is some random process independent from $w(\cdot)$.

The strategy (portfolio strategy) is a process $(\beta(t), \gamma(t))$ with values in $\mathbf{R} \times \mathbf{R}^N$, $\gamma(t) = (\gamma_1(t), \ldots, \gamma_N(t))$, where $\gamma_i(t)$ is the quantity of the ith stock, and $\beta(t)$ is the quantity of the bond. The total wealth is $X(t) = \beta(t)B(t) + \sum_i \gamma_i(t)S_i(t)$. A strategy $(\beta(\cdot), \gamma(\cdot))$ is said to be *self-financing* if there is no income from or outflow to external sources. In that case,

$$dX(t) = \sum_{i=1}^{N} \gamma_i(t)dS_i(t) + \beta(t)dB(t).$$

To ensure that $S_i(t)$ and $X(t)$ are well defined as Ito processes, some restrictions on measurability and integrability must be imposed for the processes $a, \sigma, \gamma,$ and β.

It can be seen that

$$\beta(t) = \frac{X(t) - \sum_{i=1}^{N} \gamma_i(t)S_i(t)}{B(t)}.$$

Let $\widetilde{S}_i(t) = \{\widetilde{S}_i(t)\} \triangleq B(t)^{-1}S_i(t)$ be the discounted stock price. Similarly to Theorem 2.13, it can be shown that

$$d\widetilde{X}(t) = \sum_{i=1}^{N} \gamma_i(t)d\widetilde{S}_i(t),$$

where $\widetilde{X}(t) \triangleq B(t)^{-1}X(t)$ is the discounted wealth.

Then absence of arbitrage for this model can be described loosely as the condition where a risk-free gain cannot be achieved with a self-financing strategy.

The following example shows that absence of arbitrage for single stock markets defined for isolated stocks does not guarantee that the corresponding multi-stock market with the same stocks is arbitrage free.

Example 2.42 *Let $N = 2$, $n = 1$, and let*

$$d\widetilde{S}_i(t) = \widetilde{S}_i(t)[\widetilde{a}_i(t)dt + \sigma_i(t)dw(t)],$$

where $\widetilde{a}_i(t)$ and $\sigma_i(t)$ are some pathwise continuous processes, $\sigma_2(t) \geq \text{const} > 0$. Let $\psi(t) \triangleq \sigma_1(t)\widetilde{S}_1(t)[\widetilde{S}_2(t)\sigma_2(t)]^{-1}$ and $I(t) \triangleq \mathbb{I}_{\{a_1(t) > \psi(t)a_2(t)\}}$,

$$\gamma_1(t) \equiv I(t), \quad \gamma_2(t) \equiv -I(t)\psi(t).$$

Then

$$d\widetilde{X}(t) = \gamma_1(t)d\widetilde{S}_1(t) + \gamma_1(t)d\widetilde{S}_2(t) = I(t)[d\widetilde{S}_1(t) - \psi(t)d\widetilde{S}_2(t)]$$
$$= I(t)[\widetilde{a}_1(t) - \psi(t)\widetilde{a}_2(t)]dt.$$

Hence

$$\widetilde{X}(T) = X(0) + \int_0^T (\widetilde{a}_1(t) - \psi(t)\widetilde{a}_2(t))^+ dt.$$

Clearly, this two stock market model allows arbitrage for some $\widetilde{a}_i(\cdot)$. □

In the last example, the model was such that $N > n$. However, it is possible that $N > n$ and the market is still arbitrage free.

Similar to the case when $n = N = 1$, it can be shown that the market is arbitrage free if there exists an equivalent risk-neutral measure such that the discounted stock price vector $\widetilde{S}(t)$ is a martingale.

Let us show that there is no arbitrage if there exists a process $\theta(t)$ with values in $\mathbf{R}^n$ such that

$$\sigma(t)\theta(t) = \widetilde{a}(t) \tag{2.19}$$

and such that some conditions of integrability of θ are satisfied. (These conditions are always satisfied if the process $\theta(t)$ is bounded.) This process $\theta(t)$ is called *the market price of risk process*. Here $\widetilde{a}(t) \stackrel{\Delta}{=} \widetilde{a}(t) - r(t)\mathbf{1}$, where $\mathbf{1} = (1, \ldots, 1)^\top \in \mathbf{R}^N$.

Let us show that the existence of the process $\theta(\cdot)$ implies existence of an equivalent risk-neutral measure. It suffices to show that the measure $\mathbf{P}_*$ defined in Girsanov Theorem 1.78 for this θ is an equivalent risk-neutral measure. Set

$$w_*(t) \stackrel{\Delta}{=} w(t) + \int_0^t \theta(s)ds.$$

By Girsanov Theorem 1.78, $w_*(t)$ is a Wiener process under $\mathbf{P}_*$. Clearly,

$$\sigma(t)dw_*(t) = \widetilde{a}(t)dt + \sigma(t)dw(t).$$

In addition, $d\widetilde{S}(t) = \widetilde{\mathbf{S}}(t)[\widetilde{a}(t)dt + \sigma(t)dw(t)]$, where $\widetilde{\mathbf{S}}(t)$ is a diagonal matrix with the main diagonal $(\widetilde{S}_1(t), \ldots, \widetilde{S}_N(t))$. Hence

$$d\widetilde{S}(t) = \widetilde{\mathbf{S}}(t)\sigma(t)dw_*(t).$$

It follows that $\widetilde{S}(t)$ is a martingale under $\mathbf{P}_*$.

For instance, if $n = N$ and the matrix σ is non-degenerate, then $\theta(t) = \sigma(t)^{-1}\widetilde{a}(t)$. If this process is bounded, then the market is arbitrage free.

3

Some special market models

In this chapter, we consider special models that are not covered by the assumptions of Chapter 2. For example, this will include models with an unbounded appreciation rate and models with random bond prices described by Ito's equations.

3.1 Mean-reverting market model

In this section, we consider a so-called mean-reverting single stock market model with constant risk-free rate $r > 0$ model with stock price $S(t) = e^{rt}\widetilde{S}(t)$, where the discounted stock price $\widetilde{S}(t)$ evolves as

$$
\begin{aligned}
\widetilde{S}(t) &= s_0 e^{\widetilde{R}(t)}, \\
d\widetilde{R}(t) &= (\alpha - \lambda \widetilde{R}(t))dt + \sigma dw(t),
\end{aligned}
\tag{3.1}
$$

and where $\sigma > 0$, α, $\lambda > 0$ are deterministic, $s_0 = S(0)e^{-\widetilde{R}(0)}$. We assume that $\widetilde{R}(0)$ is a random number independent from $\{w(t)\}_{t>0}$, and $\mathbf{E}\widetilde{R}(0)^2 < +\infty$. (If $\alpha = 0$, then $\widetilde{R}(t)$ is an Ornstein–Uhlenbek process.)

3.1.1 Basic properties of a mean-reverting model

From Itô's formula, we obtain

$$
d\widetilde{S}(t) = \widetilde{S}(t)[\alpha - \lambda \hat{R}(t) + \sigma^2/2]dt + \sigma S(t)dw(t).
$$

Clearly, this is a modification of our model from Chapter 2 where

$$
\widetilde{a}(t) = \alpha - \lambda \widetilde{R}(t) + \frac{\sigma^2}{2}, \quad \sigma(t) = \sigma.
$$

It can be noted that this $\widetilde{a}(t)$ is not a bounded process and therefore it does not satisfy conditions imposed in Chapter 2.

Let $\widetilde{w}(\cdot)$ be a standard Wiener process independent from $w(\cdot)$. Set

$$
\widetilde{R}_0(t) = \int_{-\infty}^{t} e^{-\lambda(t-s)}\alpha ds + \int_{-\infty}^{t} e^{-\lambda(t-s)}\sigma dw(s) = \frac{\alpha}{\lambda} + \int_{-\infty}^{t} e^{-\lambda(t-s)}\sigma dw(s).
$$

We use here a small modification of the stochastic integral to define $\int_{-\infty}^{0} e^{-\lambda(t-s)}\sigma dw(s)$. More precisely, we use the following definition: $\int_{-\infty}^{0} e^{-\lambda(t-s)}\sigma dw(s) \overset{\Delta}{=} \int_{0}^{\infty} e^{-\lambda(t+s)}\sigma d\widetilde{w}(s)$. (Normally, a Wiener process is defined on time interval $[t_0, +\infty)$, where $t_0 \in \mathbf{R}$ is initial time.)

Note that the results below are valid for any choice of the process $\widetilde{w}(\cdot)$.

Clearly, the process $\widetilde{R}_0(t)$ is Gaussian, and $\widetilde{R}(t)$ is conditionally Gaussian given $\widetilde{R}(0)$.

The following proposition is well known.

Proposition 3.1 $\mathbf{E}|R(t) - R_0(t)|^2 \to 0$ *and* $\widetilde{R}(t) \to \widetilde{R}_0(t)$ *a.s. as* $t \to +\infty$.

For this mean-reverting market model, arbitrage opportunities and other speculative opportunities are studied below.

3.1.2 Absence of arbitrage and the Novikov condition

We say that the Novikov condition is satisfied for a time interval $[0, T]$ if

$$\mathbf{E} \exp \frac{1}{2} \int_0^T \widetilde{a}(t)^2 \sigma(t)^{-2} dt < +\infty. \tag{3.2}$$

(See Condition 2.16.)

It is well known that if the Novikov condition is satisfied, then $\mathbf{E}\mathcal{Z} = 1$, where

$$\mathcal{Z} = \mathcal{Z}(T) \overset{\Delta}{=} \exp\left(-\int_0^T \widetilde{a}(t)\sigma(t)^{-1}dw(t) - \frac{1}{2}\int_0^T \widetilde{a}(t)^2\sigma(t)^{-2}dt\right). \tag{3.3}$$

In that case, one can define the (equivalent) probability measure $\mathbf{P}_{*,T}$ by $d\mathbf{P}_{*,T}/d\mathbf{P} = \mathcal{Z}(T)$. By Girsanov's theorem, it follows that $w_*(t) \overset{\Delta}{=} w(t) + \int_0^t \sigma^{-1}\widetilde{a}(s)ds$ is a Wiener process under $\mathbf{P}_{*,T}$ for $t \in [0, T]$.

Furthermore, if the Novikov condition is satisfied for some $T > 0$. Then the market model does not allow arbitrage for the time interval $[0, T]$.

Theorem 3.2 *Let* $\widetilde{R}(0)$ *be non-random. Then, for any* $\kappa \in \mathbf{R}$, $T > 0$,

$$\mathbf{E} \exp\left(\frac{1}{2\sigma^2}\int_0^T [\kappa - \lambda\widetilde{R}(t)]^2 dt\right) < +\infty. \tag{3.4}$$

In particular, the Novikov condition (3.2) holds for any $T > 0$ *for the mean-reverting stock price model with* $\widetilde{a}(t) = \alpha - \lambda\widetilde{R}(t) + \sigma^2/2$.

As is known, $\mathbf{E}\exp(T\xi^2) = +\infty$ for any Gaussian random variable ξ and any $T > 1/(2\mathrm{Var}\,\xi)$. For the mean-reverting model, $\widetilde{a}(t)$ is Gaussian, and $\widetilde{a}(t) \to$ const $- \lambda\widetilde{R}_0(t)$ as $t \to +\infty$, where $\widetilde{R}_0(t)$ is a stationary Gaussian process. Therefore, the fact that the Novikov condition holds for the mean-reverting model for large T is non-trivial, and it is even counterintuitive. The proof uses certain properties of the mean-reverting process; in particular, constant in time Gaussian processes are excluded.

Corollary 3.3 *For any $T > 0$, there exists an equivalent probability measure $\mathbf{P}_* = \mathbf{P}_{*,T}$ such that the process $\widetilde{R}(t)$ is a martingale under $\mathbf{P}_*$ in $t \in [0,T]$ with respect to the filtration $\mathcal{F}_t$. If $\widetilde{R}(0) = 0$, then the process $\widetilde{R}(t)/\sigma$ is a Wiener process under $\mathbf{P}_*$ in $t \in [0,T]$.*

It will be useful to consider the log-normal processes.

Corollary 3.4 *For any $T > 0$, there exists an equivalent probability measure $\hat{\mathbf{P}}_* = \hat{\mathbf{P}}_{*,T}$ such that the process $\widetilde{S}(t) \triangleq s_0 e^{\widetilde{R}(t)}$ is a martingale under $\hat{\mathbf{P}}_*$ in $t \in [0,T]$ with respect to the filtration $\mathcal{F}_t$.*

Remark 3.5 *Corollary 3.3 and Corollary 3.4 hold if $\widetilde{R}(0)$ is random, independent from $\{w(t)\}_{t \geq 0}$, and such that $\mathbf{E}\widetilde{R}(0)^2 < +\infty$.*

Existence of an equivalent martingale measure is a sufficient condition of absence of arbitrage for a given finite time interval $[0,T]$.

Corollary 3.6 *The mean-reverting model does not allow arbitrage for the time interval $[0,T]$ for any $T > 0$.*

Non-robustness of the Novikov condition for large time intervals

Lemma 3.7 *Let $\widetilde{R}(0)$ be such as described in Remark 3.5. Let $\alpha = 0$. Then, for any $\varepsilon > 0$, there exists $T > 0$ such that*

$$\mathbf{E} \exp\left(\left[\frac{\lambda^2}{2\sigma^2} + \varepsilon \right] \int_0^T \widetilde{R}(t)^2 dt \right) = \infty. \tag{3.5}$$

Corollary 3.8 *Let the stock price evolution be described by equation (3.1) with $(a(t), \sigma(t))$ such that $\widetilde{a}(t) \equiv \lambda_1 \widetilde{R}(t)$, $\sigma(t) \equiv \sigma$, where $\lambda_1, \sigma \in \mathbf{R}$ are given, and where $\widetilde{R}(t)$ is defined by (3.1) with the same σ, with $\alpha = 0$, and with some $\lambda > 0$. It follows from the results above that if $|\lambda_1| \leq \lambda$, then the Novikov condition holds for any finite time interval $[0,T]$. If $|\lambda_1| > \lambda$, then there exists time $T > 0$ such that the Novikov condition does not hold for time interval $[0,T]$. (Note that the case when $\lambda_1 = \lambda$ corresponds to the mean reverting model.)*

Lemma 3.7 and Corollary 3.8 mean that the mean-reverting model is on the "edge" of the area where the Novikov condition holds for all time intervals.

3.1.3 Proofs

Proof of Proposition 3.1. We have that

$$d\widetilde{R}_0(t) = (\alpha - \lambda\widetilde{R}_0(t))dt + \sigma dw(t), \tag{3.6}$$

$$\widetilde{R}_0(0) = \frac{\alpha}{\lambda} + \int_{-\infty}^0 e^{\lambda s}\sigma dw(s) = \frac{\alpha}{\lambda} + \int_0^\infty e^{-\lambda s}\sigma d\widetilde{w}(s), \tag{3.7}$$

and $Y(t) \triangleq \widetilde{R}_0(t) - \widetilde{R}(t)$ satisfies

$$dY(t) = -\lambda Y(t)dt, \quad Y(0) = \widetilde{R}_0(0) - \widetilde{R}(0), \tag{3.8}$$

i.e., $Y(t) = \widetilde{R}_0(t) - \widetilde{R}(t) = e^{-\lambda t}[\widetilde{R}_0(0) - \widetilde{R}(0)]$. Clearly, this process converges to zero in mean square and with probability 1. $\square$

Proof of Theorem 3.2. Let $e_m(\cdot) : \mathbf{R}^n \to [0, 1]$ be continuous functions such that $e_m(x) = 1$ if $|x| \leq m$, $e_m(x) = 0$ if $|x| > m$, and $e_m(x) \leq e_{m+1}(x)$ for all x, $m = 1, 2, 3, \dots$. Let $u_m(x, t, T)$ be the solution of the Cauchy problem for the following backward parabolic equation

$$\frac{\partial u_m}{\partial t}(x, t, T) + (\alpha - \lambda x)\frac{\partial u_m}{\partial x}(x, t, T) + \frac{\sigma^2}{2}\frac{\partial^2 u_m}{\partial x^2}(x, t, T)$$
$$+ \frac{1}{2\sigma^2}(\kappa - \lambda x)^2 e_m(x)u_m(x, t, T) = 0, \quad t < T, \ x \in \mathbf{R},$$
$$u_m(x, T, T) = 1. \tag{3.9}$$

Clearly, (3.9) is the Kolmogorov's equation for the process $R(t)$, and

$$u_m(x, t, T) = \mathbf{E}\left\{\exp\frac{1}{2\sigma^2}\int_t^T (\kappa - \lambda\widetilde{R}(s))^2 e_m(\widetilde{R}(s))ds \,\Big|\, \widetilde{R}(t) = x\right\}, \quad t < T.$$

Let $\widetilde{b}(t) \triangleq \kappa - \lambda\widetilde{R}(s)$,

$$\zeta_m \triangleq \exp\frac{1}{2\sigma^2}\int_0^T \widetilde{b}(s)^2 e_m(\widetilde{R}(s))ds, \quad \zeta \triangleq \exp\frac{1}{2\sigma^2}\int_0^T \widetilde{b}(s)^2 ds.$$

By the definitions, it follows that

$$u_m(\widetilde{R}(0), 0, T) = \mathbf{E}\zeta_m.$$

Clearly, $\zeta_m \to \zeta$ as $m \to +\infty$ a.s., and the convergence is monotonic, i.e., $\zeta_m > 0$ is non-decreasing in m a.s. It follows that the function $u_m(\widetilde{R}(0), 0, T)$ is non-negative, and it is monotonic and non-decreasing in m for any T. In addition, $\mathbf{E}\zeta_m \to \mathbf{E}\zeta$ as $m \to +\infty$ (even if $\mathbf{E}\zeta = +\infty$), i.e., $u_m(\widetilde{R}(0), 0, T) = \mathbf{E}\zeta_m \to \mathbf{E}\zeta$.

To prove the theorem, it suffices to show that

$$\sup_{m>0} u_m(x, 0, T) < +\infty \quad \forall T > 0, \forall x \in \mathbf{R}. \tag{3.10}$$

Let us prove (3.10). Let $y \triangleq \frac{\lambda}{2\sigma^2}$, and let $v_m(x, t) \triangleq u_m(x, t, T)e^{-x^2y}$. We have

$$u_m(x, t, T) = v_m(x, t)e^{x^2y}, \quad \frac{\partial u_m}{\partial t} = \frac{\partial v_m}{\partial t}e^{x^2y},$$

$$\frac{\partial u_m}{\partial x} = \frac{\partial v_m}{\partial x}e^{x^2y} + v_m e^{x^2y} \cdot 2xy = e^{x^2y}\left[\frac{\partial v_m}{\partial x} + 2xyv_m\right],$$

$$\frac{\partial^2 u_m}{\partial x^2} = \frac{\partial^2 v_m}{\partial x^2}e^{x^2y} + \frac{\partial v_m}{\partial x}e^{x^2y} \cdot 4xy + v_m e^{x^2y}(2xy)^2 + v_m e^{x^2y}2y$$

$$= e^{x^2y}\left[\frac{\partial^2 v_m}{\partial x^2} + 4xy\frac{\partial v_m}{\partial x} + (2xy)^2 v_m + 2yv_m\right].$$

Using these formulas, equation (3.9) can be transformed into an equation for v_m:

$$\frac{\partial v_m}{\partial t}(x,t) + (\alpha - \lambda x)\left[\frac{\partial v_m}{\partial x}(x,t) + 2xyv_m(x,t)\right]$$

$$+ \frac{\sigma^2}{2}\left[\frac{\partial^2 v_m}{\partial x^2}(x,t) + 4xy\frac{\partial v_m}{\partial x}(x,t) + (2xy)^2 v_m(x,t) + 2yv_m(x,t)\right]$$

$$+ \frac{1}{2\sigma^2}(\kappa - \lambda x)^2 e_m(x)v_m(x,t) = 0, \qquad t < T,\ x \in \mathbf{R},$$

$$v_m(x,T) = e^{-x^2 y}. \tag{3.11}$$

We have that

$$-\lambda x + \frac{\sigma^2}{2}\cdot 4xy = 0, \quad -2\lambda y + 2\sigma^2 y^2 + \frac{\lambda^2}{2\sigma^2} = 2\sigma^2\left(y - \frac{\lambda}{2\sigma^2}\right)^2 = 0. \tag{3.12}$$

By (3.12), equation (3.11) can be rewritten as

$$\frac{\partial v_m}{\partial t}(x,t) + \alpha\frac{\partial v_m}{\partial x}(x,t) + 2\alpha xyv_m(x,t) + \frac{\sigma^2}{2}\left[\frac{\partial^2 v_m}{\partial x^2}(x,t) + 2yv_m(x,t)\right]$$

$$+ \frac{1}{2\sigma^2}(\kappa^2 - 2\kappa\lambda x)e_m(x)v_m(x,t) - yx^2(1 - e_m(x))v_m(x,t) = 0, \tag{3.13}$$

$$t < T,\ x \in \mathbf{R},$$

$$v_m(x,T) = e^{-x^2 y}. \tag{3.14}$$

Clearly, this equation has a unique solution that can be presented as

$$v_m(x,t) = \mathbf{E}e^{-y\xi(T)^2}\exp\int_t^T\left(\sigma^2 y + 2\alpha y\xi(s) + \frac{1}{2\sigma^2}[\kappa^2 - 2\kappa\lambda\xi(s)]e_m(\xi(s))\right.$$

$$\left. - y[1 - e_m(\xi(s))]\xi(s)^2\right)ds,$$

and where $\xi(s) = \xi^{x,t}(s)$ is the solution of the following linear Ito equation:

$$d\xi(s) = \alpha ds + \sigma dw(s), \quad s > t, \qquad \xi(t) = x.$$

Note that

$$0 \le v_m(x,t) \le \mathbf{E}\exp\int_t^T(c_1 + c_2|\xi(s)|)ds \le c_3\mathbf{E}\exp\int_t^T c_2|\xi(s)|ds,$$

where $c_i > 0$ are constants that do not depend on m. By Jensen's inequality, it follows from convexity of the exponent that

$$\mathbf{E}\exp\int_t^T c_2|\xi(s)|ds = \mathbf{E}\exp\left(\frac{1}{T-t}\int_t^T(T-t)c_2|\xi(s)|ds\right)$$

$$\le \mathbf{E}\frac{1}{T-t}\int_t^T\exp((T-t)c_2|\xi(s)|)ds = \frac{1}{T-t}\int_t^T\mathbf{E}\exp((T-t)c_2|\xi(s)|)ds.$$

Hence

$$v_m(x,t) \leq \frac{c_3}{T-t} \int_t^T \mathbf{E} \exp((T-t)c_2|\xi(s)|)ds.$$

The process $\xi(s) = \xi^{x,t}(s)$ is Gaussian, and its mean and variance are bounded on $[t,T]$ for any given (x,t). In addition, $\xi(\cdot)$, c_2, and c_3, do not depend on m. It follows that $\sup_{m>0} v_m(x,0) < +\infty$ for any $x \in \mathbf{R}$. We have then $u_m(x,t,T) = v_m(x,t)e^{x^2 y}$ as the solution of (3.9). Hence

$$\sup_{m>0} u_m(x,0,T) = \sup_{m>0} v_m(x,0) < +\infty \quad \forall x \in \mathbf{R}.$$

By (3.10), the proof follows. $\square$

The proof of Corollaries 3.3 and 3.4 is given under the assumptions of Remark 3.5.

Proof of Corollary 3.3. It is well known that if the Novikov condition is satisfied, then $\mathbf{E}\mathcal{Z}(T)^{-1} = 1$, where $\mathcal{Z}(T)$ is defined by (3.3). It was proven above that the Novikov condition is satisfied on the conditional probability space given $\widetilde{R}(0)$, i.e., under the measure $\mathbf{P}(\cdot|\widetilde{R}(0))$. Hence $\mathbf{E}\{\mathcal{Z}(T)^{-1}|\widetilde{R}(0)\} = 1$ a.s., where $\mathcal{Z}(T)$ is defined by (3.3). It follows that $\mathbf{E}\mathcal{Z}(T)^{-1} = \mathbf{E}(\mathbf{E}\{\mathcal{Z}(T)^{-1}|\widetilde{R}(0)\}) = 1$. By the Girsanov theorem, it follows that $w_*(t) \stackrel{\Delta}{=} w(t) + \int_0^t \sigma^{-1}\widetilde{b}(s)ds$ is a Wiener process under $\mathbf{P}_{*,T}$ for $t \in [0,T]$, where $\widetilde{b}(t) \stackrel{\Delta}{=} \alpha - \lambda\widetilde{R}(t)$, and where $\mathbf{P}_{*,T}$ is a measure defined by $d\mathbf{P}_{*,T}/d\mathbf{P} = \mathcal{Z}(T)^{-1}$. Hence it is an equivalent martingale measure. This completes the proof. $\square$

Proof of Corollary 3.4. By Ito's formula,

$$d\widetilde{S}(t) = \widetilde{S}(t)\left[\frac{\sigma^2}{2}dt + d\widetilde{R}(t)\right] = \widetilde{S}(t)[\widetilde{a}(t)dt + \sigma dw(t)],$$

where $\widetilde{a}(t) = \alpha - \lambda\widetilde{R}(t) + \sigma^2/2$. By Theorem 3.2, (3.4) holds for $\kappa = \alpha + \sigma^2/2$ on the conditional probability space given $\widetilde{R}(0)$, i.e., under the measure $\mathbf{P}(\cdot|\widetilde{R}(0))$. Hence $\mathbf{E}\{\hat{\mathcal{Z}}(T)^{-1}|\widetilde{R}(0)\} = 1$ a.s., where $\hat{\mathcal{Z}}(T)$ is defined by

$$\hat{Z}(T) \stackrel{\Delta}{=} \exp\left(\int_0^T \hat{a}(t)\sigma(t)^{-1}dw(t) + \frac{1}{2}\int_0^T \hat{a}(t)^2\sigma(t)^{-2}dt\right) \quad (3.15)$$

(3.3). It follows that $\mathbf{E}\hat{\mathcal{Z}}(T)^{-1} = \mathbf{E}(\mathbf{E}\{\hat{\mathcal{Z}}(T)^{-1}|\widetilde{R}(0)\}) = 1$. By the Girsanov theorem again, it follows that $w_*(t) \stackrel{\Delta}{=} w(t) + \int_0^t \sigma^{-1}\widetilde{a}(s)ds$ is a Wiener process under $\hat{\mathbf{P}}_{*,T}$ for $t \in [0,T]$, where $\hat{\mathbf{P}}_{*,T}$ is a measure defined by $d\hat{\mathbf{P}}_{*,T}/d\mathbf{P} = \hat{\mathcal{Z}}(T)^{-1}$. This completes the proof. $\square$

Proof of Lemma 3.7. It suffices to consider non-random $\widetilde{R}(0)$. (It is the same as proving (3.5) for the conditional space given $\widetilde{R}(0)$). Let $k \stackrel{\Delta}{=} \lambda^2/\sigma^2 + 2\varepsilon$. We have

$$\mathbf{E}\exp\left(\frac{k}{2}\int_0^T \widetilde{R}(t)^2dt\right) \geq \mathbf{E}\exp\frac{k}{2T}\left(\int_0^T \widetilde{R}(t)dt\right)^2 = \mathbf{E}e^{\frac{1}{2}\xi^2},$$

where $\xi \triangleq \sqrt{\frac{k}{T}}\eta$, $\eta \triangleq \int_0^T \widetilde{R}(t)dt$. It suffices to show that, for large T,

$$\text{Var }\xi > 1, \quad \text{i.e.,} \quad \text{Var }\eta > \frac{T}{k}. \tag{3.16}$$

We then have

$$\text{Var }\eta = \mathbf{E}\left(\int_0^T \widetilde{R}(t)dt - \mathbf{E}\int_0^T \widetilde{R}(t)dt\right)^2 = \mathbf{E}\left(\int_0^T [\widetilde{R}(t) - \mathbf{E}\widetilde{R}(t)]dt\right)^2$$

$$= \mathbf{E}\left(\int_0^T dt \int_0^t e^{-\lambda(t-s)}\sigma dw(s)\right)^2$$

$$= \mathbf{E}\int_0^T \int_0^T dt dq \int_0^t e^{-\lambda(t-s)}\sigma dw(s) \int_0^q e^{-\lambda(q-p)}\sigma dw(p)$$

$$= 2\mathbf{E}\int_0^T dt \int_t^T dq \int_0^t e^{-\lambda(t-s)}\sigma dw(s) \int_0^q e^{-\lambda(q-p)}\sigma dw(p).$$

Hence

$$\text{Var }\eta =$$

$$= 2\int_0^T dt \int_t^T dq \int_0^t e^{-\lambda(t-s)}e^{-\lambda(q-s)}\sigma^2 ds$$

$$= 2\sigma^2 \int_0^T dt \int_t^T dq\, e^{-\lambda(q-t)} \int_0^t e^{-2\lambda(t-s)}ds$$

$$= 2\sigma^2 \int_0^T dt \int_t^T dq\, e^{-\lambda(q-t)}\frac{1-e^{-2\lambda t}}{2\lambda}$$

$$= 2\sigma^2 \int_0^T dt \int_t^T dq\, \frac{e^{-\lambda(q-t)} - e^{-\lambda(q-t)}e^{-2\lambda t}}{2\lambda}$$

$$= \sigma^2 \int_0^T dt \int_t^T dq\, \frac{e^{-\lambda(q-t)} - e^{-\lambda q - \lambda t}}{\lambda}$$

$$= \frac{\sigma^2}{\lambda^2} \int_0^T dt \left[e^{\lambda t}(e^{-\lambda t} - e^{-\lambda T}) - e^{-\lambda t}(e^{-\lambda t} - e^{-\lambda T})\right]$$

$$= \frac{\sigma^2}{\lambda^2}\left[T - \frac{e^{\lambda T}-1}{\lambda}e^{-\lambda T} - \frac{1-e^{-2\lambda T}}{2\lambda} + \frac{1-e^{-\lambda T}}{\lambda}e^{-\lambda T}\right] \geq \frac{1}{k-2\varepsilon}(T-c),$$

where

$$c \triangleq \max_{T>0}\left(\frac{e^{\lambda T}-1}{\lambda}e^{-\lambda T} + \frac{1-e^{-2\lambda T}}{2\lambda} - \frac{1-e^{-\lambda T}}{\lambda}e^{-\lambda T}\right)$$

does not depend on T. (Remember that $k - 2\varepsilon = \lambda^2/\sigma^2$.) Then (3.16) follows for large T. This completes the proof. $\square$

3.2   A market model with delay in coefficients

We suggest modeling the continuous time stock price process $S(t)$ via the following stochastic delay differential equation

$$dR(t) = -\lambda(R(t) - R(t - \varrho))dt + \sigma dw(t), \quad t \in [0, +\infty),$$
$$S(t) = e^{R(t)}. \tag{3.17}$$

Here $R(t)$ is the return, $\lambda \in \mathbf{R}$, $\varrho > 0$, and $\sigma > 0$ are some constants, and $w(t)$ is a standard Wiener process [116].

From Ito's formula, we obtain

$$dS(t) = S(t)[-\lambda(\log S(t) - \log S(t - \varrho)) + \sigma^2/2]dt + \sigma S(t)dw(t). \tag{3.18}$$

Clearly, this is a modification of our model from Chapter 2 where

$$a(t) = -\lambda[\widetilde{R}(t) - \widetilde{R}(t - \varrho)] + \frac{\sigma^2}{2}, \quad \sigma(t) = \sigma.$$

It can be noted that this $a(t)$ is not a bounded process and therefore it does not satisfy conditions imposed in Chapter 2.

The choice of this particular model was based on the rationale that the presence of the mean reversion reduces the variance for the mean-reverting process [116]. This feature is used in financial modeling; see, e.g., [117]. Under the mean-reverting settings, the return at time t tends to reverse to the long-term average of returns, and the variance of the process is lower than for a martingale with the same volatility. However, we found that the mean-reverting model is not particularly useful for the purpose of this section, since it was difficult to justify a selection of a particular long-term return value. To overcome this, we considered model (3.17) with a delay term. One may say that, for the case where $\lambda > 0$, the process is pushed back to its past values at selected with fixed delay and currently changing times. Respectively, for the case where $\lambda < 0$, the process is pushed away from its past values; it appears that the case in which $\lambda < 0$ is also significant.

3.2.1   Existence, regularity, and non-arbitrage properties

Model (3.17) features the existence, regularity, and non-arbitrage properties. This can be shown as the following.

Assume that $R(t)|_{[-\varrho,0]}$ is a Gaussian process independent on $w(t)|_{t>0}$ and such that its second moment is bounded. Then equation (3.17) has a unique strong solution on the time interval $(0, +\infty)$. This solution can be obtained consequently on the intervals $[0, \varrho], [\varrho, 2\varrho], \ldots, [(k - 1)\varrho, k\varrho], \ldots$. This procedure produces a Gaussian process such that there exists a sequence $\{C_k\}_{k=1}^{\infty}$ such that $\mathbf{E}R(t)^2 \leq C_k$ if $t \in [(k - 1)\varrho, \varrho]$.

We assumed below that $R(t) = 0$ for $t < 0$. In this case, the appreciation rate for the stock price $S(t)$ is

$$a(t) = -\lambda(R(t) - R(t - \varrho)\mathbb{I}_{\{t \geq \varrho\}} + \sigma^2/2.$$

This is a Gaussian process.

It follows that the Novikov condition holds for any sufficiently small interval $[\theta, \theta + \varepsilon]$, i.e.,

$$\mathbf{E} \exp\left(\frac{1}{2} \int_\theta^{\theta+\varepsilon} |a(s)|^2 \sigma^{-2} ds\right) < +\infty$$

if $\varepsilon > 0$ is sufficiently small. Therefore, by Girsanov theorem, the process $S(t)$ can be transformed by a probability measure change into a Black–Scholes price process, with the volatility σ, "locally", i.e., on any sufficiently small time interval. This means that the "true" volatility of S is σ, the same as for the Black–Scholes model; in the theory, this volatility can be restored without error from the continuous time observations of the entire path of $S(t)|_{t\in[\theta,\theta+\varepsilon]}$ or $R(t)|_{t\in[\theta,\theta+\varepsilon]}$. In particular, this implies that the standard estimates for volatility converges to σ as sampling frequency converges to infinity (i.e., the sampling interval converges to zero). Therefore, in the limit case of infinite sampling frequency, or continuous time measurements, our price process is indistinguishable from the price process for the classical Black–Scholes market model. However, it appears that, for any given finite sampling frequency, the volatility estimates behave differently for our delay equations and for the Black–Scholes price process. We show below that, for any given finite sampling frequency, the presence of the delay term makes the volatility systematically underestimated if $\lambda > 0$ (respectively, overestimated if $\lambda < 0$); in both cases, the dependence of this systematic bias on the sampling frequency appears to be monotonic.

Further, it can be noted that it is not possible to use Girsanov theorem for an arbitrarily selected time interval $[0, T]$, because it is unclear if the Novikov condition holds if T is not small enough. A similar but simpler case where $S(t) = e^{G(t)}$, where G was a Gaussian Ornstein–Uhlenbek process, was studied in [43], where it was proved that the Novikov condition holds for an arbitrarily large interval, for this case. The method [43] relied on the Markov properties of the process and cannot be extended to our case of the equation with delay. Therefore, the existence of an equivalent martingale measure for an arbitrarily selected time interval $[0, T]$ is still an open question. However, it appears that the market with the suggested stock price $S(t)$ is arbitrage free with respect to the standard class of the self-financing strategies. More precisely, it appears that there is no a strategy such that

$$\mathbf{P}(X(T) \geq 0) = 1, \quad \mathbf{P}(X(T) > 0) > 0, \quad X(0) = 0,$$

where $X(t)$ is the corresponding wealth generated by a self-financing strategy. It can be shown as the following. Suppose that such a process exists. Let $N > 0$ be such that, for $\varepsilon = T/N$,

$$\mathbf{E} \exp\left(\frac{1}{2} \int_{k\varepsilon}^{(k+1)\varepsilon} |a(s)|^2 \sigma^{-2} ds\right) < +\infty$$

for all $k = 0, 1, .., N - 1$. The market is equivalent to the Black–Scholes market on any time interval $[k\varepsilon, (k + 1)\varepsilon]$. From the absence of an arbitrage for the market defined for $t \in [T - \varepsilon, T]$ considered on the conditional probability space given $\mathcal{F}_{T-\varepsilon}$, it follows that $\mathbf{P}(X(T - \varepsilon) > 0 \,|\, \mathcal{F}_{T-\varepsilon}) = 1$. Taking backward steps, we obtain

$$\mathbf{P}(X(T - k\varepsilon) > 0 \,|\, \mathcal{F}_{T-k\varepsilon}) = 1$$

for all $k = 2, ..., N$. Hence $X(0) > 0$, and the process $X(t)$ required to demonstrate the presence of arbitrage does not exists. This makes our price model applicable for derivative pricing models.

3.2.2   Time discretization and restrictions on growth

In the Monte Carlo simulation, we have to replace stochastic delay differential equation (3.17) by the following stochastic delay difference equation with a given delay τ

$$R(t_k) = R(t_{k-1}) - \lambda(R(t_{k-1}) - R(t_{k-\tau}))\delta + \sigma\sqrt{\delta}\xi_k,$$
$$S(t_k) = e^{R(t_k)}. \tag{3.19}$$

Here, $k = 1, 2, ..., \delta = t_k - t_{k-1}$, and ξ_k are independent and identically distributed random variables from the standard normal distribution; $\tau > 0$ is an integer.

For the purposes of this section, it will be sufficient to study the sample paths of solutions of (3.19) created by Monte Carlo simulation as a substitution of (3.17), where $\varrho = \tau\delta$.

It can be noted that equation (3.19) represents a linear autoregression AR(τ) with the characteristic polynomial

$$z^\tau = z^{\tau-1} - \lambda\delta(z^{\tau-1} - 1).$$

This polynomial has a root $z = 1$. Therefore, the time series $\{R(t_k)\}$ does not converge to a stationary process as $k \to +\infty$. Let $\{z_1, ..., z_\tau\}$ be the roots of this polynomial, and let as selected $z_1 = 1$. We will be using model (3.19) for the pairs (τ, λ) such that all other roots $\{z_2, ..., z_\tau\}$ are inside of the open disc $\mathbb{D} \overset{\Delta}{=} \{z \in \mathbf{C} : |z| < 1\}$, i.e.,

$$\{z_k\}_{k=2}^\tau \subset \mathbb{D}. \tag{3.20}$$

In this case, the series $R(t_k)$ features a moderate growth rate similar to the one for the returns in the Black–Scholes model. It can be noted that if $\lambda > 0$, then (3.20) holds for all $\tau \geq 2$. If $\lambda < 0$, then it may happen that (3.20) does not hold for some τ. However, it appears that (3.20) holds for small enough $|\lambda|$ and small enough τ. In particular, we found that:

- For $\tau \leq 11$, (3.20) holds for all $\kappa = \lambda\delta \geq -0.111$.

- For $\tau \leq 15$, (3.20) holds for all $\kappa = \lambda\delta \geq -0.075$.

- For $\tau \leq 20$, (3.20) holds for all $\kappa = \lambda\delta \geq -0.055$.

- For $\tau \leq 150$, (3.20) holds for all $\kappa = \lambda\delta \geq -0.006$.

It appears that this range for the parameters allows replicating the volatilities depending on the sampling frequencies similar to the ones observed for the historical data.

3.3 A market model with stochastic numéraire

This section addresses the stock option pricing problem in a continuous time market model where there are two stochastic tradable assets, and one of them is selected as a numéraire. An equivalent martingale measure is not unique for this market, and there are non-replicable claims. Some rational choices of the equivalent martingale measures are suggested and discussed, including implied measures calculated from bond prices constructed as a risk-free investment with deterministic payoff at the terminal time. This leads to the possibility of inferring an implied market price of risk process from observed historical bond prices.

3.3.1 Model setting

It is known that various models in the financial market lead to different properties with respect to pricing methods, replicability of claims, and arbitrage opportunities. In this section, we describe a market model consisting of two tradable assets with random continuous in time prices representing a modification of the classical Black–Scholes model where one of the assets is non-random.

We consider a continuous time model of a securities market consisting of two tradable assets with the prices $S(t)$ and $B(t)$, $t \geq 0$. The prices evolve as

$$dS(t) = S(t)\big(a(t)dt + \sigma(t)dw(t) + \hat{\sigma}(t)d\hat{w}(t)\big), \quad t > 0, \tag{3.21}$$

and

$$dB(t) = B(t)\big(\alpha(t)dt + \rho(t)dw(t) + \hat{\rho}(t)d\hat{w}(t)\big). \tag{3.22}$$

We assume that $W(t) = (w(t), \hat{w}(t))$ is a standard Wiener process with independent components on a given standard probability space $(\Omega, \mathcal{F}, \mathbf{P})$, where Ω is a set of elementary events, $\mathbf{P}$ is a probability measure, and $\mathcal{F}$ is a $\mathbf{P}$-complete σ-algebra of events. The initial prices $S(0) > 0$ and $B(0) > 0$ are given constants.

We consider this model as an extension of the classic bond and stock market model, where a bond with the price $B(t)$ evolving as

$$\frac{dB}{dt}(t) = \alpha(t)B(t). \tag{3.23}$$

is used as a numéraire. Equation (3.22) for stochastic numéraire is a generalization of (3.23); one may say that equation (3.22) represents a modification of the equation for a risk-free asset that takes into account possibility of stochastic disturbances in the return rate. In this setting, $B(t)$ is not exactly a risk-free asset. However, if the processes $\rho(t)$ and $\hat{\rho}(t)$ are small in some norm, then (3.22) can be considered as the equation for the money market account with small deviations (see an example in Section 3.3.2 below). In particular, the conditions on the coefficients imposed below allow the case where $\rho(t) \equiv \hat{\rho}(t) = \varepsilon$ for an arbitrarily small $\varepsilon > 0$. It appears that the presence of arbitrarily small deviations in (3.22) changes dramatically the properties of the market model (see Section 3.3.2).

Definition 3.9

(i) If $\operatorname{ess\,sup}_{t,\omega}(|\hat{\sigma}(t,\omega)| + |\hat{\rho}(t,\omega)|) > 0$, we denote by $\{\mathcal{F}_t\}_{t \geq 0}$ the filtration generated by the process $W = (w, \hat{w})$.

(ii) If $\operatorname{ess\,sup}_{t,\omega}(|\hat{\sigma}(t,\omega)| + |\hat{\rho}(t,\omega)|) = 0$, we denote by $\{\mathcal{F}_t\}_{t \geq 0}$ the filtration generated by the process w only.

*In both cases, $\mathcal{F}_0$ is trivial, i.e., it is the **P**-augmentation of the set $\{\emptyset, \Omega\}$.*

We assume that the process $\mu(t) = (a(t), \sigma(t), \hat{\sigma}(t), \alpha(t), \rho(t), \hat{\rho}(t))$ is **F**-adapted and bounded.

Let

$$\tilde{\sigma} \overset{\Delta}{=} \sigma - \rho, \quad \tilde{\rho} \overset{\Delta}{=} \hat{\sigma} - \hat{\rho}.$$

We assume that there exists $c > 0$ such that either $|\tilde{\sigma}(t,\omega)| \geq c$ a.e. or $|\tilde{\rho}(t,\omega)| \geq c$ a.e.

Discounted stock price and equivalent martingale measures

Let $\tilde{S}(t) \overset{\Delta}{=} S(t)/B(t)$. By the Itô formula, it follows that this process evolves as

$$d\tilde{S}(t) = \tilde{S}(t)\big(\tilde{a}(t)dt + \tilde{\sigma}(t)dw(t) + \tilde{\rho}(t)d\hat{w}(t)\big), \tag{3.24}$$

$$\tilde{S}(0) = S(0)/B(0), \tag{3.25}$$

where

$$\tilde{a} \overset{\Delta}{=} a - \alpha + \rho^2 + \hat{\rho}^2 - \sigma\rho - \hat{\sigma}\hat{\rho}.$$

Let $V(t) = (V_1(t), V_2(t))^\top = (\tilde{\sigma}(t), \tilde{\rho}(t))^\top$ and $\hat{V}(t) = (\hat{V}_1(t), \hat{V}_2(t))^\top = (\rho(t), \hat{\rho}(t))^\top$. These processes take values in $\mathbf{R}^2$. By the assumptions, $\operatorname{ess\,inf}_{t,\omega} |V(t,\omega)| > 0$.

Definition 3.10 *Let Θ be the set of $\mathcal{F}_t$-adapted processes $\theta(t) = (\theta_1(t), \theta_2(t))^\top$ with values in $\mathbf{R}^2$ such that $\theta_1(t)\tilde{\sigma}(t) + \theta_2(t)\tilde{\rho}(t) = \tilde{a}(t)$, i.e., $V(t)^\top \theta(t) = \tilde{a}(t)$, and such that*

$$\operatorname{ess\,sup}_{\omega} \int_0^T |\theta(t)|^2 dt < +\infty.$$

Here and below $|\cdot|$ is the Euclidean norm of vectors.

Up to the end of this section, we use notation θ for elements of the set Θ only. For $\theta \in \Theta$, set

$$Z_\theta = \exp\left(-\int_0^T \theta(s)^\top dW(s) - \frac{1}{2}\int_0^T |\theta(s)|^2 ds\right). \tag{3.26}$$

Our standing assumptions imply that $\mathbf{E}Z_\theta = 1$. Define the probability measure $\mathbf{P}_\theta$ by $d\mathbf{P}_\theta/d\mathbf{P} = Z_\theta$; this measure is equivalent to the measure $\mathbf{P}$. Let $\mathbf{E}_\theta$ be the corresponding expectation.

Let

$$W_\theta(t) = \left(\begin{array}{c} W_{\theta 1}(t) \\ W_{\theta 2}(t) \end{array}\right) = \int_0^t \theta(s)ds + W(t). \tag{3.27}$$

By Girsanov's theorem, W_θ is a standard Wiener process in $\mathbf{R}^2$ under $\mathbf{P}_\theta$.

For $\theta \in \Theta$, equation (3.25) can be rewritten as

$$d\widetilde{S}(t) = \widetilde{S}(t)V(t)^\top dW_\theta(t).$$

Remark 3.11 *Clearly, the set Θ has more than one element; it is a linear manifold. Therefore, the selection of the process $\theta(t)$ and the measure $\mathbf{P}_\theta$, is not unique.*

Example 3.12

(i) If $\widetilde{\rho} \equiv 0$, then the process $\theta_1(t)$ is uniquely defined as $\theta_1(t) = \widetilde{\sigma}(t)^{-1}\widetilde{a}(t)$ and is called the *marked price of risk process*. If, in addition, the process $\widetilde{\sigma}(t)$ is non-random, then the process $\widetilde{S}(t)$ has the same distribution under $\mathbf{P}_\theta$ for all $\theta \in \Theta$.

(ii) If $\widetilde{\sigma} \equiv 0$, then the process $\theta_2(t)$ is uniquely defined as $\theta_2(t) = \widetilde{\rho}(t)^{-1}\widetilde{a}(t)$.

(iii) Under some special requirements for the equivalent martingale imposed in [20], there exists a unique equivalent martingale measure such that the process $\left(\widetilde{S}(t), \exp\left(\int_0^t k(s)ds\right)B(t)^{-1}\right)$ is a martingale, for a given process $k(t) \geq 0$.

(iv) Let $z(t) = \int_0^t |V(s)|^{-1}V(s)^\top dW(s)$; by Lévy's characterization theorem, it is a one-dimensional Wiener process. Let $q(t) = |V(s)|^{-1}\widetilde{a}(t)$. By the assumptions, it is a bounded process. Let $\hat{z}(t) = \int_0^t q(s)ds + z(t)$. We have that $V(t)^\top dW(t) = |V(t)|dz(t)$ and

$$d\widetilde{S}(t) = \widetilde{S}(t)\big(\widetilde{a}(t)dt + |V(t)|dz(t)\big) = \widetilde{S}(t)|V(t)|d\hat{z}(t).$$

By the Girsanov theorem, there is an equivalent martingale measure $\hat{\mathbf{P}}$ such that $\hat{z}(t)$ is a Wiener process under $\hat{\mathbf{P}}$; in this case, $\widetilde{S}(t)$ is a martingale under $\hat{\mathbf{P}}$. This martingale measure was studied in [34].

Let $\mathcal{Y}$ be the set of all $\mathcal{F}_t$-adapted measurable processes with values in $\mathbf{R}^2$ that are square integrable on $[0, T] \times \Omega$ with respect to $\ell_1 \times \mathbf{P}_\theta$, where ℓ_1 is the Lebesgue measure.

Let $\mathcal{H}_\theta$ be the Hilbert space formed as the completion of the set of $\mathcal{F}_t$-adapted measurable processes $y(t)$ such that $\|y\|_{\mathcal{H}_\theta} = \left(\mathbf{E}_\theta \int_0^T |\widetilde{S}(t) y(t)|^2 dt \right)^{1/2} < +\infty$.

Wealth and discounted wealth

Let $X(0) > 0$ be the initial wealth at time $t = 0$ and let $X(t)$ be the wealth at time $t > 0$.

We assume that the wealth $X(t)$ at time $t \geq 0$ is

$$X(t) = \beta(t)B(t) + \gamma(t)S(t). \tag{3.28}$$

Here $\beta(t)$ is the quantity of the numéraire portfolio, $\gamma(t)$ is the quantity of the stock portfolio, $t \geq 0$. The pair $(\beta(\cdot), \gamma(\cdot))$ describes the state of the securities portfolio at time t. Each of these pairs is called a *strategy*.

Definition 3.13 *Let $\theta \in \Theta$. A pair $(\beta(\cdot), \gamma(\cdot))$ is said to be an admissible strategy under $\mathbf{P}_\theta$ if the processes $\beta(t)$ and $\gamma(t)$ are progressively measurable with respect to the filtration $\{\mathcal{F}_t\}_{t \geq 0}$ and such that*

$$\mathbf{E}_\theta \int_0^T \widetilde{S}(t)^2 \gamma(t)^2 dt < +\infty. \tag{3.29}$$

Definition 3.14 *Let $\theta \in \Theta$ be given. A pair $(\beta(\cdot), \gamma(\cdot))$ that is an admissible strategy under $\mathbf{P}_\theta$ is said to be a self-financing strategy, if there exists a sequence of Markov times $\{T_k\}_{k=1}^\infty$ with respect to $\{\mathcal{F}_t\}_{t \geq 0}t$ such that $0 \leq T_k \leq T_{k+1} \leq T$ for all k, $T_k \to T$ as $k \to +\infty$ a.s. and that the following holds:*

(i) For $k = 1, 2, ..,$

$$\mathbf{E}_\theta \int_0^{T_k} \left(\beta(t)^2 B(t)^2 + S(t)^2 \gamma(t)^2 \right) dt < +\infty.$$

(ii) The corresponding wealth $X(t) = \gamma(t)S(t) + \beta(t)B(t)$ is such that

$$dX(t) = \gamma(t)dS(t) + \beta(t)dB(t).$$

It should be noted that condition (i) in Definition 3.14 ensures that the stochastic differentials in condition (ii) here are well defined.

The process $\widetilde{X}(t) \triangleq X(t)/B(t)$ is said to be the discounted wealth.

Lemma 3.15 *If a strategy $(\beta(t), \gamma(t))$ is self-financing and admissible under $\mathbf{P}_\theta$ for some $\theta \in \Theta$, then, for the corresponding discounted wealth,*

$$d\widetilde{X}(t) = \gamma(t)d\widetilde{S}(t). \tag{3.30}$$

Remark 3.16 *Since we assume that the coefficients for the equations for $S(t)$ and $B(t)$ are bounded, it follows from Lemma 3.15 that if (3.29) holds for some θ then $\mathbf{E}_\theta \widetilde{X}(T)^2 < +\infty$ for this θ.*

Lemma 3.17 *For every $\theta \in \Theta$, the processes $\widetilde{X}(t)$ and $\widetilde{S}(t)$ are martingales under Q with respect to $\{\mathcal{F}_t\}_{t \geq 0}t$, i.e., $\mathbf{E}_\theta\{\widetilde{S}(T)\,|\mathcal{F}_t\} = \widetilde{S}(t)$ and $\mathbf{E}_\theta\{\widetilde{X}(T)\,|\mathcal{F}_t\} = \widetilde{X}(t)$.*

Remark 3.18 *Consider a European option with the payoff $B(T)\xi$, where ξ is an $\mathcal{F}_T$-measurable random variable. For any $\theta \in \Theta$ such that $\mathbf{E}_\theta \xi^2 < +\infty$, the option price $\mathbf{E}_\theta \xi$ is an arbitrage-free price.*

3.3.2 Replication of claims: Strategies and hedging errors

In financial mathematics, the most common approach to option pricing is representing the prices as the minimal initial wealth that can be raised, via self-financing strategies, into the wealth allowing fulfilment of option obligations. This reduces the pricing problem to analysis of replicability of random contingent claims. This section describes certain features of the claim replication for the model described above.

For $\theta \in \Theta$, let $\mathcal{X}$ be the subspace of $L_2(\Omega, \mathcal{F}_T, \mathbf{P}_\theta)$ consisting of all $\zeta \in L_2(\Omega, \mathcal{F}_T, \mathbf{P}_\theta)$ such that there exists an admissible self-financing strategy $(\beta(\cdot), \gamma(\cdot))$ under $\mathbf{P}_\theta$ and the corresponding wealth process $X(t)$ such that $X(0) = 0$ and $X(T) = B(T)\zeta$.

Let

$$\mathcal{X}^\perp \triangleq \{\eta \in L_2(\Omega, \mathcal{F}_T, \mathbf{P}_\theta) : \ \mathbf{E}_\theta \eta = 0, \ \mathbf{E}_\theta[\zeta\eta] = 0 \text{ for all } \zeta \in \mathcal{X}\}.$$

Let $\xi \in L_2(\Omega, \mathcal{F}_T, \mathbf{P}_\theta)$. By the martingale representation theorem, we have that, for some uniquely defined $U_\theta \in \mathcal{Y}$ and $c_\theta \in \mathbf{R}$,

$$\xi = c_\theta + \int_0^T U_\theta(t)^\top dW_\theta(t). \tag{3.31}$$

In addition, it follows from the properties of closed subspaces in Hilbert spaces that ξ can be represented via Föllmer–Schweizer decomposition

$$\xi = c_\theta + I_\theta + R_\theta. \tag{3.32}$$

Here $c_\theta = \mathbf{E}_\theta \xi$, $R_\theta \in \mathcal{X}^\perp$, and

$$I_\theta = \int_0^T \gamma_\theta(t) d\widetilde{S}(t) \in \mathcal{X} \tag{3.33}$$

for some $\gamma_\theta \in \mathcal{H}_\theta$, i.e., it is the terminal discounted wealth $\widetilde{X}(T)$ for some admissible self-financing strategy $(\beta_\theta(\cdot), \gamma_\theta(\cdot))$ under $\mathbf{P}_\theta$ and for the initial wealth $X(0) = 0$. Therefore, a contingent claim $B(T)\xi$ can be decomposed as $B(T)(\widetilde{\xi}_\theta + R_\theta)$, where $B(T)\widetilde{\xi}_\theta$ is a replicable part such that $\widetilde{\xi}_\theta = c_\theta + \int_0^T \gamma_\theta(t) d\widetilde{S}(t)$.

We regard the process $\gamma_\theta(t)$ here as the hedging strategy, we regard $B(T)R_\theta$ as the hedging error, and we regard R_θ as the discounted hedging error. We could regard the value $c_\theta = \mathbf{E}_\theta \xi$ as the price of an option with the payoff $B(T)\xi$ given that either this price does not depend on the choice of θ or it is calculated under some reasonable choice of θ; some choices of θ are discussed in Example 3.12 and in Section 3.3.3 below.

Let us express γ_θ via U_θ.

Proposition 3.19 *Let $\xi \in L_2(\Omega, \mathcal{F}_T, \mathbf{P}_\theta)$, and let U_θ be defined by (3.31). Let*

$$\nu_\theta(t) = \frac{U_\theta(t)^\top V(t)}{|V(t)|^2}, \qquad \eta_\theta(t) = U_\theta(t) - \nu_\theta(t)V(t). \tag{3.34}$$

Then (3.32) holds with

$$I_\theta = \int_0^T \nu_\theta(t)V(t)^\top dW_\theta(t), \qquad R_\theta = \int_0^T \eta_\theta(t)^\top dW_\theta(t). \tag{3.35}$$

In addition,

$$|U_\theta(t)|^2 \equiv |\nu_\theta(t)V(t)|^2 + |\eta_\theta(t)|^2, \qquad \eta_\theta(t)^\top V(t) \equiv 0, \qquad \mathbf{E}_\theta I_\theta R_\theta = 0,$$

and (3.33) holds with

$$\gamma_\theta(t) = \nu_\theta(t)\widetilde{S}(t)^{-1}. \tag{3.36}$$

Proof. It suffices to observe that $\eta_\theta \in \mathcal{Y}$, and that $\nu_\theta V$ is the projection of U_θ on V. In particular, it follows that $R_\theta \in \mathcal{X}^\perp$. The uniqueness follows from the properties of orthogonal subspaces of a Hilbert space. $\square$

Some cases of non-replicability and replicability

The following statement follows from the non-uniqueness of the equivalent martingale measures and the second fundamental theorem of asset pricing.

Proposition 3.20 *Assume that $\widetilde{\rho}(\cdot) \neq 0$, i.e., it is not an identically zero process. Then the set $\mathcal{X}^\perp$ contains non-zero elements, i.e., $\sup_{\eta \in \mathcal{X}^\perp} \mathbf{E}_\theta |\eta| > 0$.*

By this proposition, the discounted hedging error R_θ is non-zero in the general case. In other words, a contingent claim of a general type is not replicable.

Let us describe some cases of replicability.

Let $\{\mathcal{F}_t^w\}_{t \geq 0}$ be the filtration generated by the process $w(t)$, and let $\{\mathcal{F}_t^{\widetilde{S}}\}_{t \geq 0}$ be the filtration generated by the process $\widetilde{S}(t)$.

Theorem 3.21 *Assume that the processes $\widetilde{\sigma}(t)$ and $\widetilde{\rho}(t)$ are non-random. Then the claims $B(T)\xi$ are replicable for $\xi \in L_2(\Omega, \mathcal{F}_T^{\widetilde{S}}, \mathbf{P}_\theta)$ for any $\theta \in \Theta$. More precisely, there exists an $\mathcal{F}_t$-adapted process $\gamma(t)$ such that $\mathbf{E}_\theta \int_0^T \gamma(t)^2 \widetilde{S}(t)^2 dt < +\infty$ and $\xi = \mathbf{E}_\theta \xi + \int_0^T \gamma(t) d\widetilde{S}(t)$.*

The case of a complete market

Note that the market described in Theorem 3.21 is incomplete since there are claims that cannot be replicated. The following theorem describes an important special case when the market is complete.

Theorem 3.22 *Assume that the processes $\widetilde{a}(t)$ and $\widetilde{\sigma}(t)$ are adapted to the filtration $\{\mathcal{F}_t^w\}_{t \geq 0}$ generated by the process $w(t)$, and that $\widetilde{\rho}(t) \equiv 0$, i.e., it is an identically zero process. Then $\theta_1(t) = \widetilde{a}(t)\widetilde{\sigma}(t)^{-1}$ for any $\theta \in \Theta$, and the claims $B(T)\xi$ are replicable for $\xi \in L_2(\Omega, \mathcal{F}_T^w, \mathbf{P}_\theta)$.*

Corollary 3.23 *Assume that either conditions of Theorem 3.21 or the conditions of Theorem 3.22 hold. Then the choice of the hedging (replicating) strategy γ is unique, i.e., it is the same for all $\theta \in \Theta$ such that $\mathbf{E}_\theta \xi^2 < +\infty$; the expectation $\mathbf{E}_\theta \xi$ is also the same for all these θ.*

On the relativity of the price and the hedging error

The following theorems demonstrate that this price $c_\theta = \mathbf{E}_\theta \xi$ can be selected quite arbitrarily even for the case of arbitrarily small stochastic deviations in (3.22), i.e., for arbitrarily small processes $\rho(t)$ and $\hat{\rho}(t)$. For instance, we can select $\rho(t) \equiv \hat{\rho}(t) = \varepsilon$ for an arbitrarily small $\varepsilon > 0$. This means that the presence of small deviations in (3.22) changes dramatically the properties of the market model.

We denote $x^+ = \max(0, x)$ for $x \in \mathbf{R}$.

Theorem 3.24 *Assume that*

$$\operatorname*{ess\,inf}_{t,\omega} |\widetilde{\sigma}(t,\omega)\hat{\rho}(t,\omega) - \rho(t,\omega)\widetilde{\rho}(t,\omega)| > 0. \tag{3.37}$$

Let $\kappa \in (0, +\infty)$ be given, and let $\xi = B(T)^{-1}(\kappa - S(T))^+$. Then the following holds.

 (i) For any $\varepsilon > 0$, there exists $\theta \in \Theta$ such that $c_\theta = \mathbf{E}_\theta \xi \in [0, \varepsilon]$.

 (ii) For any $M > 0$, there exists $\theta \in \Theta$ such that $c_\theta = \mathbf{E}_\theta \xi \geq M$.

Theorem 3.25 *Assume that (3.37) holds. Let $\kappa \in (0, +\infty)$ be given, and let $\xi = B(T)^{-1}(S(T) - \kappa)^+$. Then the following holds.*

 (i) For any $\varepsilon > 0$, there exists $\theta \in \Theta$ such that $c_\theta = \mathbf{E}_\theta \xi \in [0, \varepsilon]$.

 (ii) For any $\varepsilon > 0$, there exists $\theta \in \Theta$ such that $c_\theta = \mathbf{E}_\theta \xi \in [S(0) - \varepsilon, S(0)]$.

Consider a strategy that replicates the claim $B(T)(c_\theta + I_\theta)$, where $c_\theta \in \mathbf{R}$ and $I_\theta \in \mathcal{X}$ are such that (3.32) holds with the discounted hedging error $R_\theta \in \mathcal{X}^\perp$.

The following theorems show that, for any given θ, the value of the second moment of R_θ is varying widely if it is calculated with respect to other equivalent martingale measures and can take extreme values.

Theorem 3.26 *Let ξ be a random claim such that (3.32) holds for some $\theta \in \Theta$, $c_\theta \in \mathbf{R}$, $I_\theta \in \mathcal{X}$, and $R_\theta \in \mathcal{X}^\perp$ such that $\mathbf{E}_\theta R_\theta^2 > 0$. Assume that (3.31) holds for $U_\theta \in \mathcal{Y}$ such that*

$$\operatorname{ess\,sup}_\omega \int_0^T |U_\theta(t,\omega)|^2 dt < +\infty, \qquad \operatorname{ess\,inf}_\omega \int_0^T |\eta_\theta(t,\omega)| dt > 0,$$

where η_θ is defined by (3.34). Then, for any $M > 0$, there exists $\vartheta \in \Theta$ such that $\mathbf{E}_\vartheta R_\theta^2 \geq M$.

Theorem 3.27 *Let ξ be a random claim such that (3.32) holds for some $\theta \in \Theta$, $c_\theta \in \mathbf{R}$, $I_\theta \in \mathcal{X}$, and $R_\theta \in \mathcal{X}^\perp$ such that $\mathbf{E}_\theta R_\theta^2 > 0$. Assume that (3.31) holds for $U_\theta \in \mathcal{Y}$ such that*

$$\operatorname{ess\,sup}_{t,\omega} |U_\theta(t,\omega)| < +\infty, \qquad \operatorname{ess\,inf}_{t,\omega} |\eta_\theta(t,\omega)| > 0,$$

where η_θ is defined by (3.34). Then, for any $\varepsilon > 0$, there exists $\vartheta \in \Theta$ such that $\mathbf{E}_\vartheta R_\theta^2 \leq \varepsilon$.

It can be noted that it is not uncommon to observe pricing abnormalities in meaningful market models; see e.g., [105] and the references therein.

To overcome relativity of pricing outlined in Theorems 3.24–3.26, we need to investigate reasonable choices for the martingale measures.

3.3.3   On selection of θ and the equivalent martingale measure

We have found that, for any $\theta \in \Theta$, the discounted price process $\widetilde{S}$ is a martingale under the measure $\mathbf{P}_\theta$. Therefore, in a general case, there are many equivalent martingale measures. A question arises of which particular θ should be used for pricing of options. In the literature, there are many methods developed for this problem, mainly for the incomplete market models with random volatility and appreciation rate.

One may look for "optimal" θ and c_θ in the spirit of mean-variance pricing, such that $\mathbf{E}R_\theta^2$ is minimal; see, e.g., [108]. A generalization of this approach leads to minimization of $\mathbf{E}|R_\theta|^q$ for $q \geq 1$. An alternative approach is to define the price as $\sup_{\theta \in \Theta_0} c_\theta$ for some reasonably selected set $\Theta_0 \subset \Theta$. In particular, this pricing rule leads to a corrected volatility smile in the case of an incomplete market with random volatility [45].

The following theorem provides a unifying approach for selection of θ. In particular, this approach allows including models listed in Example 3.12.

Theorem 3.28 *Let $\theta = (\theta_1, \theta_2)^\top \in \Theta$ be given, and let $\varrho(t) = \hat{V}(t)^\top \theta(t)$, i.e.,*

$$\tilde{\sigma}\theta_1 + \tilde{\rho}\theta_2 = \tilde{a},$$
$$\rho\theta_1 + \hat{\rho}\theta_2 = \varrho. \tag{3.38}$$

Then

$$dS(t) = S(t)\big([a(t) - \tilde{a}(t) - \varrho(t)]dt + \sigma(t)dW_{1\theta}(t) + \hat{\sigma}(t)dW_{2\theta}(t)\big),$$
$$dB(t) = B(t)([\alpha(t) - \varrho(t)]dt + \rho(t)dW_{1\theta}(t)) + \hat{\rho}(t)dW_{2\theta}(t)). \tag{3.39}$$

Theorem 3.28 gives a parametrization of the set Θ via ϱ.

Examples 3.29–3.32 below demonstrate how this parametrization helps to identify some reasonable choices of θ. For these examples, we assume that

$$\operatorname*{ess\,inf}_{t,\omega} |\tilde{\sigma}(t,\omega)\hat{\rho}(t,\omega) - \rho(t,\omega)\tilde{\rho}(t,\omega)| > 0. \tag{3.40}$$

This condition ensures that system (3.38) defines a unique $\theta \in \Theta$ for any $\mathcal{F}_t$-adapted process ϱ such that $\operatorname{ess\,sup}_\omega \int_0^T \varrho(t)^2 dt < +\infty$.

Example 3.29 For θ from Theorem 3.28 with $\varrho \equiv 0$, the process $(S(t), B(t))$ evolves as

$$dS(t) = S(t)\big([a(t) - \tilde{a}(t)]dt + \sigma(t)dW_{1\theta}(t) + \hat{\sigma}(t)dW_{2\theta}(t)\big),$$
$$dB(t) = B(t)\big(\alpha(t)dt + \rho(t)dW_{1\theta}(t)\big) + \hat{\rho}(t)dW_{2\theta}(t)\big).$$

In this case, the equation for B has the same coefficients as the equation for $B(t)$ under $\mathbf{P}$, with replacement of $W(t)$ by $W_\theta(t)$. Respectively, the distribution of $B(t)$ under $\mathbf{P}_\theta$ and under the historical measure $\mathbf{P}$ is the same if the coefficients $\alpha(t)$, $\rho(t)$, and $\hat{\rho}(t)$, are non-random. In addition, if $\hat{\sigma} \equiv 0$ then the choice $\varrho \equiv 0$ ensures that $\theta_1 = \hat{a}/\sigma$.

Example 3.30 For θ from Theorem 3.28 with $\varrho = -\tilde{a}$, the evolution of S under $\mathbf{P}_\theta$ is described by an Ito equation with the same coefficients as the equation for $S(t)$ under $\mathbf{P}$, with replacement of $W(t)$ by $W_\theta(t)$. In this case, the distribution of $S(t)$ under $\mathbf{P}_\theta$ and under the historical measure $\mathbf{P}$ is the same if the coefficients $a(t), \sigma(t)$, and $\hat{\sigma}(t)$, are non-random.

Example 3.31 Let $k \in (0,1)$ and $r_B \in \mathbf{R}$ be given. Let us calculate θ from (3.38) with $\varrho = k(\alpha - r_B)$ i.e.,

$$\tilde{\sigma}\theta_1 + \tilde{\rho}\theta_2 = \tilde{a},$$
$$\rho\theta_1 + \hat{\rho}\theta_2 = k(\alpha - r_B). \tag{3.41}$$

By (3.39), this leads to the equation

$$dB(t) = B(t)\big([kr_B + (1-k)\alpha]dt + \rho dW_{1\theta}(t)\big) + \hat{\rho}dW_{2\theta}(t)\big),$$

i.e., the appreciation rate coefficient for B under $\mathbf{P}_\theta$ is $kr_B + (1-k)\alpha$. Therefore, there exists a choice of θ that ensures that the appreciation rate for B under $\mathbf{P}_\theta$ can be arbitrarily close to r_B. This can be achieved with selection of k close to 1 in (3.41).

In is also possible to consider a modification of this approach with r_B depending on θ. For example, an important example considered in Section 6.7 below is where $r_B = \alpha - r_\theta$, where

$$r_\theta = \alpha - \rho^2 - \hat{\rho}^2 - \rho\theta_1 - \hat{\rho}\theta_2.$$

This gives equations

$$\tilde{\sigma}\theta_1 + \tilde{\rho}\theta_2 = \tilde{a},$$
$$\rho\theta_1 + \hat{\rho}\theta_2 = \frac{k}{1-k}(\rho^2 + \hat{\rho}^2). \tag{3.42}$$

In particular, this r_B can be selected as the expected average risk-free rate associated with the zero coupon bond under the measure $\mathbf{P}_\theta$, as is described in Section 6.7 below.

Example 3.32 An important example of the selection of θ is

$$\theta(t) = \tilde{a}(t)V(t)/|V(t)|^2. \tag{3.43}$$

The following theorem shows that this corresponds to the choice of θ with the minimal norm.

Theorem 3.33 *Let $\theta(t)$ be defined by (3.43). Then, for every t, ω, the value of $|\theta(t, \omega)|$ is minimal among all $\theta \in \Theta$. In addition, if $\xi = c_\theta + \int_0^T \gamma_\theta(t)d\tilde{S}(t) + R_\theta$ for some $R_\theta \in \mathcal{X}^T$ and γ_θ is an adapted process such that $\gamma_\theta \sigma \in \mathcal{H}_\theta$, then $\mathbf{E}(R_\theta \mathcal{M}(T)) = 0$, where $\mathcal{M}(T) = \int_0^T \gamma_\theta(t)\tilde{S}(t)V(t)^\top dW(t)$ represents the "martingale" part of the integral*

$$\int_0^T \gamma_\theta(t)d\tilde{S}(t) = \int_0^T \gamma_\theta(t)\tilde{S}(t)\hat{a}(t)dt + \mathcal{M}(T).$$

The selection of θ described in Theorem 3.33 ensures that the corresponding self-financing strategy with the quantity of shares $\gamma(t)$ is a so-called *locally risk minimizing strategy*; see, e.g., [57, 21].

Let us reconsider Example 3.12 (iv). We will be using the measure $\hat{\mathbf{P}}$ and the processes $q(t)$, $z(t)$, and $\hat{z}(t)$ defined in this example.

Set $\mathcal{V}(t) = \hat{V}(t) - k(t)V$, where

$$k(t) = \hat{V}(t)^\top V(t)/|V(t)|^2.$$

Clearly, we have that $\mathcal{V}(t)^\top V(t) = 0$.

Further, there exists a one-dimensional Wiener process $z_1(t)$ such that $\int_0^t \mathcal{V}(s)^\top dW(s) = \int_0^t |\mathcal{V}(s)|dz_1(s)$ and

$$
\begin{aligned}
dB(t) &= B(t)\big(\alpha(t)dt + k(t)V(t)^\top dW(t) + \mathcal{V}(t)^\top dW(t)\big) \\
&= B(t)\big(\alpha(t)dt + k(t)|V(t)|dz(t) + |\mathcal{V}(t)|dz_1(t)\big).
\end{aligned}
$$

For $q(t) = \tilde{a}(t)/|V(t)|$, we have

$$dB(t) = B(t)\big(\alpha(t)dt + k(t)|V(t)|(d\hat{z}(t) - q(t)dt) + |\mathcal{V}(t)|dz_1(t)\big).$$

On the other hand,

$$dB(t) = B(t)\big(\alpha(t)dt + \hat{V}(t)^\top dW(t)\big) = B(t)\big(\alpha(t)dt + \hat{V}(t)^\top(dW_\theta(t) - \theta(t)dt)\big).$$

This means that, in our notation, $\hat{\mathbf{P}} = \mathbf{P}_\theta$, where $\theta \in \Theta$ is such that

$$k(t)q(t)|V(t)| = k(t)\tilde{a}(t) = \hat{V}(t)^\top \theta(t).$$

The only $\theta \in \Theta$ satisfying this is $\theta(t) = \tilde{a}(t)V(t)/|V(t)|^2$ from Theorem 3.33.

3.3.4 Markov case

The values for the prices, errors, and hedging strategies obtained above were expressed via integrands the existence of which is ensured by the martingale representation theorem. In this section, we suggests some representations via solutions of deterministic partial differential equations (PDEs) which could be more convenient.

Assume that $\theta \in \Theta$ is given.

We will be using the processes $s(t) = \log \tilde{S}(t)$ and $b(t) = \log B(t)$. By Ito's formula, it follows that

$$\begin{aligned} ds(t) &= (\tilde{a} - \tilde{\sigma}^2/2 - \tilde{\rho}^2/2)dt + \tilde{\sigma}dw(t) + \tilde{\rho}d\hat{w}(t) \\ &= (-\tilde{\sigma}^2/2 - \tilde{\rho}^2/2)dt + V(t)^\top dW_\theta(t), \end{aligned} \qquad (3.44)$$

where $V^\top = (\tilde{\sigma}, \tilde{\rho})^\top$, and

$$\begin{aligned} d\lambda(t) &= (\alpha - \rho^2/2 - \hat{\rho}^2/2)dt + \rho dw(t) + \hat{\rho}d\hat{w}(t) \\ &= (\alpha - \rho^2/2 - \hat{\rho}^2/2)dt + \hat{V}(t)^\top dW_\theta(t), \end{aligned} \qquad (3.45)$$

where $\hat{V}^\top = (\rho, \hat{\rho})^\top$.

Up to the end of this section, we assume that there exists a measurable function $\Theta : \mathbf{R}^2 \times [0, T] \to \mathbf{R}^8$ such that

$$(\tilde{a}(t), \tilde{\sigma}(t), \tilde{\rho}(t), \alpha(t), \rho(t), \hat{\rho}(t), \theta_1(t), \theta_2(t))^\top = \Theta(s(t), \lambda(t), t).$$

To simplify notation, we will describe it as the following: we assume that the processes $\tilde{a}(t)$, $\tilde{\sigma}(t)$, $\tilde{\rho}(t)$, $\alpha(t)$, $\rho(t)$, $\hat{\rho}(t)$, $\theta(t)$, $V(t)$, $\hat{V}(t)$ (defined on $[0, T] \times \Omega$) are replaced by the processes $\tilde{a}(s(t), \lambda(t), t)$, $\tilde{\sigma}(s(t), \lambda(t), t)$, $\tilde{\rho}(s(t), \lambda(t), t)$, $\alpha(s(t), \lambda(t), t)$, $\rho(s(t), \lambda(t), t)$, $\hat{\rho}(s(t), \lambda(t), t)$, $\theta(s(t), \lambda(t), t)$, $V(s(t), \lambda(t), t)$, and $\hat{V}(s(t), \lambda(t), t)$, respectively, for some measurable functions $\tilde{a}(s, b, t)$, $\tilde{\sigma}(s, b, t)$, $\tilde{\rho}(s, b, t)$, $\alpha(s, b, t)$, $\rho(s, b, t)$, $\hat{\rho}(s, b, t)$, $\theta(s, b, t)$, $V(s, b, t)$, $\hat{V}(s, b, t)$, defined on $\mathbf{R}^2 \times [0, T]$.

Let $H = H_\theta = H_\theta(s, b, t)$ be the solution of the following Cauchy problem for a linear parabolic equation in $\mathbf{R}^2 \times [0, T]$

$$\begin{aligned} &H'_t + H'_s(-\tilde{\sigma}^2/2 - \tilde{\rho}^2/2) + H'_b(\alpha - \rho^2/2 - \hat{\rho}^2/2) + \mathcal{L}H = 0, \\ &H(s, b, T) = F(e^s, e^b), \quad (s, b) \in \mathbf{R}^2. \end{aligned} \qquad (3.46)$$

Here

$$\mathcal{L}H = \frac{1}{2}\begin{pmatrix} \tilde{\sigma} \\ \rho \end{pmatrix}^\top H'' \begin{pmatrix} \tilde{\sigma} \\ \rho \end{pmatrix} + \frac{1}{2}\begin{pmatrix} \tilde{\rho} \\ \hat{\rho} \end{pmatrix}^\top H'' \begin{pmatrix} \tilde{\rho} \\ \hat{\rho} \end{pmatrix},$$

$$H'' = \begin{pmatrix} H''_{ss} & H''_{sb} \\ H''_{bs} & H''_{bb} \end{pmatrix}.$$

In this section, we assume that there exists a generalized solution $H(s, b, t)$ of the Cauchy problem (3.46) such that its gradient with respect to (s, b) is bounded.

Under these assumptions, consider the pricing problem for the claim $B(T)\xi$, where $\xi = F(\widetilde{S}(T), B(T))$ for some measurable function $F : (0, +\infty)^2 \to \mathbf{R}$ such that $\mathbf{E}_\theta \xi^2 < +\infty$ for some $\theta \in \Theta$.

Proposition 3.34 *The price in (3.32) can be represented as*

$$c_\theta = \mathbf{E}_\theta \xi = H(s(0), \lambda(0), 0) = H(\log \widetilde{S}(0), \log B(0), 0).$$

Furthermore, the hedging strategy in (3.35) can be represented as

$$\gamma(t) = f_\theta(s(t), \lambda(t), t)e^{-s(t)} = f_\theta(\log \widetilde{S}(t), \log B(t), t)\widetilde{S}(t)^{-1},$$

where

$$f_\theta(s, b, t) = H_s'(s(t), \lambda(t), t)$$

$$+ H_b'(s(t), \lambda(t), t)\frac{\widetilde{\sigma}(s, b, t)\rho(s, b, t) + \widetilde{\rho}(s, b, t)\hat{\rho}(s, b, t)}{\widetilde{\sigma}(s, b, t)^2 + \widetilde{\rho}(s, b, t)^2}.$$

Further, let us consider the problem of estimation of $\mathbf{E}R_\theta^2$ for the discounted hedging error R_θ. This error is represented in (3.32)–(3.35) as

$$R_\theta = \mathbf{E} \int_0^T \eta_\theta(t)^\top dW_\theta(t), \qquad \eta_\theta(t) = U_\theta(t) - \nu_\theta(t)V(t).$$

Consider a function $g_\theta = g_\theta(s, b, t) : \mathbf{R}^2 \times [0, T] \to \mathbf{R}$ defined as

$$g_\theta(s, b, t) = H_s'(s, b, t) \begin{pmatrix} \widetilde{\sigma}(s, b, t) \\ \widetilde{\rho}(s, b, t) \end{pmatrix} + H_b'(s, b, t) \begin{pmatrix} \rho(s, b, t) \\ \hat{\rho}(s, b, t) \end{pmatrix}$$

$$- f_\theta(s, b, t) \begin{pmatrix} \widetilde{\sigma}(s, b, t) \\ \widetilde{\rho}(s, b, t) \end{pmatrix}.$$

Let $J = J(y, s, b, t)$ be the solution of the following Cauchy problem for a linear parabolic equation in $\mathbf{R}^3 \times [0, T]$

$$J_t' + J_y' g_\theta^\top \theta + J_s'(\widetilde{a} - \widetilde{\sigma}^2/2 - \widetilde{\rho}^2/2) + J_b'(r - \rho^2/2 - \hat{\rho}^2/2) + \mathcal{D}J,$$

$$J(y, s, b, T) = y^2, \quad (y, s, b) \in \mathbf{R}^3. \tag{3.47}$$

Here $g_\theta = (g_{\theta,1}, g_{\theta,2})^\top$,

$$\mathcal{D}J = \frac{1}{2} \begin{pmatrix} g_{\theta,1} \\ \widetilde{\sigma} \\ \rho \end{pmatrix}^\top J'' \begin{pmatrix} g_{\theta,1} \\ \widetilde{\sigma} \\ \rho \end{pmatrix} + \frac{1}{2} \begin{pmatrix} g_{\theta,2} \\ \widetilde{\rho} \\ \hat{\rho} \end{pmatrix}^\top J'' \begin{pmatrix} g_{\theta,2} \\ \widetilde{\rho} \\ \hat{\rho} \end{pmatrix},$$

$$J'' = \begin{pmatrix} J_{xx}'' & J_{xy}'' & J_{xz}'' \\ J_{yx}'' & J_{yy}'' & J_{yz}'' \\ J_{zx}'' & J_{zy}'' & J_{zz}'' \end{pmatrix}.$$

We assume that there exists a generalized solution $J(y, s, b, t)$ of the Cauchy problem (3.47) such that its gradient with respect to (y, s, b) is bounded.

Proposition 3.35 *We have that*

$$\mathbf{E}R_\theta^2 = J(0, \log s(0), \lambda(0), 0) = J(0, \log \widetilde{S}(0), \log B(0), 0).$$

3.3.5 Proofs

Proof of Lemma 3.15. Let $(\widetilde{X}(t), \gamma(t))$ be a process such that (2.6) holds. Then it suffices to prove that $X(t) \triangleq B(t)\widetilde{X}(t)$ is the wealth corresponding to the self-financing strategy $(\beta(\cdot), \gamma(\cdot))$, where $\beta(t) = (X(t) - \gamma(t)S(t))B(t)^{-1} = \widetilde{X}(t) - \gamma(t)\widetilde{S}(t)$. Clearly, the process $(\widetilde{X}(t), \widetilde{S}(t), \widetilde{B}(t))$ is pathwise continuous. Let Markov times $\{T_k\}_{k=1}^\infty$ be selected as $T_k = \inf\{t \in [0, T] : |X(t)| + |S(t)| + |B(t)| \geq k\}$. It follows that $0 \leq T_k \leq T_{k+1} \leq T$ for all k, $T_k \to T$ as $k \to +\infty$ a.s., and condition (i) in Definition 3.14 holds.

By Ito's formula applied to the product $B(t)\widetilde{X}(t)$ and by (3.30), we have that

$$dX(t) = B(t)d\widetilde{X}(t) + \widetilde{X}(t)dB(t) + \gamma(t)\widetilde{S}(t)B(t)[\widetilde{\sigma}\rho + \widetilde{\rho}\hat{\rho}]dt$$
$$= B(t)d\widetilde{X}(t) + rX(t)dt + \widetilde{X}(t)B(t)(\rho dw(t) + \hat{\rho}d\hat{w}(t))$$
$$+\gamma(t)\widetilde{S}(t)B(t)[\widetilde{\sigma}\rho + \widetilde{\rho}\hat{\rho}]dt.$$

By (3.30), it can be extended as

$$dX(t) = B(t)\gamma(t)d\widetilde{S}(t) + rX(t)dt + \widetilde{X}(t)B(t)(\rho dw(t) + \hat{\rho}d\hat{w}(t))$$
$$+\gamma(t)\widetilde{S}(t)B(t)[\widetilde{\sigma}\rho + \widetilde{\rho}\hat{\rho}]dt$$
$$= \gamma(t)S(t)[(\widetilde{a} + \widetilde{\sigma}\rho + \widetilde{\rho}\hat{\rho})dt + \widetilde{\sigma}dw(t) + \widetilde{\rho}d\hat{w}(t)]$$
$$+rX(t)dt + X(t)(\rho dw(t) + \hat{\rho}d\hat{w}(t)).$$

By the definitions, we have that $\widetilde{a} + \widetilde{\sigma}\rho + \widetilde{\rho}\hat{\rho} + r = a$. It follows that

$$dX(t) = \gamma(t)S(t)[(\widetilde{a} + \widetilde{\sigma}\rho + \widetilde{\rho}\hat{\rho})dt + \widetilde{\sigma}dw(t) - \hat{\rho}d\hat{w}(t)]$$
$$+r[\gamma(t)S(t) + \beta(t)B(t)]dt + [\gamma(t)S(t) + \beta(t)B(t)](\rho dw(t) + \hat{\rho}d\hat{w}(t))$$
$$= \gamma(t)S(t)[adt + \sigma dw(t) + \hat{\sigma}d\hat{w}(t)]$$
$$+\beta(t)B(t)[rdt + \rho dw(t) + \hat{\rho}d\hat{w}(t)] = \gamma(t)dS(t) + \beta(t)dB(t).$$

This completes the proof of Lemma 3.15. $\square$

Proof of Lemma 3.17 follows immediately from equation (3.30) and from the fact that $d\widetilde{S}(t) = \widetilde{S}(t)V(t)^\top dW_\theta(t)$. $\square$

Proof of Proposition 3.20. By Lemma 3.15 and 3.17, the set $\mathcal{X}$ contains random variables

$$\int_0^T \gamma(t)d\widetilde{S}(t) = \int_0^T \gamma(t)\widetilde{S}(t)V(t)^\top dW_\theta(t),$$

where $\gamma \in \mathcal{H}_\theta$ and where W_θ is defined by (3.27).

For any $\zeta \in \mathcal{X}^\perp$, there exists $U(t) = (U_1(t), U_2(t))^\top \in \mathcal{Y}$ such that

$$\zeta = \int_0^T U(t)^\top dW_\theta(t).$$

Let us show that if $\zeta \in \mathcal{X}^\perp$ then $U(t)^\top V(t) = 0$. For this ζ, we have that

$$\mathbf{E}_\theta \zeta \int_0^T \gamma(t) d\widetilde{S}(t) = \mathbf{E}_\theta \int_0^T \gamma(t) \widetilde{S}(t) V(t)^\top U(t) dt = 0 \quad \forall \gamma \in \mathcal{H}_\theta.$$

Hence $\widetilde{S}(t) V(t)^\top U(t) = 0$ a.e. Hence $V(t)^\top U(t) = 0$ a.e.

To show that the set $\mathcal{X}^\perp$ contains non-zero elements, it suffices to take $U_1(t) = \psi(t)\hat{\rho}(t)$ and $U_2(t) = \psi(t)\widetilde{\sigma}(t)$, with an arbitrary $\psi \in \mathcal{Y}$, i.e.,

$$\zeta = \int_0^T \psi(t)[\hat{\rho}(t) dW_{\theta 1}(t) + \widetilde{\sigma}(t) dW_{\theta 2}(t)]. \tag{3.48}$$

This completes the proof. $\square$

Proof of Theorem 3.21. Under the assumptions, $d\widetilde{S}(t) = \widetilde{S}(t)|V(t)|dz_\theta(t)$, where $z_\theta(t)$ is a one-dimensional Wiener process such that $\int_0^t V(s)^\top dW_\theta(s) = \int_0^t |V(s)|dz_\theta(s)$. Hence the filtration $\{\mathcal{F}_t^{z_\theta}\}_{t \geq 0}$ generated by $z_\theta(t)$ is such that $\mathcal{F}_T^{z_\theta} = \mathcal{F}_T^{\widetilde{S}}$. Hence any $\xi \in L_2(\Omega, \mathcal{F}_T^{\widetilde{S}}, \mathbf{P}_\theta)$ belongs to $L_2(\Omega, \mathcal{F}_T^{z_\theta}, \mathbf{P}_\theta)$. By the martingale representation theorem, it follows that there exists an $\mathcal{F}_t^{z_\theta}$-adapted process $u_\theta(t)$ such that $\mathbf{E}_\theta \int_0^T u_\theta(t)^2 dt < +\infty$ and $\xi = \mathbf{E}_\theta \xi + \int_0^T u_\theta(t) dz_\theta(t)$. It suffices to select $\gamma(t) = u_\theta(t) \widetilde{S}(t)^{-1}$. This completes the proof of Theorem 3.21. $\square$

Proof of Theorem 3.22. The proof follows from the martingale representation theorem applied on the probability space $(\Omega, \mathcal{F}_T^w, \mathbf{P}_\theta)$. Let give this proof for the sake of completeness.

By the martingale representation theorem, it suffices to show that the set $\mathcal{X}^\perp$ is trivial, i.e., $\sup_{\eta \in \mathcal{X}^\perp} \mathbf{E}_\theta |\zeta| = 0$. By Lemma 3.15 and Lemma 3.17, the set $\mathcal{X}$ contains random variables

$$\xi = \int_0^T \gamma(t) \widetilde{S}(t) \widetilde{\sigma}(t) dW_{\theta 1}(t).$$

Assume that

$$\zeta = c + \int_0^T \varphi(t) dW_{\theta 1}(t) \in \mathcal{X}^\perp,$$

where $c \in \mathbf{R}$ and where $\varphi(t)$ is an $\mathcal{F}_t$-adapted process that is square integrable under $\mathbf{P}_\theta$. By the definition of $\mathcal{X}^\perp$, it follows that, for all γ,

$$\mathbf{E}_\theta(\xi\zeta) = \mathbf{E}_\theta \int_0^T \gamma(t) \widetilde{S}(t) \widetilde{\sigma}(t) \varphi(t) dt = 0.$$

It follows that $\varphi(\cdot) = 0$. Hence $\zeta = 0$ (i.e., $\mathbf{E}_\theta |\zeta|^2 = 0$). This completes the proof of Theorem 3.22. $\square$

Proof of Corollary 3.23. Let the initial wealth $c_{\theta i} = X^{(i)}(0)$ and the strategy $(\beta^{(i)}(\cdot), \gamma^{(i)}(\cdot))$ be such that $\widetilde{X}^{(i)}(T) = \xi$ a.s. for the corresponding discounted wealth $X^{(i)}(t)$, $i = 1, 2$.

Set
$$g(t) \triangleq \gamma^{(1)}(t) - \gamma^{(2)}(t), \qquad Y(t) \triangleq \widetilde{X}^{(1)}(t) - \widetilde{X}^{(2)}(t).$$

We have that $Y(T) = 0$ a.s. Hence

$$Y(T) = Y(0) + \int_0^T g(t) d\widetilde{S}(t) = 0.$$

For $K > 0$, consider first exit times $T_K = T \wedge \inf\{t : \int_0^t (|\gamma^{(1)}(s)| + |\gamma^{(2)}(s)|^2 ds \geq K\}$; they are Markov times with respect to $\{\mathcal{F}_t\}_{t \geq 0} t$. We have that

$$Y(T_K) = Y(0) + \int_0^{T_K} g(t) d\widetilde{S}(t) = \mathbf{E}_{\theta i}\{Y(T) \mid \mathcal{F}_{T_K}\} = 0, \quad i = 1, 2.$$

Hence

$$0 = Y(0)^2 + \mathbf{E}_{\theta_i} \int_0^{T_K} g(s)^2 \widetilde{S}(s)^2 |V(s)|^2 dt = 0.$$

It follows that $Y(0) = 0$, and $g(t)|_{[0,T_K]} = 0$ for any $K > 0$. In addition, $T_K \to T$ a.s. as $T_K \to +\infty$. Hence $g = 0$. This completes the proof of Corollary 3.23. $\square$

Proof of Theorem 3.24. By (3.37), for any $K \in \mathbf{R}$, there exists $\theta = \theta_K \in \Theta$ such that

$$\theta_1 \widetilde{\sigma} + \theta_2 \widetilde{\rho} = \widetilde{a},$$
$$\theta_1 \rho + \theta_2 \hat{\rho} = \hat{V}(t)^\top \theta(t) = K - \alpha + \rho^2 + \hat{\rho}^2.$$

By Girsanov's theorem, $W_\theta(t) = W(t) + \int_0^t \theta(s) ds$ is a standard Wiener process in $\mathbf{R}^2$ under $\mathbf{P}_\theta$. We have $d\widetilde{S}(t) = \widetilde{S}(t) V(t)^\top dW_\theta(t)$ and

$$
\begin{aligned}
dB(t)^{-1} &= B(t)^{-1}([-\alpha + \rho^2 + \hat{\rho}^2] dt - \rho dw(t) - \hat{\rho} d\hat{w}(t)) \\
&= B(t)^{-1}([-\alpha + \rho^2 + \hat{\rho}^2] dt - \hat{V}(t)^\top dW(t)) \\
&= B(t)^{-1}([-\alpha + \rho^2 + \hat{\rho}^2] dt - \hat{V}(t)^\top \theta(t) dt + \hat{V}(t)^\top dW_\theta(t)) \\
&= B(t)^{-1}([-\alpha + \rho^2 + \hat{\rho}^2] dt - (K - \alpha + \rho^2 + \hat{\rho}^2) dt + \hat{V}(t)^\top dW_\theta(t)) \\
&= B(t)^{-1}(-K dt + \hat{V}(t)^\top dW_\theta(t)).
\end{aligned}
$$

Let $\hat{B}(t)^{-1} = e^{Kt} B(t)^{-1}$. We have

$$d\hat{B}(t)^{-1} = \hat{B}(t)^{-1} \hat{V}(t)^\top dW_\theta(t).$$

It follows that $\hat{B}(t)^{-1}$ is a martingale under $\mathbf{P}_\theta$.

Let us prove statement (i) of Theorem 3.24. Let $K > 0$. Then, we have

$$
\begin{aligned}
\mathbf{E}_\theta \xi = \mathbf{E}_\theta B(T)^{-1} (\kappa - S(T))^+ &\leq \mathbf{E}_\theta B(T)^{-1} \kappa = e^{-KT} \kappa \mathbf{E}_\theta \hat{B}(T)^{-1} \\
&= e^{-KT} \kappa \hat{B}(0)^{-1} \to 0 \quad \text{as} \quad K \to +\infty.
\end{aligned}
$$

Let us prove statement (ii) of Theorem 3.24. Let $K < 0$. We have

$$\mathbf{E}_\theta \xi = \mathbf{E}_\theta B(T)^{-1}(\kappa - S(T))^+ = \mathbf{E}_\theta (B(T)^{-1}\kappa - \widetilde{S}(T))^+$$
$$\geq \mathbf{E}_\theta B(T)^{-1}\kappa - \mathbf{E}_\theta \widetilde{S}(T) \kappa e^{-KT} \mathbf{E}_\theta \hat{B}(T)^{-1} - \widetilde{S}(0)$$
$$= \kappa e^{-KT} \hat{B}(0)^{-1} - \widetilde{S}(0) \to +\infty \quad \text{as} \quad K \to -\infty.$$

This completes the proof of Theorem 3.24. $\square$

Proof of Theorem 3.25. Let $\theta = \theta_K$ and $\hat{B}(t)$ be such as defined in the proof of Theorem 3.24.

Let us prove statement (i) of Theorem 3.25. We have

$$\mathbf{E}_\theta \xi = \mathbf{E}_\theta B(T)^{-1}(S(T) - \kappa)^+ = \mathbf{E}_\theta \mathbb{I}_{\{S(T)>\kappa\}} B(T)^{-1}(S(T) - \kappa)$$
$$\leq \mathbf{E}_\theta \mathbb{I}_{\{S(T)>\kappa\}} B(T)^{-1} S(T).$$

Let $\hat{S}(t) = e^{Kt} S(t)$. By the definitions, we have

$$\begin{aligned}
dS(t) &= S(t)\left(adt + V^\top dW(t) + \hat{V}^\top d\hat{W}(t)\right) \\
&= S(t)\left(adt - V^\top \theta dt - \hat{V}^\top \theta dt + V^\top dW_\theta(t) + \hat{V}^\top d\hat{W}_\theta(t)\right) \\
&= S(t)\left((a - \tilde{a})dt - (K - \alpha + \rho + \hat{\rho}^2)dt + V^\top dW_\theta(t) + \hat{V}^\top d\hat{W}_\theta(t)\right)
\end{aligned}$$

and

$$d\hat{S}(t) = \hat{S}(t)\left((a - \tilde{a})dt + (\alpha - \rho - \hat{\rho}^2)dt + V^\top dW_\theta(t) + \hat{V}^\top d\hat{W}_\theta(t)\right).$$

It follows from the standard estimates for stochastic differential equations that

$$\sup_K \mathbf{E}_{\theta_K} |\hat{S}(T)| < +\infty,$$

for $\theta = \theta_K$; see, e.g., Chapter 2 in [79]. Hence $\mathbb{I}_{\{S(T)>\kappa\}} = \mathbb{I}_{\{\hat{S}(T)>\kappa e^{KT}\}} \to 0$ a.s. as $K \to +\infty$. By the Lebesgue dominated convergence theorem, it follows that $\mathbf{E}_\theta \xi \to 0$ as $K \to +\infty$. Hence statement (i) of Theorem 3.25 follows.

Let us prove statement (ii) of Theorem 3.25. For $K > 0$, we have that

$$\begin{aligned}
\mathbf{E}_\theta \xi &= \mathbf{E}_\theta B(T)^{-1}(S(T) - \kappa)^+ = \mathbf{E}_\theta (\widetilde{S}(T) - B(T)^{-1}\kappa)^+ \\
&\geq \mathbf{E}_\theta \widetilde{S}(T) - e^{-KT} \mathbf{E}_\theta \hat{B}(T)^{-1}\kappa = \widetilde{S}(0) - e^{-KT} \hat{B}(0)^{-1}\kappa \to S(0) \\
&\quad \text{as} \quad K \to +\infty.
\end{aligned}$$

In addition, we have

$$\mathbf{E}_\theta \xi = \mathbf{E}_\theta B(T)^{-1}(S(T) - \kappa)^+ = \mathbf{E}_\theta (\widetilde{S}(T) - B(T)^{-1}\kappa)^+ \leq \mathbf{E}_\theta \widetilde{S}(T) = S(0).$$

This completes the proof of Theorem 3.25. $\square$

Proof of Theorem 3.26. Let $K > 0$, and $\vartheta = \vartheta_K = -K\eta_\theta + \theta$. By Girsanov's theorem,

$$W_\vartheta(t) = W_\theta(t) - K \int_0^t \eta_\theta(s) ds$$

is a Wiener process under $\mathbf{P}_\vartheta$. By the definitions,

$$R_\theta = \int_0^T \eta_\theta(t)^\top dW_\theta(t) = K \int_0^T |\eta_\theta(t)|^2 dt + \int_0^T \eta_\theta(t)^\top dW_\vartheta(t).$$

Let

$$N_1(K) \triangleq \mathbf{E}_\vartheta \left(\int_0^T |\eta_\theta(t)|^2 dt \right)^2,$$

$$N_2(K) \triangleq \mathbf{E}_\vartheta \left(\int_0^T \eta_\theta(t)^\top dW_\vartheta(t) \right)^2.$$

We have

$$\inf_{K>0} N_1(K) = \inf_{K>0} \mathbf{E}_\vartheta \left(\int_0^T (|\eta_\theta(t)|^2 dt \right)^2 \geq \operatorname{ess\,inf}_\omega \left(\int_0^T |\eta_\theta(t,\omega)|^2 dt \right)^2 > 0.$$

By Proposition 3.20 and by assumptions on U_θ, it follows that $\operatorname{ess\,sup}_{t,\omega} \int_0^T |\eta_\theta(t,\omega)|^2 dt < +\infty$. Hence

$$\sup_{K>0} N_2(K) = \sup_{K>0} \mathbf{E}_\vartheta \int_0^T |\eta_\theta(t)|^2 dt \leq \operatorname{ess\,sup}_\omega \int_0^T |\eta_\theta(t,\omega)|^2 dt < +\infty.$$

Hence $\mathbf{E}_\vartheta R_\theta^2 \geq K^2 N_1(K) - 2K\sqrt{N_1(K)N_2(K)} + N_2(K) \to +\infty$ as $K \to +\infty$. This completes the proof of Theorem 3.26. $\square$

Proof of Theorem 3.27. Let $K > 1$, and let $y(t)$ evolves as

$$dy(t) = \eta_\theta(t)^\top dW_\theta(t), \quad y(0) = 0.$$

Let $T_K = T \wedge \inf\{t > 0: \int_0^t y(s)^2 ds \geq K\}$. Let

$$q(t) = q_K(t) \triangleq -Ky(t)\frac{\eta_\theta(t)}{|\eta_\theta(t)|} \mathbb{I}_{\{t \leq T_K\}}.$$

This is an $\mathcal{F}_t$-adapted process such that $q(t)^\top V(t) = 0$, $q(t)^\top \eta_\theta(t) = -Ky(t)$ for $t \leq T_K$, $q(t) = 0$ for $t > T_K$, $|q(t)| \leq K|y(t)|$, and

$$\int_0^T |q(s)|^2 ds = \int_0^{T_K} |q(s)|^2 ds \leq K^2 \int_0^{T_K} y(s)^2 ds \leq K^3.$$

In particular, it follows that $\theta - q \in \Theta$.

Let $\vartheta = \vartheta_K \triangleq \theta - q$. Then

$$W_\vartheta(t) = W_{\theta-q}(t) = W_\theta(t) - \int_0^t q(s)ds$$

is a Wiener process under $\mathbf{P}_\vartheta$. By the definitions, it follows that

$$dy(t) = -Ky(t)dt + \eta_\theta(t)^\top dW_\vartheta(t), \quad y(0) = 0.$$

By the Girsanov theorem, the measure $\mathbf{P}_\vartheta$ is equivalent to $\mathbf{P}_\theta$. We have

$$
\begin{aligned}
R_\theta &= \int_0^T \eta_\theta(t)^\top dW_\theta(t) = \int_0^T [\eta_\theta(t)^\top q(t)dt + \eta_\theta(t)^\top dW_\vartheta(t)] \\
&= -\int_0^{T_K} Ky(t)dt + \int_0^T \eta_\theta(t)^\top dW_\vartheta(t) = y(T_K) + \int_{T_K}^T \eta_\theta(t)^\top dW_\vartheta(t).
\end{aligned}
$$

By the assumptions on U_θ and by (3.34), it follows that

$$\mathbf{C}_\eta \stackrel{\Delta}{=} \operatorname{ess\,sup}_{t,\omega} |\eta_\theta(t,\omega)| < +\infty. \tag{3.49}$$

Clearly,

$$
\begin{aligned}
\mathbf{E}_\vartheta y(T_K)^2 &= \mathbf{E}_\vartheta \int_0^{T_K} e^{-2K(T_K-s)}|\eta_\theta(s)|^2 ds \\
&\leq \mathbf{C}_\eta^2 \mathbf{E}_\vartheta \int_0^{T_K} e^{-2K(T_K-s)} ds = \mathbf{C}_\eta^2 \mathbf{E}_\vartheta \frac{1-e^{-2KT_K}}{2K} \leq \frac{1-e^{-2KT}}{2K} \to 0
\end{aligned}
$$

$$\text{as} \quad K \to +\infty. \tag{3.50}$$

Consider events $A_K = \{\int_0^T y(t)^2 dt > K\} = \{T_K < T\}$. We have

$$\mathbf{E}_\vartheta \left(\int_{T_K}^T \eta_\theta(t)^\top dW_\vartheta(t) \right)^2 \leq \mathbf{E}_\vartheta \int_{T_K}^T |\eta_\theta(t)|^2 dt \leq \mathbf{E}_\vartheta \mathbb{I}_{A_K} \int_0^T |\eta_\theta(t)|^2 dt$$

$$\leq T\mathbf{C}_\eta^2 \mathbf{P}_\vartheta(A_K).$$

Thus, we have

$$\mathbf{E}_\vartheta y(t)^2 dt \leq \mathbf{C}_\eta^2 \int_0^t e^{-2K(t-s)} ds \leq \mathbf{C}_\eta^2 T.$$

By the Markov inequality and by (3.49), it follows that

$$\mathbf{P}_\vartheta(A_K) \leq \frac{1}{K}\mathbf{E}_\vartheta \int_0^T y(t)^2 dt \leq \frac{1}{K}\mathbf{C}_\eta^2 T^2 \to 0 \quad \text{as} \quad K \to +\infty. \tag{3.51}$$

By (3.50)–(3.51), $\mathbf{E}_\vartheta R_\theta^2 \to 0$ as $K \to +\infty$. This completes the proof of Theorem 3.27. $\square$

Proof of Theorem 3.28. It can be verified directly that the equations for S and B have the desired form. $\square$

Proof of Theorem 3.33. First, the standard Lagrange optimization techniques give immediately that the selected θ is such that $|\theta(t, \omega)|$ is minimal over all $\theta \in \Theta$ and it is a unique solution of the problem

$$\text{Minimize} \quad |\theta| \quad \text{subject to} \quad V(t, \omega)^{\top}\theta = \widetilde{a}(t, \omega).$$

Further, by martingale representation theorem, we see that, for some $U_\theta \in \mathcal{Y}$, presentation (3.31) holds. It was shown in Section 3.3.2 that (3.33) and (3.34) hold. By Proposition 3.19, we see $V(t)^{\top}\eta_\theta(t) \equiv 0$. For our choice of θ, this gives $\theta(t)^{\top}\eta_\theta(t) \equiv 0$. It follows that

$$R_\theta = \int_0^T \eta_\theta(t)^{\top} dW_\theta(t) = \int_0^T \eta_\theta(t)^{\top} dW(t).$$

Hence

$$\mathbf{E}R_\theta \int_0^T \gamma(t)\widetilde{S}(t)V(t)^{\top} dW(t)$$

$$= \mathbf{E}\int_0^T \eta_\theta(t)^{\top} dW(t) \int_0^T \gamma(t)\widetilde{S}(t)V(t)^{\top} dW(t)$$

$$= \mathbf{E}\int_0^T \gamma(t)\widetilde{S}(t)\eta_\theta(t)^{\top}V(t)dt = 0.$$

This completes the proof of Theorem 3.33. $\square$

Proof of Proposition 3.34. By the Ito formula, it follows that

$$\xi - H(s(0), \lambda(0), 0) = H(s(T), \lambda(T), T) - H(s(0), \lambda(0), 0)$$

$$= \int_0^T (H_t' + H_s'(-\widetilde{\sigma}^2/2 - \widetilde{\rho}^2/2) + H_b'(\alpha - \rho^2/2 - \hat{\rho}^2/2) + \mathcal{L}H)dt$$

$$+ \int_0^T U_\theta(t)^{\top}W_\theta(t),$$

where

$$U_\theta(t) = H_s'(s(t), \lambda(t), t)V(t) + H_b'(s(t), \lambda(t), t)\hat{V}(t),$$

and where $V(t) = V(s(t), \lambda(t), t)$, $\hat{V}(t) = \hat{V}(s(t), \lambda(t), t)$. By the choice of H, it follows that

$$\xi - H(s(0), \lambda(0), 0) = \int_0^T U_\theta(t)^{\top}W_\theta(t).$$

Hence $\mathbf{E}_\theta\xi = H(s(0), \lambda(0), 0)$ and (3.31) holds with this $U_\theta(t)$. We have established in Section 3.3.2 that

$$\gamma(t) = \nu_\theta(t)\widetilde{S}(t)^{-1}, \quad \nu_\theta(t) = \frac{U_\theta(t)^{\top}V(t)}{|V(t)|^2}.$$

In addition, we have that

$$\frac{U_\theta(t)^\top V(t)}{|V(t)|^2} = H'_s(\mathrm{s}(t), \lambda(t), t) + H'_b(\mathrm{s}(t), \lambda(t), t)\frac{\hat{V}(t)^\top V(t)}{|V(t)|^2}.$$

This gives (3.47), since, by the assumptions of this section, we have that

$$\frac{\hat{V}(t)^\top V(t)}{|V(t)|^2} = \frac{(\widetilde{\sigma}(s, b, t)\rho(s, b, t) + \widetilde{\rho}(s, b, t)\hat{\rho}(s, b, t))}{\widetilde{\sigma}(s, b, t)^2 + \widetilde{\rho}(s, b, t)^2}.$$

This completes the proof of Proposition 3.34. □

Proof of Proposition 3.35. We have that

$$\mathbf{E}R_\theta^2 = \mathbf{E}\int_0^T |\eta_\theta(t)|^2 dt, \qquad \eta_\theta(t) = U_\theta(t) - \nu_\theta(t)V(t).$$

Under the assumptions of this section, $\eta_\theta(t) = g_\theta(\mathrm{s}(t), \lambda(t), t)$. Further, under the assumption of Proposition 3.19, $R_\theta = y(T)$, where the process $x(t)$ evolves as

$$dy(t) = \eta_\theta(t)^\top dW_\theta(t) = \eta_\theta(t)^\top \theta(t)dt + \eta_\theta(t)^\top dW(t), \quad y(0) = 0.$$

This can be rewritten as

$$dy(t) = g_\theta(\mathrm{s}(t), \lambda(t), t)^\top \theta(\mathrm{s}(t), \lambda(t), t)dt + g_\theta(\mathrm{s}(t), \lambda(t), t)^\top dW(t).$$

This equation coupled with equations (3.44) and (3.45) describes the evolution of a diffusion Markov process $(y(t), \mathrm{s}(t), \lambda(t))$ such that $J(y, s, b, t)$ is the solution in $\mathbf{R}^3 \times [0, T]$ of the corresponding backward Kolmogorov–Fokker–Planck parabolic equation for the Markov diffusion process $(y(t), \mathrm{s}(t), \lambda(t))$.

By the Ito formula, it follows that

$$\begin{aligned}
&\mathbf{E}[y(T)^2 - J(y(0), \mathrm{s}(0), \lambda(0), 0)]\\
&= \mathbf{E}[J(y(T), \mathrm{s}(T), \lambda(T), T) - J(0, \mathrm{s}(0), \lambda(0), 0)]\\
&= \mathbf{E}\int_0^T (J'_t + J'_y g_\theta^\top \theta + J'_s(\widetilde{a} - \widetilde{\sigma}^2/2 - \widetilde{\rho}^2/2)\\
&\quad + J'_b(r - \rho^2/2 - \hat{\rho}^2/2) + \mathcal{D}J)dt.
\end{aligned}$$

By the choice of J, it follows that the right-hand part of this equality is zero, i.e.,

$$\mathbf{E}[y(T)^2 - J(0, \mathrm{s}(0), \lambda(0), 0)] = 0.$$

This completes the proof of Proposition 3.35. □

3.4   Bibliographic notes and literature review

Section 3.1

Section 3.1 is based on the results from [43].

Section 3.2

Stochastic delay differential equations have been widely studied, including quite general models with nonlinear dependence on the delayed term were allowed; see, e.g., [13, 70, 92, 93, 108], and the bibliographies therein. The first market model with a stochastic delay equation for the prices was introduced and investigated in [113], where no-arbitrage properties were established. Unfortunately, the results from [13, 70, 92, 93, 113] cannot be used for our relatively simple model (3.17), because the coefficients in (3.18) with log functions do not satisfy the conditions on regularity imposed therein. For this reason, non-arbitrage properties established in [13, 113] cannot be applied directly to our model (3.17).

Section 3.3

Incomplete modifications of the Black–Scholes model, where $B(t)$ was an Ito process, were considered in [34, 77, 20, 16]. These works considered the martingale pricing method where the option price is calculated as the expectation of the discounted claim under some equivalent risk-neutral measure (martingale measure) such that the discounted stock price $S(t)/B(t)$ is a martingale on a given time interval $[0, T]$ under this measure. A particular equivalent martingale measure corresponding to Example 3.12 (iv) was considered in [34]. A multi-stock market under requirements that make the choice of an equivalent martingale measure unique in the case of a single stock and stochastic bond was considered in [20]; see Example 3.12 (iii). Pricing of replicable claims was considered in [58]. Asymptotic properties of this price with respect to a particular equivalent martingale measure were studied in [77]. In [118], an applied random numéraire was used to reduce the computational dimension for Asian options and general semimartingales. In [69], random numéraire was applied to reduce the computational dimension for variance swap options and models based on both Brownian motion and jump processes. Some related results and more references can be found [71], Ch.2, and [107, 119].

The model described in Section 3.3 was introduced in [50] and is close to the model from [34] with a modification that ensures the existence of many equivalent martingale measures. In [34], an impact of the absence of an equivalent martingale measure was studied in the setting with a stochastic bond price $B(t)$ such that $B(T) = 1$. In this case, the appreciation rate of the discounted stock price is imploding when terminal time is approached, and the Novikov condition of existence of an equivalent martingale measure is not satisfied. The setting described in Section 3.3 removes this feature; it uses a stochastic numéraire without restrictions on the terminal price. This could be close to the model from [34] if one considers a stochastic bond with the price $B(t)$ maturing at $T + \varepsilon$, i.e., such that $B(T + \varepsilon) = 1$, for an arbitrarily small $\varepsilon > 0$.

For portfolio selection problems, related questions arise in the setting with a random numéraire; see, e.g., [18, 73, 74, 123]. In addition, a similar setting covers portfolio selection models without riskless assets, including discrete time market models and even single period market models; see, e.g. [123]. Lemma 3.15 represents a special case of Proposition 1 in [58].

4

Pathwise inference for the parameters of market models

Summary

In this chapter, some pathwise methods for inferring the parameters of the diffusion coefficient of continuous time market models are presented. These methods are applicable to a sole path of the observed prices and do not require the observation of an ensemble of such paths. In addition, a Kalman filter approach estimating the appreciation rate of stock prices is discussed.

4.1   Estimation of volatility

In this section, we consider estimation of the historical volatility of stock prices.

4.1.1   Representation theorems for the volatility

The underlying continuous time model

It is assumed that the stock prices are represented as time series formed as samples of the solution of a stochastic differential equation with random and time-varying parameters; these parameters are not observable directly and have an unknown evolution law. The price samples are available with limited frequency only. We describe below the standard summation formula for the volatility and two modifications. In addition, a linear transformation eliminating the appreciation rate and preserving the volatility is suggested.

Consider a risky asset (stock, foreign currency unit, etc.,) with time series of the prices $S_1, S_2, S_3, \ldots$, for example, daily prices.

We consider the mainstream diffusion model for stock prices described above. For this model, $S_k = S(t_k)$, where $S(t)$ is a continuous time random process such that

$$dS(t) = S(t)[a(t)dt + \sigma(t)dw(t)]. \tag{4.1}$$

Here $w(t)$ is a Wiener process, $a(t)$ is the appreciation rate, $\sigma(t)$ is the volatility, $t > 0$. We assume that a and σ are some scalar random processes such that $(a(t), \sigma(t))$

73

is independent from $w(\tau) - w(\theta)$ for all θ, τ such that $\theta > \tau \geq t$. We assume that the process $(a(t), \sigma(t))$ belongs to $L_2(0, T)$ with probability 1 (i.e., $\int_0^T [a(s)^2 + \sigma(s)^2]ds < +\infty$ with probability 1), for a given $T > 0$.

As was discussed above, this model has many financial applications, including pricing of derivatives and optimal portfolio selection. Practical implementation of this model requires estimating (a, σ) from the historical data. For constant a and σ, satisfactory estimates can be obtained from sufficiently large samples. For financial models, estimation of a is challenging since the trend for financial time series is usually relatively small and unstable. Estimation of σ gives more robust results.

The significance of the precision of the volatility estimation can be illustrated as the following. For instance, assume that some volatility estimate is applied for option pricing as a parameter for the Black–Scholes formula. Consider, for example, calculation of a call option price with the exercise time $T = 1$, the initial stock price $S(0) = 1$, the risk-free short-term rate 0.03, and with the strike price 1.2. The option price calculated for constant volatility $\sigma = 0.4$ is 0.1016, and the option price calculated for constant volatility $1.05\sigma = 0.42$ is 0.1095. This means that the 5% error for volatility estimate gives 7% error for the option price which is quite significant.

It is convenient to characterize the volatility process via its running mean

$$v(t) = \frac{1}{\Delta t} \int_{t-\Delta t}^{t} \sigma(s)^2 ds.$$

In fact, the process $v(t)$ is the one that is usually estimated from the historical data, rather than the underlying process $\sigma(t)$ itself. Up to the end of this section, we consider estimation of the process $v(t)$.

We give two useful theorems below for representation of the running average $v(t)$.

Theorem 4.1 *Model (4.1) implies that the value* $v(t) = \frac{1}{\Delta t} \int_{t-\Delta t}^{t} \sigma(s)^2 ds$ *can be found explicitly as*

$$v(t) = \frac{2}{\Delta t} \left(\int_{t-\Delta t}^{t} \frac{dS(s)}{S(s)} - \log S(t) + \log S(t - \Delta t) \right). \tag{4.2}$$

Proof. It is well known that any solution of equation (4.1) is such that

$$S(t) = S(t - \Delta t) \exp \left(\int_{t-\Delta t}^{t} a(s)ds - \frac{1}{2} \int_{t-\Delta t}^{t} \sigma(s)^2 ds + \int_{t-\Delta t}^{t} \sigma(s)dw(s) \right).$$

It follows that

$$\log S(t) - \log S(t - \Delta t) = \int_{t-\Delta t}^{t} a(s)ds - \frac{1}{2} \int_{t-\Delta t}^{t} \sigma(s)^2 ds + \int_{t-\Delta t}^{t} \sigma(s)dw(s).$$

In addition,

$$\int_{t-\Delta t}^{t} \frac{dS(t)}{S(t)} = \int_{t-\Delta t}^{t} [a(s)ds + \sigma(s)dw(s)].$$

Hence

$$\frac{1}{2}\int_{t-\Delta t}^{t}\sigma(s)^2 ds = \int_{t-\Delta t}^{t}\frac{dS(t)}{S(t)} - \log S(t) + \log S(t - \Delta t). \qquad (4.3)$$

Then equation (4.3) follows. □
    Let us give one more estimate of $v(t)$.

Theorem 4.2 *Model (4.1) implies that*

$$\int_{t-\Delta t}^{t}\sigma(s)^2 ds = 2\log|X(t)|, \qquad (4.4)$$

where $X(s)$ is a complex-valued process defined for $s \in [t - \Delta t, t]$ such that

$$dX(s) = iX(s)\frac{dS(s)}{S(s)}, \quad s \in (t - \Delta t, t),$$

$$X(t - \Delta t) = 1. \qquad (4.5)$$

Here $i = \sqrt{-1}$ is the imaginary unit.

   Proof. By the Ito formula, the solution $X(t)$ of (4.5) is

$$\begin{aligned}
X(t) &= X(t - \Delta t)\exp\left(i\int_{t-\Delta t}^{t}a(s)ds - \frac{i^2}{2}\int_{t-\Delta t}^{t}\sigma(s)^2 ds\right.\\
&\quad \left.+i\int_{t-\Delta t}^{t}\sigma(s)dw(s)\right)\\
&= \exp\left(i\int_{t-\Delta t}^{t}a(s)ds + \frac{1}{2}\int_{t-\Delta t}^{t}\sigma(s)^2 ds + i\int_{t-\Delta t}^{t}\sigma(s)dw(s)\right).
\end{aligned}$$

Hence

$$|X(t)| = \exp\left(\frac{1}{2}\int_{t-\Delta t}^{t}\sigma(s)^2 ds\right).$$

Then (4.4) follows. □

4.1.2   Estimation of discrete time samples

In the continuous time setting, the process $\sigma(t)$ is always adapted to the filtration generated by the historical prices $S(s)$, $s \leq t$. This implies that, in theory, $\sigma(t)$ can be estimated without error from the observation of the continuous path on the time interval $[t - \varepsilon, t]$ for an arbitrarily small $\varepsilon > 0$. In practice, only finite time series of the prices observed with limited frequency are available. This generates the error in matching the statistical estimates with the value of $\sigma(t)$ in the continuous time model. The problem of reducing this error is the main focus of this section. We consider estimates for random and time variable volatility at time t based on statistics collected at time $[t - \Delta t, t]$, where $\Delta t > 0$ is given.

Our goal is to construct an estimate $v(t)$ from available samples $S(t_k)$, where $t_k \in [t - \Delta t, t]$, $k = m_0, m_0 + 1, ..., m$. We assume that the time points t_k are equally spaced with a sampling interval $\delta = t_k - t_{k-1}$. Furthermore, we assume that $t_{m_0} = t - \Delta t$ and $t_m = t$. This means that $\Delta t = (m - m_0)\delta$.

The traditional estimate

The traditional estimate of $v(t)$ is represented by the sample variance of the series $\delta^{-1/2} \log(S(t_k)/S(t_{k-1}))$. This traditional estimate can be calculated as

$$\hat{v}(t_m) = \frac{1}{\Delta t} \sum_{k=m_0+1}^{m} (A_m - Z_k)^2, \tag{4.6}$$

where $\Delta t = (m - m_0)\delta$,

$$A_m = \frac{1}{m - m_0} \sum_{k=m_0+1}^{m} Z_k,$$

and where

$$Z_k = \log S(t_k) - \log S(t_{k-1}).$$

(See, e.g., estimate (9.1) in [43]. We suggest two modifications below of this estimate. These modifications are based on the assumptions that the underlying time series are generated by model (4.1) and on the properties of the continuous time Ito processes.)

The estimate based on (4.2)

Unfortunately, the second integral in (4.3) cannot be calculated without error but rather has to be estimated using available prices $S(t_k)$.

Let

$$\xi_k = \frac{S(t_k) - S(t_{k-1})}{S(t_{k-1})}. \tag{4.7}$$

For $t = t_m$ and $t - \Delta t = t_{m_0}$, we have to use approximation

$$\int_{t-\Delta t}^{t} \frac{dS(s)}{S(s)} \sim \sum_{k=m_0+1}^{m} \xi_k. \tag{4.8}$$

Formula (4.2) leads to the following estimate of $v(t)$:

$$\hat{v}(t) = \frac{2}{\Delta t} \left(\sum_{k=m_0+1}^{m} \xi_k - \log S(t_m) + \log S(t_{m_0}) \right), \tag{4.9}$$

where $\Delta t = (m - m_0)\delta$.

The estimate based on (4.4)

The numerical implementation of Theorem 4.1 and (4.4) suggests estimating the value $X(t)$. Again, this value cannot be calculated without error but rather has to be estimated using available prices $S(t_k)$.

The time discretization of (4.5) leads to the stochastic difference equation

$$X(t_k) - X(t_{k-1}) = iX(t_{k-1})\xi_k, \quad k \geq m_0 + 1,$$
$$X(t_{m_0}) = 1$$

where ξ_k are defined by (4.7). This equation can be rewritten as

$$X(t_k) = X(t_{k-1})(1 + i\xi_k), \quad k \geq m_0 + 1,$$
$$X(t_{m_0}) = 1.$$

Hence

$$X(t_m) = \prod_{k=m_0+1}^{m} (1 + i\xi_k).$$

Clearly,

$$|X(t_m)| = \prod_{k=m_0+1}^{m} |1 + i\xi_k| = \prod_{k=m_0+1}^{m} (1 + \xi_k^2)^{1/2},$$

and

$$\log|X(t_m)| = \sum_{k=m_0+1}^{m} \log[(1 + \xi_k^2)^{1/2}] = \frac{1}{2} \sum_{k=m_0+1}^{m} \log(1 + \xi_k^2).$$

Therefore, formula (4.4) leads to the following estimate of $v(t)$:

$$\hat{v}(t) = \frac{1}{\Delta t} \sum_{k=m_0+1}^{m} \log(1 + \xi_k^2), \tag{4.10}$$

where $\Delta t = (m - m_0)\delta$.

Note that estimate (4.6) represents the sample variance of the series $\delta^{-1/2} \log(S(t_k)/\log(S(t_{k-1}))$ and is not directly associated with model (4.1). Estimates (4.9) and (4.10) have a different origin; they were derived from the properties of model (4.1).

Figure 4.1 shows estimates (4.6) and (4.10) for applied for Monte Carlo simulated prices with $\sigma(s) = 1 + 0.25\cos(s\pi/(3\Delta t))$ and with $a(s) \equiv 0.5$, in the setting described in Section 4.1.5 below. This figure demonstrates that these estimates produce close but still different results.

Estimate (4.6) compensates for the impact of a large appreciation rate even for a low sampling frequency. The approach described in the following section helps to achieve this feature for estimates (4.9) and (4.10) as well.

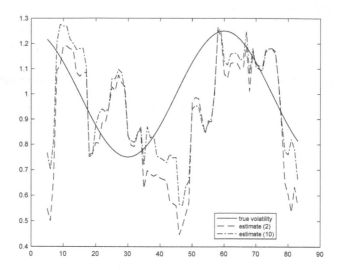

Figure 4.1: $- - -$: values of $\sigma(t)$; ——: values of $\hat{v}(t)$ defined by estimate (4.6); $- \cdot - \cdot -$: values of $\hat{v}(t)$ defined by estimate (4.10). This figure shows that these estimates produce close but still different results.

4.1.3   Reducing the impact of the appreciation rate

Since only short time series $S(t_k)$ are observable, it is not possible to separate the impact of the noise $\sigma(t)dw(t)$ from the impact of the random and time variable input $a(t)dt$ defined by the appreciation rate process.

Let $\gamma(t)$ be an adapted process, and let

$$\hat{S}(t) = S(0) + \int_0^t \gamma(s)\hat{S}(s)S(s)^{-1}dS(s).$$

It follows from the definitions that $\hat{S}(t)$ is the solution of the equation

$$d\hat{S}(t) = \gamma(t)\hat{S}(t)S(t)^{-1}dS(t), \quad t > 0,$$
$$\hat{S}(0) = S(0),$$

i.e.,

$$d\hat{S}(t) = \hat{S}(t)[\hat{a}(t)dt + \hat{\sigma}(t)dw(t)],$$

where

$$\hat{a}(t) = \gamma(t)a(t), \quad \hat{\sigma}(t) = \gamma(t)\sigma(t).$$

Lemma 4.3 *There exists a sequence of the processes* $\gamma(t) = \gamma_j(t)$ *such that* $|\gamma_j(t)| \equiv 1$ *for all* j *and that*

$$\int_0^T \gamma_j(t)f(t)dt \to 0 \quad as \quad j \to +\infty \quad for\ any \quad f(\cdot) \in L_2(0,T).$$

Proof. It suffices to take piecewise constant functions $\gamma_j(t) = (-1)^{k(j,t)}$, where $k(i,t) = 1$ if $t \in [2mT/j, (2m+1)T/j)$, $k(j,t) = -1$ if $t \in [(2m+1)T/j, (2m+2)T/j)$, $m = 0,1,2,....$ Clearly, the required limit holds for all $f_j \in C(0,T)$, and the set $C(0,T)$ is dense in $L_2(0,T)$. Since $\|\gamma_j\|_{L_2(0,T)} = \text{const}$, it follows that $\gamma_j \to 0$ as $j \to +\infty$ weakly in $L_2(0,T)$. This completes the proof of Lemma 4.3. $\square$

Let us consider the sequence $\{\gamma(\cdot)\} = \{\gamma_j(\cdot)\}$ from the proof of Lemma 4.3 and the corresponding processes $\hat{S}(t) = \hat{S}_j(t)$, $\hat{a}(t) = \hat{a}_j(t) = \gamma_j(t)a(t)$, and $\hat{\sigma}(t) = \hat{\sigma}_j(t) = \gamma_j(t)\sigma(t)$. Since

$$\hat{S}(t) = \hat{S}(0) + \int_0^t \gamma_j(s)a(s)\hat{S}(s)ds + \int_0^t \gamma_j(s)\sigma(s)\hat{S}(s)dw(s),$$

we have that $\hat{S}(t) = S(0)$ and

$$\hat{S}(t) = \hat{S}(0)\exp\left(\int_0^t \hat{a}(s)ds - \frac{1}{2}\int_0^t \hat{\gamma}(s)^2\sigma(s)^2 ds + \int_0^t \hat{\sigma}(s)dw(s)\right)$$
$$= S(0)\exp\left(\int_0^t \gamma_j(s)a(s)ds - \frac{1}{2}\int_0^t \gamma_j(s)^2\sigma(s)^2 ds + \int_0^t \gamma_j(s)\sigma(s)dw(s)\right).$$

Given the choice of $\{\gamma_j\}$, we have that $\sigma(t)^2 = \hat{\sigma}(t)^2$ and

$$\int_0^t \hat{a}(s)ds = \int_0^t \gamma_j(s)a(s)ds \to 0 \quad as \quad j \to +\infty \quad \text{a.s..}$$

Therefore, the processes $\hat{S}(t) = \hat{S}_j(t)$ can be interpreted as processes with a vanishing appreciation rate as $i \to +\infty$.

Clearly,

$$\frac{1}{\Delta t}\int_{t-\Delta t}^t \sigma(s)^2 ds = \frac{1}{\Delta t}\int_{t-\Delta t}^t \hat{\sigma}(s)^2 ds$$

Therefore, the estimate of

$$\frac{1}{\Delta t}\int_{t-\Delta t}^t \sigma(s)^2 ds$$

can be obtained via calculating the similar value for the process $\hat{S}(t) = \hat{S}_j(t)$ for which the impact of the appreciation rate $a(t)$ is eliminated in the limit case where $j \to +\infty$.

We call $\hat{S}(t)$ a process with an eliminated appreciation rate. In fact, the process $\hat{S}(t)$ converges to a martingale.

In practical calculations, the processes $\hat{S}(t) = \hat{S}_j(t)$ and $\gamma(t) = \gamma_j(t)$ are represented by discrete time processes; it is natural to select $\gamma_j(t_k) = (-1)^k$.

Figure 4.2 shows an example of the simulated processes $S(t)$ and $\hat{S}(t)$ with $\gamma(t) = \gamma_j(t)$ defined as in the proof of Lemma 4.3, with the price parameters defined as

$$a(t) \equiv 0.5, \quad \sigma(t) \equiv 0.3, \quad t \in [0, 1],$$

and with $\delta = t_k - t_{k+1} = 0.004$. Here t_k are the times where the prices were observed; the same times are used as the points of discontinuity for $\gamma(t)$. This sample represents daily prices; the plot shows the evolution on a one-year time horizon. It can be noted that the impact of the appreciation rate elimination is barely seen from the local dynamics, since the volatility dominates the appreciation rate.

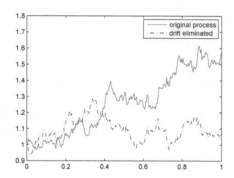

Figure 4.2: Drift elimination: $- - - -$: the original process $S(t)$ with appreciation rate; $- \cdot - \cdot -$: the process $\hat{S}(t)$ with eliminated appreciation rate, for the sample of daily prices. These processes have the same volatility.

4.1.4   The algorithm

Assume that the series of historical prices $S(t_k)$ is available, and that this is the series of data of some sufficient frequency, to justify the use of the continuous time diffusion model (4.1). We suggests the following procedure to estimate $v(t) = \frac{1}{\Delta t} \int_{t-\Delta t}^{t} \sigma(s)^2 ds$.

1. Apply the appreciation rate eliminating procedure described above with $\gamma(t_k) = (-1)^k$. Let $\hat{S}(t_k)$ be the corresponding process with eliminated appreciation rate.

2. Estimate the volatility using the series $\hat{S}(t_k)$ and one of equations (4.6), (4.9), or (4.10).

The nature of the diffusion model (4.1) is such that a precise estimate of the volatility is achievable for the high frequency data only; the error increases if the frequency decreases. Therefore, it is preferable to use the data of the highest available frequency.

On some features of pathwise estimates

The estimates for the volatility coefficients discussed above are pathwise estimates. These estimates do not rely on the distribution of the underlying process and on a particular choice of the drift.

4.1.5 Some experiments

Monte Carlo simulation

In our experiments, we used Monte Carlo simulation of the time series for $S(t_k)$ evolving as

$$S(t_{k+1}) = S(t_k) + a(t_k)S(t_k)\delta + \sigma S(t_k)\sqrt{\delta}\,\eta_{k+1},$$

with mutually independent random η_k from the standard normal distribution $N(0,1)$, and with fixed

$$\sigma = 0.3, \qquad \delta = t_k - t_{k+1} = 0.004.$$

This choice of δ corresponds to the time series of the daily prices. In the experiments, we considered only the cases of short series with $m - m_0 = 10$. Note that the selection of the constant volatility in the experiments described above does not undermine the purpose of studying processes with time variable volatility, since only short time series are used. Our choice of the short memory of $m - m_0 = 10$ periods corresponds to the case when the volatility is not expected to remain the same for longer than two weeks (or ten business days).

For $100,000$ Monte Carlo trials, we simulated a sequence of 250 daily prices for every Monte Carlo trial. For every sequence of prices, we considered subsequences of 10 consequent daily prices, for 240 possible initial times $t - \Delta t$. We estimated the volatility using these subsequences.

To compare different methods, we estimate the expected error

$$\mathbf{E}\left|\sigma - \hat{v}(t)^{1/2}\right|.$$

More precisely, we estimate the sample mean error $\mathbb{E}e$, where

$$e = \left|\sigma - \hat{v}(t_m)^{1/2}\right|.$$

Here $\mathbb{E}$ denotes the sample mean over all Monte Carlo simulation trials and over all subsequences $\{S(t_k)\}$, $k = m_0, m_0 + 1, ..., m_0 + 10$ of the sequences of the simulated stock prices, for all possible m_0. The total size of these samples was above 1,000,000. We found that enlarging the sample does not improve the results. Actually, the experiments with samples of $10,000$ Monte Carlo trials produced the same results. Table 4.1 shows the values of $\mathbb{E}e$ in the experiments described above for some choices with $a(s)|_{[t-\Delta t, t]}$.

$a(s)$	$\mathbb{E}e$ for (4.6)	$\mathbb{E}e$ for (4.9)	$\mathbb{E}e$ for (4.10)
$a(s) \equiv 0.5$; without drift elimination	0.0584477	0.0546515	0.05482599
$a(s) \equiv 0.5$; with drift elimination	0.0581887	0.0548541	0.05477097
$a(s) = 6\sin(2\pi[S(s) - S(s - \tau(s))])$; without drift elimination	0.06096515	0.06715185	0.06704281
$a(s) = 6\sin(2\pi[S(s) - S(s - \tau(s))])$; with drift elimination	0.05993648	0.0596709	0.05960616

Table 4.1: The mean error $\mathbb{E}e$ for the traditional estimate (4.6) and for estimates (4.9), and (4.10) for Monte Carlo simulated prices and for new estimates (4.9) and (4.10).

Here $\tau(s) = 0.04\lfloor s/0.04 \rfloor$; we denote by $\lfloor s \rfloor$ the integer part of s.

Further, we estimated the standard deviation σ_e of the error $e = |\sigma - \hat{v}(t_m)^{1/2}|$ as the following:

1. The error ε_t was calculated for $t = 11, ..., 250$.

2. The sample $\bar{e}$ was formed from 100,000 values $\frac{1}{240} \sum_{t=11}^{250} \varepsilon_t$ obtained in 100,000 Monte Carlo trials.

3. σ_ε was calculated as the standard deviation of the series $\bar{e}$.

Table 4.2 shows the values for σ_e in these experiments.

$a(s)$	σ_e for (4.6)	σ_e for (4.9)	σ_e for (4.10)
$a(s) \equiv 0.5$; without drift elimination	0.00675680407	0.0065980581	0.00660535275
$a(s) \equiv 0.5$; with drift elimination	0.00673944223	0.0066648496	0.00663868046
$a(s) = 6\sin(2\pi[S(s) - S(t - \tau)])$; without drift elimination	0.00720963109	0.011522603	0.0117100558
$a(s) = 6\sin(2\pi[S(s) - S(t - \tau)])$; with drift elimination	0.00724451753	0.00788851706	0.0078621931

Table 4.2: The standard deviation σ_e for the traditional estimate (4.6) and for estimates (4.9) and (4.10) for Monte Carlo simulated prices.

The values for $\mathbb{E}e$ and σ_e obtained in the experiments are very stable; the results are practically the same for far fewer Monte Carlo trials.

Note that the appreciation rate elimination does not take effect in a single term k under the sums in (4.6), (4.9), and (4.10). However, it can make the error less systematic after mixing all $m - m_0$ terms in the sum. This explains some of the improvement achieved with the drift elimination for time-dependent and random a in the experiment described above.

It can also be noted that some minor improvement in the performance was observed for an estimate constructed as the mean of estimates (4.6), (4.9), and (4.10).

Experiment with historical prices

We have carried out some experiments for the time series representing the returns for the historical stock prices. Using daily price data from 1984 to 2009 for 19 American and Australian stocks (Citibank, Coca Cola, IBM, AMC, ANZ, LEI, LLC, LLN, MAY, MLG, MMF, MWB, MIM, NAB, NBH, NCM, NCP, NFM and NPC), we generated samples of price data for one synthetic price process $S(t_k)$. In fact, the full 25 years of data was not available for all the stocks; the total number of the prices in the sample was 69,948.

For the historical prices, the "true" volatility process is not available. Moreover, it cannot even be presumed with certainty that model (4.1) is suitable for particular prices samples. Therefore, we cannot estimate the "error" in this experiment. So far, we will demonstrate only that different estimation rules produce close enough but still different distributions of random estimates.

We estimated the value of

$$\mathbf{E} \left(\frac{1}{\Delta t} \int_{t-\Delta t}^{t} \sigma(s)^2 ds \right)^{1/2} .$$

More precisely, we estimated the corresponding sample mean

$$\bar{\sigma} = \mathbb{E} \left[\hat{v}(t_m)^{1/2} \right] ,$$

with estimates $\hat{v}_k$ obtained accordingly to the different rules described above. We considered again short series consisting of 10 daily prices, with $t - \Delta t = t_{m_0}$, and $t = t_m$, $m - m_0 = 10$. The sample mean $\mathbb{E}$ used here represents the averaging over all possible initial times t_m and over different stocks; the total number of short time series was 66,590. The results of this experiment are presented in Table 4.3.

	$\bar{\sigma}$ for (4.6)	$\bar{\sigma}$ for (4.9)	$\bar{\sigma}$ for (4.10)
Without drift elimination	0.2449	0.2516	0.2511
With drift elimination	0.2446	0.2454	0.2511

Table 4.3: $\bar{\sigma}$ for estimates (4.6), (4.9), and (4.10) for the large set of historical prices

It can be seen from this table that the average values of the volatility calculated over a large number of short time series are close for different estimates. For a smaller number of short series, this effect is less noticeable. For instance, a similar experiment with 50 prices for NAB and with the averaging over 29 short time series gives the results presented in Table 4.4. Note that, in the experiment described by Table

	$\bar{\sigma}$ for (4.6)	$\bar{\sigma}$ for (4.9)	$\bar{\sigma}$ for (4.10)
Without drift elimination	0.1387	0.1570	0.1569
With drift elimination	0.1549	0.1570	0.1569

Table 4.4: $\bar{\sigma}$ for estimates (4.6), (4.9), and (4.10) for a smaller set of historical prices

4.4, the traditional estimate gives a value that is about 10% less than the values for the other estimates.

We found already that, for Monte Carlo simulation of the series generated by Ito equations, the different estimates, with or without appreciation rate elimination, produce different estimates for the same model. For historical prices, we observed again that the different estimates produced different results. Unlike the case of the Monte Carlo simulation, we cannot tell which estimate produces a smaller error, since the true volatility is unknown. This leads to the conclusion that it could be beneficial to use and compare different methods simultaneously for the estimation of the volatility from the historical prices.

4.2   Modeling the impact of the sampling frequency

Volatility estimation for time series has many applications, especially for financial time series. Due to the growth of computing power and data storage capacity, the high-frequency market data is now available for analysis. This has created new computational challenges for both academics and practitioners. It appears that the volatility calculated from different frequencies (or on the different time scales) can have different values. It was often observed that the volatility increases when the sampling frequency increases; see, e.g., [2]. It also appears that the volatility may decrease when more frequent samples are used; the summary of such observations is shown at Table 4.8. There is a demand in modeling the underlying price processes to capture these features. For the data generated from the classical Black–Scholes market model, this effect does not take place, i.e., the volatility is independent of the sampling frequency. Therefore, the observed phenomena of the dependency of the volatility on sampling frequency requires modifications to the classical model.

In this section, we readdress the problem of modeling the price process whose volatility depends on the sampling frequency, i.e., on the timescale. We describe a continuous time model for stock prices such that the volatility calculated via different samplings may take certain preselected values. It appears that Ito equations with delay feature this property. The presence of additional parameters that describe the delay enables us to capture the difference in volatility estimated at alternative sampling rates, unlike other price processes such as geometric Brownian motion, the mean-reversion model, Heston model, and others. So far, we have considered the simplest linear delay equation with only one delay term. This new term provides us with the capability of matching the volatility from the simulated price process with the volatility from the historical data, on three different sampling frequencies. More precisely, our model only allows us to replicate the bias arising for the volatility measured on different sampling frequencies.

4.2.1 Analysis of the model's parameters

In this section we discuss the Monte Carlo simulation of the price process defined by (3.19). First, we define the volatility estimator we will be using. We then discuss the experiment we performed.

Volatility estimator

We will be using the classical estimator for volatility based on samples collected within the $[t - \Delta t, t]$ interval, where $\Delta t > 0$ is given. As was mentioned above, it is convenient to characterize the volatility via the process with the integral

$$v(t) = \frac{1}{\Delta t} \int_{t-\Delta t}^{t} \sigma(s)^2 \mathrm{d}s.$$

We consider that the $v(t)$ estimation process is usually carried out with the historical data.

We will construct an estimate of $v(t)$ from samples of stock prices $S(\theta_k)$, where $\theta_k \in [t - \Delta t, t]$, $k = m_0, m_0 + 1, ..., m$.. We also assume that the time points θ_k are equally spaced with sampling interval $\hat{\delta} = \theta_k - \theta_{k-1}$. Furthermore, we assume that $\theta_{m_0} = t - \Delta t$ and $t_m = t$. Therefore, $\Delta t = (m - m_0)\hat{\delta}$.

Note that the choice of $\hat{\delta}$ could be different from δ in (3.19). In our experiments, we consider only the cases where $\hat{\delta} = N\delta$ for some integer $N \geq 1$ such that $\{\theta_k\} \subset \{t_k\}$; if $N = 1$, then $\delta = \hat{\delta}$ and $t_k = \theta_k$.

For a given choice of $\hat{\delta}$, we estimate $v(t)$ via the value

$$\hat{\sigma}(t) = \sqrt{v(t)} = \left[\frac{1}{\Delta t} \sum_{k=m_0+1}^{m} (\overline{R}_m - \hat{R}(\theta_k))^2 \right]^{1/2}, \qquad (4.11)$$

where

$$\Delta t = (m - m_0)\hat{\delta}, \quad \hat{R}(\theta_k) = \log S(\theta_k) - \log S(\theta_{k-1}),$$

$$\overline{R}_m = \frac{1}{m - m_0} \sum_{k=m_0+1}^{m} \hat{R}(\theta_k).$$

Note that the particular choice of this estimator is not crucial for our purposes;

other estimators can be used instead. For instance, the estimator proposed by Andersen [7] gives very similar estimates on different frequencies. To illustrate this, we compared the annualized daily volatility estimated by (4.11) and by Andersen's realized volatility for the S&P 500 index for 2008–2013. We estimated the mean absolute difference (MAD) between the two estimators as:

$$MAD = \frac{\Sigma|RV_{30sec}^A - RV_{30sec}^C|}{n},$$

where n is the number of observations in our sample. Here we have n = 1511 days. We found that $MAD_{30sec} = 0.000164$, $MAD_{5min} = 0.001647$, and $MAD_{15min} = 0.004826$.

4.2.2   Monte Carlo simulation of the process with delay

We first simulate the process generated by (3.19) at some high frequency (small δ). We then compute the volatility at lower frequencies from the same simulated path. For instance, assuming that there are 6.5 trading-hours per day and 252 trading-days per year, we simulate a one-year sample path at 15-second frequency, i.e., $\delta = \frac{1}{252 \times 6.5 \times 60 \times 4} \approx 2.5437 \times 10^{-6}$. To measure the volatility at lower frequencies, we sub-sample the simulated path at 5-minute and 1-hour frequencies by using the tick-aggregation technique [124]. This is done by taking the last price realized before each new grid point.

In our Monte Carlo simulation, we use the following selection criteria for the model's parameters:

1. Selection of δ: We used δ for 15-second data throughout this experiment.

2. Selection of (τ, λ): We used $\tau = 5$, 20, and 120. This corresponds to $\varrho = 75$ sec, 5 min, and 30 min, respectively. In addition, we selected a variety of $\lambda \in [-2000, 20000]$ such that (3.20) holds.

3. Selection of σ: We used $\sigma = 0.3$.

We generated 100,000 instances for each combination of σ, τ, and λ. To analyze the results statistically, we summarize the average and the standard deviation of the estimated volatility from these instances at each selected sampling frequency.

The results of the simulation experiments

For the prices simulated from the above settings, the measured volatility under different timescale are summarized in Tables 4.5 and 4.6.

$\kappa = \lambda \delta \tau$		5			20			120	
		mean	sd		mean	sd		mean	sd
$\kappa = -0.0005; \lambda = -196.56$	σ_{15sec}	0.3000	0.0003	σ_{15sec}	0.3000	0.0003	σ_{15sec}	0.3000	0.0003
	σ_{5min}	0.3004	0.0015	σ_{5min}	0.3014	0.0015	σ_{5min}	0.3015	0.0010
	σ_{hour}	0.3009	0.0052	σ_{hour}	0.3027	0.0052	σ_{hour}	0.3140	0.0030
		mean	sd		mean	sd		mean	sd
$\kappa = -0.0025; \lambda = -982.8$	σ_{15sec}	0.3000	0.0003	σ_{15sec}	0.3000	0.0003	σ_{15sec}	0.3002	0.0003
	σ_{5min}	0.3026	0.0015	σ_{5min}	0.3073	0.0015	σ_{5min}	0.3101	0.0016
	σ_{hour}	0.3029	0.0053	σ_{hour}	0.3141	0.0054	σ_{hour}	0.3892	0.0069
		mean	sd		mean	sd		mean	sd
$\kappa = -0.005; \lambda = -1965.6$	σ_{15sec}	0.3002	0.0003	σ_{15sec}	0.3006	0.0003	σ_{15sec}	0.3000	0.0003
	σ_{5min}	0.3052	0.0016	σ_{5min}	0.3153	0.0019	σ_{5min}	0.3341	0.0025
	σ_{hour}	0.3059	0.0049	σ_{hour}	0.3301	0.0057	σ_{hour}	0.5667	0.0050

Table 4.5: Simulation with $\lambda < 0$ in the delay model.

$\kappa = \lambda \delta \tau$		5			20			120	
		mean	sd		mean	sd		mean	sd
$\kappa = 0.0005; \lambda = 196.56$	σ_{15sec}	0.3000	0.0003	σ_{15sec}	0.3000	0.0003	σ_{15sec}	0.3000	0.0003
	σ_{5min}	0.3000	0.0015	σ_{5min}	0.2985	0.0015	σ_{5min}	0.2987	0.0015
	σ_{hour}	0.3000	0.0050	σ_{hour}	0.2971	0.0050	σ_{hour}	0.2873	0.0052
		mean	sd		mean	sd		mean	sd
$\kappa = 0.005; \lambda = 1965.6$	σ_{15sec}	0.3003	0.0003	σ_{15sec}	0.3004	0.0003	σ_{15sec}	0.3003	0.0003
	σ_{5min}	0.2948	0.0015	σ_{5min}	0.2866	0.0015	σ_{5min}	0.2912	0.0015
	σ_{hour}	0.2941	0.0037	σ_{hour}	0.2749	0.0047	σ_{hour}	0.2123	0.0037
		mean	sd		mean	sd		mean	sd
$\kappa = 0.05; \lambda = 19656$	σ_{15sec}	0.3003	0.0003	σ_{15sec}	0.3038	0.0004	σ_{15sec}	0.3067	0.0004
	σ_{5min}	0.2559	0.0012	σ_{5min}	0.2148	0.0007	σ_{5min}	0.2848	0.0021
	σ_{hour}	0.2503	0.0020	σ_{hour}	0.1594	0.0019	σ_{hour}	0.1045	0.0020

Table 4.6: Simulations with $\lambda > 0$ in the delay model.

Let us discuss the results presented in Tables 4.5 and 4.6. Due the randomness of the data, short samples produces random estimates of the volatilities featuring significant variance. To decrease this variance for demonstration purposes, we selected longer time series using observations within a 1-year window from the simulated data. In our experiments, we used samples with 393,120 observations, 19,656 observations, and 1,638 observations for 15-sec, 5-min, and 1-hour data, respectively. Tables 4.5 and 4.6 show the impact of the standard deviation on the estimated volatility. Shorter time series would produce the same estimates for the volatilities but with higher variance; see Table 4.7.

Annual Volatility	Mean	Standard Deviation
1-day windows	$\sigma_{15sec}= 0.3002$ $\sigma_{5min} = 0.2923$ $\sigma_{hour} = 0.2111$	$\sigma_{15sec}= 0.0054$ $\sigma_{5min} = 0.0254$ $\sigma_{hour} = 0.0792$
5-day windows	$\sigma_{15sec}= 0.3002$ $\sigma_{5min} = 0.2926$ $\sigma_{hour} = 0.2143$	$\sigma_{15sec}= 0.0022$ $\sigma_{5min} = 0.0098$ $\sigma_{hour} = 0.02482$
22-day windows	$\sigma_{15sec}= 0.3002$ $\sigma_{5min} = 0.2926$ $\sigma_{hour} = 0.2134$	$\sigma_{15sec}= 0.0010$ $\sigma_{5min} = 0.0045$ $\sigma_{hour} = 0.0113$

Table 4.7: The average and standard deviation of the annual volatility using different windows for $\kappa = 0.005$ ($\lambda = 1965.6$) and $\tau = 120$.

For the case where $\lambda < 0$, we observe that the estimated volatility increases as the sampling frequency decreases. Given a fixed τ, the larger λ reduces the difference in the estimated volatility per sampling frequency. On the other hand, given a fixed λ, bigger τ results in larger gaps in the estimated volatility at each sampling frequency. For example, with $\tau = 120$, we have that

$$\left.\frac{\sigma_{hour}}{\sigma_{5min}}\right|_{\kappa=-0.005} = 1.696 > \left.\frac{\sigma_{hour}}{\sigma_{5min}}\right|_{\kappa=-0.0025} = 1.255,$$

whereas with $\kappa = -0.005$,

$$\left.\frac{\sigma_{hour}}{\sigma_{5min}}\right|_{\tau=20} = 1.047 < \left.\frac{\sigma_{hour}}{\sigma_{5min}}\right|_{\tau=120} = 1.696.$$

For $\lambda > 0$, we have the opposite results, i.e., the estimated volatility decreases as the sampling frequency decreases. Given a fixed τ (or a fixed λ), the larger λ (or τ) results in larger gaps in the estimated volatility as the sampling frequency decreases. For instance, $\tau = 120$,

$$\left.\frac{\sigma_{hour}}{\sigma_{5min}}\right|_{\kappa=0.0005} = 0.962 > \left.\frac{\sigma_{hour}}{\sigma_{5min}}\right|_{\kappa=0.005} = 0.729,$$

while with $\kappa = 0.005$,

$$\left.\frac{\sigma_{hour}}{\sigma_{5min}}\right|_{\tau=20} = 0.959 > \left.\frac{\sigma_{hour}}{\sigma_{5min}}\right|_{\tau=120} = 0.729.$$

These experiments demonstrate that, with the proposed model, we can replicate the volatility and timescale dependence characteristic of financial timeseries. The parameters within the model can be used to control the changes of the estimated volatility at different sampling frequencies.

It is noted that we observed that the estimated volatility $\hat{\sigma}_{15sec}$ is fairly close to the "true" volatility $\sigma = 0.3$ for all choices of (λ, τ).

As was mentioned above, the results of the experiments are robust with respect to the choice of the volatility estimator. It appears that the results for numerical simulation are also robust with respect to variations in all other parameters. The average values for estimated volatilities do not change significantly when we increased the number of Monte Carlo trials, and they change very smoothly and systematically for different choices of $(\kappa, \tau, \lambda, \sigma, \delta)$ given that condition (3.20) on (τ, λ, δ) is satisfied. This was established as a result of a much larger series of numerical experiments than we reported in herein. Condition (3.20) is essential to ensure that the simulated process has a moderate growth.

4.2.3 Examples for dependence of volatility on sampling frequency for historical data

In this section, we discuss how to to obtain the parameters λ and τ in the proposed model for matching the volatility behavior for some sets of historical data.

Analysis of historical data

In practice, the highest frequency financial time series available for analysis is the tick-data. Tick-data is recorded in discrete time that is not necessary equally spaced. Some data services however provide financial data which are sampled on given equispaced frequency (i.e., every 15 seconds or 5 minutes). These samples are sub-samples of the tick-data and are aggregated using different weighting schemes.

For our empirical study, we estimate the volatility of financial time series from different markets. As was discussed in Section 4.2.1, our statistical volatility estimator requires the data to be collected at equispace. Hence, we use the tick-data as the baseline data for each underlying asset. We then perform the data cleaning process to obtain samples at equal intervals. We used the data sets obtained from SIRCA, the Securities Industry Research Centre of Asia-Pacific [110], for the period 2008–2010.

We selected the top most-traded indices and US stocks listed on Reuters Finance, including the following.

Category 1: Stock indexes: DAX (Deutsche Boerse AG German Stock Index, trading between 9:00 am and 5:45 pm CET), FTSE 100 (a share index of the 100 companies listed on the London Stock Exchange, trading between 8:00 am and 4:30 pm GMT); IBEX 35 (an index of the Spanish Continuous Market, opening between 9:00 am and 05:30 pm); SMI (Switzerland's blue-chip stock market index, between 9:00 am and 5:30 pm CET); S&P 500 (a stock market index based on the market capitalizations of the 500 largest companies having common stock listed on the NYSE, between 9:30 am and 4:00 pm); and S&P 200 (a stock market index based on the market capitalizations of 200 large companies having common stock listed on the Australian Stock Exchange, operating between 10:00 am and 4:00 pm); and TSX 60 (stock market index of 60 large companies listed on the Toronto Stock Exchange).

Category 2: Individual company stock symbols including AAPL, IBM, JPM, GE,

GOOG, MSFT, and XOM. These stocks are traded between 9:30 am and 4:00 pm on the NYSE.

Methodology

We took the following steps to analyze our datasets.

Step 1. Obtained the tick-data for each stock/index from SIRCA.

Step 2. Performed data cleaning. By using the tick-data, we force these asynchronously and irregularly recorded series to a synchronized and equispaced time grid using the previous tick aggregation to obtain samples at different frequencies.

Step 3. Estimated the volatility using the formula discussed in Section 4.11 for the entire year to obtain the annualized volatility.

Step 4. Repeated Steps 2 and 3 for each sampling frequency and each stock/index.

Tables 4.8 and 4.9 show how the volatility varies when it is measured at different sampling frequencies for the selected stocks and indexes. In this table, the volatility was calculated by applying estimation (4.11) for the observations collected over a whole year. This time window was selected quite randomly; similar experiments could be done for data collected arbitrarily over shorter time periods.

Some additional results for volatility calculated for different historical data with different frequencies can be found in [98], where sampling periods of 5 minutes, 1 hour, and 4 days were used. Overall, for the examples given above and for examples in [98], the volatility increases with increased sampling frequency for individual stocks and decreases for indexes.

Matching the volatilities for the delay equations and the historical prices

In this section, we demonstrate that it is possible to calibrate the proposed model with the historical data such that the volatilities match for three different sampling frequencies.

In our experiments, we considered data available at three sampling frequencies: 15 seconds, 5 minutes, and 1 hour.

The volatility of historical 15 second data was accepted as σ in (3.19); this is the data sampled at the highest available frequency.

The proposed model has parameters λ, τ, and σ. Our purpose is to select (σ, λ, τ) to ensure matching of the volatilities for the simulated process and for a set of historical data for 15 second, 5 minutes and 1-hour samplings. For this, we used the following simple and straightforward heuristic algorithm.

Step 1. Estimate the volatility at the selected sampling frequencies σ_{15sec}, σ_{5min}, and σ_{1hour} using the historical data.

Step 2. Select a finite set of $(\lambda_i, \tau_j, \sigma_k)$ for the search space.

Stock Index	2008		2009		2010	
DAX	$\sigma_{15sec} =$	0.3569	$\sigma_{15sec} =$	0.2476	$\sigma_{15sec} =$	0.1755
	$\sigma_{5min} =$	0.3928	$\sigma_{5min} =$	0.2682	$\sigma_{5min} =$	0.1854
	$\sigma_{hourly} =$	0.3959	$\sigma_{hourly} =$	0.2777	$\sigma_{hourly} =$	0.1888
FTSE 100	$\sigma_{15sec} =$	0.2660	$\sigma_{15sec} =$	0.1822	$\sigma_{15sec} =$	0.1346
	$\sigma_{5min} =$	0.3462	$\sigma_{5min} =$	0.2313	$\sigma_{5min} =$	0.1686
	$\sigma_{hourly} =$	0.3606	$\sigma_{hourly} =$	0.2344	$\sigma_{hourly} =$	0.1708
IBEX 35	$\sigma_{15sec} =$	0.3289	$\sigma_{15sec} =$	0.2428	$\sigma_{15sec} =$	0.2549
	$\sigma_{5min} =$	0.3569	$\sigma_{5min} =$	0.2446	$\sigma_{5min} =$	0.2772
	$\sigma_{hourly} =$	0.3620	$\sigma_{hourly} =$	0.2571	$\sigma_{hourly} =$	0.2847
SMI	$\sigma_{15sec} =$	0.3265	$\sigma_{15sec} =$	0.2106	$\sigma_{15sec} =$	0.1603
	$\sigma_{5min} =$	0.3421	$\sigma_{5min} =$	0.2171	$\sigma_{5min} =$	0.1469
	$\sigma_{hourly} =$	0.3513	$\sigma_{hourly} =$	0.2278	$\sigma_{hourly} =$	0.1564
S&P 500	$\sigma_{15sec} =$	0.2881	$\sigma_{15sec} =$	0.1952	$\sigma_{15sec} =$	0.1300
	$\sigma_{5min} =$	0.3654	$\sigma_{5min} =$	0.2507	$\sigma_{5min} =$	0.1771
	$\sigma_{hourly} =$	0.3747	$\sigma_{hourly} =$	0.2601	$\sigma_{hourly} =$	0.1808
S&P 200	$\sigma_{15sec} =$	0.2124	$\sigma_{15sec} =$	0.1464	$\sigma_{15sec} =$	0.1055
	$\sigma_{5min} =$	0.2796	$\sigma_{5min} =$	0.1805	$\sigma_{5min} =$	0.1288
	$\sigma_{hourly} =$	0.3367	$\sigma_{hourly} =$	0.2100	$\sigma_{hourly} =$	0.1530
TSX 60	σ_{15sec}	0.3345	$\sigma_{15sec} =$	0.2256	$\sigma_{15sec} =$	0.1246
	σ_{5min}	0.3884	$\sigma_{5min} =$	0.2602	$\sigma_{5min} =$	0.1341
	σ_{hourly}	0.4207	$\sigma_{hourly} =$	0.3884	$\sigma_{hourly} =$	0.1352

Table 4.8: Volatility of stock indexes under different sampling frequencies.

Stock Symbol	2008		2009		2010	
AAPL	$\sigma_{15sec} =$	0.7032	$\sigma_{15sec} =$	0.3508	$\sigma_{15sec} =$	0.3180
	$\sigma_{5min} =$	0.6347	$\sigma_{5min} =$	0.3373	$\sigma_{5min} =$	0.2915
	$\sigma_{hourly} =$	0.5832	$\sigma_{hourly} =$	0.3267	$\sigma_{hourly} =$	0.2738
IBM	$\sigma_{15sec} =$	0.5245	$\sigma_{15sec} =$	0.3085	$\sigma_{15sec} =$	0.2179
	$\sigma_{5min} =$	0.4507	$\sigma_{5min} =$	0.2766	$\sigma_{5min} =$	0.2005
	$\sigma_{hourly} =$	0.3932	$\sigma_{hourly} =$	0.2622	$\sigma_{hourly} =$	0.1812
JPM	$\sigma_{15sec} =$	0.9068	$\sigma_{15sec} =$	0.7274	$\sigma_{15sec} =$	0.3238
	$\sigma_{5min} =$	0.8217	$\sigma_{5min} =$	0.6741	$\sigma_{5min} =$	0.3039
	$\sigma_{hourly} =$	0.7487	$\sigma_{hourly} =$	0.6586	$\sigma_{hourly} =$	0.2873
GE	$\sigma_{15sec} =$	0.7163	$\sigma_{15sec} =$	0.7137	$\sigma_{15sec} =$	0.4021
	$\sigma_{5min} =$	0.6220	$\sigma_{5min} =$	0.6040	$\sigma_{5min} =$	0.3124
	$\sigma_{hourly} =$	0.5790	$\sigma_{hourly} =$	0.5844	$\sigma_{hourly} =$	0.2919
GOOG	$\sigma_{15sec} =$	0.8114	$\sigma_{15sec} =$	0.3543	$\sigma_{15sec} =$	0.3353
	$\sigma_{5min} =$	0.6276	$\sigma_{5min} =$	0.3053	$\sigma_{5min} =$	0.2820
	$\sigma_{hourly} =$	0.5937	$\sigma_{hourly} =$	0.2944	$\sigma_{hourly} =$	0.2587
MSFT	$\sigma_{15sec} =$	0.7032	$\sigma_{15sec} =$	0.3908	$\sigma_{15sec} =$	0.3180
	$\sigma_{5min} =$	0.6347	$\sigma_{5min} =$	0.3373	$\sigma_{5min} =$	0.2915
	$\sigma_{hourly} =$	0.5832	$\sigma_{hourly} =$	0.3267	$\sigma_{hourly} =$	0.2738
XOM	$\sigma_{15sec} =$	0.5071	$\sigma_{15sec} =$	0.2911	$\sigma_{15sec} =$	0.2156
	$\sigma_{5min} =$	0.4908	$\sigma_{5min} =$	0.2708	$\sigma_{5min} =$	0.2083
	$\sigma_{hourly} =$	0.4876	$\sigma_{hourly} =$	0.2620	$\sigma_{hourly} =$	0.1834

Table 4.9: Volatility of company stocks under different sampling frequencies.

Step 3. For each triplet $(\lambda_i, \tau_j, \sigma_k)$, generate a path using (3.19) and estimate the volatilities $\hat{\sigma}_{15sec}(\lambda_i, \tau_j)$, $\hat{\sigma}_{5min}(\lambda_i, \tau_j)$ and $\hat{\sigma}_{1hour}(\lambda_i, \tau_j)$, calculated for the corresponding sampling frequencies.

Step 4. Among these outputs, find the values of $(\lambda_i, \tau_j, \sigma_k)$ that generate $\hat{\sigma}_{15sec}$, $\hat{\sigma}_{5min}$, and $\hat{\sigma}_{1hour}$ matching with the volatilities estimated from the historical data.

Step 5. If there are no matching values, extend the set $\{(\lambda_i, \tau_j, \sigma_k)\}$ and repeat steps 2–4.

It appears that selecting $\sigma = \sigma_{15sec}$ allows us to find satisfactory (λ, τ) for all our experiments. We suggest this initial selection to reduce the search.

It can be noted that we do not try to find the best matching (λ, τ, σ) via minimization of the fitting errors (residuals) for the paths of historical series as is usually done by the least square estimators. We match only the volatilities of the historical series and simulated series on the given set of sampling frequencies. Therefore, our simulation does not replicate other characteristics of the price evolution such as the rate of growth.

4.2.4 Matching delay parameters for historical data

We present some examples for both cases $\lambda > 0$ and $\lambda < 0$. The table below is extracted from Table 4.8 where we only selected observations for the S&P 500 Index and Google stock in 2008. We use $\lambda < 0$ to reproduce the dependence on the sampling frequency for S&P 500 index and we use $\lambda > 0$ to replicate that characteristic for Google stock prices.

Underlying Assets	σ_{15sec}	σ_{5min}	σ_{1hour}
S&P 500 index	0.2881	0.3654	0.3747
Google stock	0.8114	0.6276	0.5937

In this sample, we considered the volatility calculated for the time window of an entire year. Shorter time periods price their own volatility values that could also be implemented.

For $\lambda < 0$, the volatility of the S&P 500 measured at different sampling frequencies was $\sigma_{15sec} = 0.2881$, $\sigma_{5min} = 0.3654$, and $\sigma_{hour} = 0.3747$. Using the steps discussed above, we obtained the following parameters:

$$\begin{cases} \sigma = 0.2881 \\ \tau = 9 \\ \delta = \dfrac{1}{252 \times 6.5 \times 60 \times 4} \\ \kappa = \lambda\delta = -0.0325, \lambda = -12776.4 \end{cases}$$

We substituted these parameters into equation (3.19), simulated 100,000 instances, and obtained the following average volatilities at each sampling frequency:

S&P 500	$mean_{sim}$	sd_{sim}
σ_{15sec}	0.2881	0.2882	0.0003
σ_{5min}	0.3654	0.3615	0.0018
σ_{hour}	0.3747	0.3751	0.0065

Similarly, for $\lambda > 0$, the volatility of Google stock on the NYSE in 2008 measured on alternative timescales was $\sigma_{15sec} = 0.8114$, $\sigma_{5min} = 0.6276$, and $\sigma_{hour} = 0.5937$. We obtained the following parameters

$$\begin{cases} \sigma = 0.8114 \\ \tau = 10 \\ \delta = \dfrac{1}{252 \times 6.5 \times 60 \times 4} \\ \kappa = \lambda\delta = 0.0445, \lambda = 17493.84 \end{cases}$$

which provides

GOOG	$mean_{sim}$	sd_{sim}
σ_{15sec}	0.8144	0.8147	0.0010
σ_{5min}	0.6276	0.6302	0.0017
σ_{hour}	0.5937	0.5892	0.0101

To illustrate the search process for the matching values, we show in Tables 4.10 and 4.12 the errors corresponding to (λ_i, τ_j) used in calibrating the S&P 500 index and Google stock, respectively, and calculated as

$$\epsilon_{(\lambda_i, \tau_j)} = \sqrt{(\sigma_{15sec} - \hat{\sigma}_{15sec})^2 + (\sigma_{5min} - \hat{\sigma}_{5min})^2 + (\sigma_{1hour} - \hat{\sigma}_{1hour})^2}.$$
(4.12)

For both tables, we selected $\sigma = \sigma_{15sec}$.

$\lambda\tau$	5	6	7	8	9	10	11	12	13
-16707.6	0.0396	0.0120	0.0147	0.0500	0.0716	0.0845	0.1260	0.1637	0.2335
-12776.4	0.0593	0.0557	0.0268	0.0108	**0.0098**	0.0306	0.0554	0.0674	0.1214
-8845.2	0.0736	0.0645	0.0507	0.0427	0.0378	0.0291	0.0246	0.0217	0.0251
-4914.0	0.0852	0.0878	0.0813	0.0739	0.0716	0.0691	0.0698	0.0612	0.0626
-982.8	0.1108	0.1080	0.1091	0.1063	0.1001	0.1051	0.1040	0.1049	0.1017

Table 4.10: The values of error (4.12) in calibrating S&P 500 historical data for some (λ_i, τ_j).

$\lambda\tau$	5	6	7	8	9	10	11	12	13
-16707.6	0.0012	0.0015	0.0019	0.0026	0.0028	0.0038	0.0042	0.0052	0.0065
-12776.4	0.0007	0.0010	0.0007	0.0012	0.0001	0.0017	0.0018	0.0020	0.0031
-8845.2	0.0003	0.0008	-0.0001	0.0003	0.0007	0.0009	0.0003	0.0013	0.0011
-4914.0	-0.0001	-0.0004	0.0002	0.0002	0.0005	0.0004	0.0004	-0.0004	0.0008
-982.8	-0.0003	0.0002	0.0006	0.0003	0.0001	-0.0001	-0.0001	-0.0002	0.0003

Table 4.11: Measures of $\hat{\sigma}_{15sec} - \sigma_{15sec}$ in calibrating S&P 500 historical data.

$\lambda\tau$	5	6	7	8	9	10	11	12	13
17886.96	0.0919	0.0661	0.0456	0.0307	0.0154	0.0052	0.0128	0.0137	0.0231
17690.40	0.0797	0.0681	0.0482	0.0317	0.0114	0.0145	0.0081	0.0119	0.0254
17493.84	0.0946	0.0651	0.0504	0.0332	0.0238	**0.0030**	0.0079	0.0173	0.0246
17297.28	0.0967	0.0633	0.0460	0.0369	0.0199	0.0113	0.0076	0.0102	0.0210
17100.72	0.0840	0.0678	0.0448	0.0349	0.0228	0.0141	0.0089	0.0105	0.0196

Table 4.12: The values of error (4.12) in calibrating GOOG historical data for some (λ_i, τ_j).

$\lambda\tau$	5	6	7	8	9	10	11	12	13
17886.96	0.0015	0.0034	0.0047	0.0039	0.0044	0.0051	0.0067	0.0053	0.0076
17690.40	0.0015	0.0024	0.0046	0.0055	0.0049	0.0062	0.0065	0.0066	0.0055
17493.84	0.0031	-0.0001	0.0059	0.0054	0.0053	0.0002	0.0049	0.0057	0.0073
17297.28	0.0021	0.0018	0.0044	0.0033	0.0046	0.0053	0.0048	0.0056	0.0070
17100.72	0.0025	0.0036	0.0036	0.0051	0.0051	0.0066	0.0055	0.0051	0.0055

Table 4.13: Measures of $\hat{\sigma}_{15sec} - \sigma_{15sec}$ in calibrating GOOG historical data.

These tables show that selection of $\sigma = \sigma_{15sec}$ allows us to find the values according to λ and τ to match the simulated volatility with sufficient accuracy on the given set of given sampling frequencies.

General properties of the simulated processes

As was mentioned above, we have to select the parameters (λ, τ) only among the pairs such that (3.20) holds, i.e., the characteristic polynomial for autoregression (3.19) does not have roots outside the unit circle. It appears that, under this restriction, the behavior of the simulated process with delay does not demonstrate any unusual and undesirable features such as excessive growth, and is quite similar to the behavior of the underlying processes, as well as to the behavior of standard Ito processes and autoregressions used to model the financial time series.

Time-varying volatility and shorter time windows

To analyze the time-varying volatility for historical prices, one could match the volatility for shorter time windows. In this case, we have to select (λ, τ, σ) separately for the corresponding time windows. For example, assume that we wish to match the quarterly data for Google stock prices during one year, and we calculate $\sigma_{15se}^{(k)}$, $\sigma_{5min}^{(k)}$, and $\sigma_{1hour}^{(k)}$ for each quarter, where $k \in \{1, 2, 3, 4\}$ represents a quarter. In this case, we have to select for each quarter $(\lambda^{(k)}, \tau^{(k)}, \sigma^{(k)})$ to match the corresponding volatilities. To replicate quarterly time $(\sigma_{15se}^{(k)}, \sigma_{5min}^{(k)}, \sigma_{1hour}^{(k)})$ depending, it suffices to accept equation (3.19) with a time dependent set of parameters $(\lambda, \tau, \sigma) = (\lambda^{(k)}, \tau^{(k)}, \sigma^{(k)})$ that depends on the particular quarter. The same approach can be used for other time windows, for instance, for weekly or daily volatilities. Again, we have to select the parameters (λ, τ, σ) for each time window separately, and accept equation (3.19) with piecewise constant parameters.

4.3 Inference for diffusion parameters for CIR-type models

In this section, we consider inference , of the parameters of the diffusion term for continuous time stochastic processes with a power type dependence of the diffusion coefficient from the underlying process such as Cox–Ingersoll–Ross, Chan–Karolyi–Longstaff–Sanders (CKLS), and similar processes. These processes are important for applications; in particular, they are used for interest rate models and for volatility models in finance.

4.3.1 The underlying continuous time model

Let $\theta \in \mathbf{R}$ and $T \in (\theta, +\infty)$ be given. We are also given a standard complete probability space $(\Omega, \mathcal{F}, \mathbf{P})$ and a right-continuous filtration $\{\mathcal{F}_t\}_{t \in [\theta, T]}$ of complete σ-algebras of events. In addition, we are given a one-dimensional Wiener process $w(t)|_{t \in [\theta, T]}$, that is a Wiener process adapted to $\{\mathcal{F}_t\}$ and such that $w(\theta) = 0$ and that $\mathcal{F}_t$ is independent from $w(s) - w(q)$ if $t \geq s > q \geq \theta$.

Consider a continuous time one-dimensional random process $y(t)|_{t \geq \theta}$ such that $y(\theta) > 0$ and

$$dy(t) = f(y(\cdot), t)dt + \sigma(t)y(t)^\gamma dw(t), \qquad t \in (\theta, T). \tag{4.13}$$

Here $\gamma \in [0, 1]$, $\sigma(t)$ is a bounded $\mathcal{F}_t$-adapted process, $f(s, t) : C([\theta, T]) \times [\theta, T] \to \mathbf{R}$ is a measurable function such that $f(s, t)$ is $\mathcal{F}_t$-adapted for any $s \in C([\theta, T])$, and that $f(s_1, t) = f(s_2, t)$ if $(s_1 - s_2)|_{[\theta, t]} \equiv 0$. In addition, we assume that, for any $\delta > 0$, $|f(s_1, t) - f(s_2, t)| \leq c_1 \|s_1 - s_2\|_{C([0,T])}$ and $|f(s, t)| \leq c_2 (\|s\|_{C([\theta, T])} + 1)$ for some constants $c_k = c_k(\delta) > 0$ a.s. (almost surely) for all $t \in [\theta, T]$, $s_1, s_2 \in C([\theta, T])$, such that $\inf_{t \in [\theta, T], k=1,2} s_k(t) > \delta$.

Under these assumptions, there exists a Markov (with respect to $\{\mathcal{F}_t\}$) time

τ with values in $(\theta, T]$ such that there exists a unique a.s. continuous solution $y(s)|_{s \in [\theta, \tau]}$ such that $\inf_{t \in [\theta, \tau]} y(s) > 0$.

Examples from financial modeling

The assumptions for the diffusion coefficient allow us to cover many important financial models. In particular, it covers the so-called Cox–Ingersoll–Ross process with equation

$$dy(t) = a[b - y(t)]dt + \sigma y(t)^{1/2} dw(t), \qquad t > 0, \tag{4.14}$$

where $a > 0$, $b > 0$, and $\sigma > 0$ are some constants.

An extension of this model

$$dy(t) = a[b - y(t)]dt + \sigma y(t)^{\gamma} dw(t), \qquad t > 0, \tag{4.15}$$

where $\gamma \neq 1/2$, is called a *Chan–Karolyi–Longstaff–Sanders (CKLS) model*; see e.g., [68].

4.3.2 A representation theorem for the diffusion coefficient

Up to the end of this section, we assume that the conditions for f and σ formulated above hold. We assume below that τ is a Markov time with respect to $\{\mathcal{F}_t\}$ such that $\tau \in (\theta, T]$ a.s. and that $\inf_{s \in [\theta, \tau]} y(s) > 0$ a.s. In particular, one can select $\tau = T \wedge \inf\{s > \theta : y(s) \leq M\}$ for any given $M \in (0, y(\theta))$.

The following representation theorem for the diffusion coefficient provides the main tool for estimation of the pair (σ, γ) that will be used below.

Theorem 4.4 *For any $h \in [0, 1]$,*

$$\int_{\theta}^{\tau} y(s)^{2(\gamma - h)} \sigma(s)^2 ds = 2 \log |Y_h(\tau)| \quad a.s., \tag{4.16}$$

where $Y_h(s)$ is a complex-valued process defined for $s \in [\theta, \tau]$ such that

$$dY_h(s) = iY_h(s) \frac{dy(s)}{y(s)^h}, \qquad s \in (\theta, \tau),$$

$$Y_h(\theta) = 1. \tag{4.17}$$

In (4.17), $i = \sqrt{-1}$ is the imaginary unit.

Corollary 4.5 *(a) We have that*

$$\int_{\theta}^{\tau} \sigma(s)^2 ds = 2 \log |Y_{\gamma}(\tau)| \quad a.s.. \tag{4.18}$$

(b) If $\sigma(t) = \sigma$ is constant, then, for any $h \in [0, 1]$,

$$\sigma^2 = 2 \left(\int_{\theta}^{\tau} y(s)^{2(\gamma - h)} ds \right)^{-1} \log |Y_h(\tau)| \quad a.s.. \tag{4.19}$$

Proof of Theorem 4.4. Let $\widetilde{a}(t) = f(y(t),t)y(t)^{-h}$. We have, for any $M \in (0, y(\theta))$,

$$dY_h(t) = iY_h(t)[\widetilde{a}(t)dt + y(t)^{\gamma-h}\sigma(t)dw(t)], \quad t \in (\theta, \tau_M).$$

According to the Ito formula again, for any $M \in (0, y(\theta))$,

$$
\begin{aligned}
Y_h(\tau_M) &= Y_h(\theta)\exp\left(i\int_\theta^{\tau_M}\widetilde{a}(s)ds - \frac{i^2}{2}\int_\theta^{\tau_M}y(s)^{2(\gamma-h)}\sigma(s)^2ds\right.\\
&\quad + \left. i\int_\theta^{\tau_M}y(s)^{\gamma-h}\sigma(s)dw(s)\right)\\
&= \exp\left(i\int_\theta^{\tau_M}\widetilde{a}(s)ds + \frac{1}{2}\int_\theta^{\tau_M}y(s)^{2(\gamma-h)}\sigma(s)^2ds\right.\\
&\quad + \left. i\int_\theta^{\tau_M}y(s)^{\gamma-h}\sigma(s)dw(s)\right) \quad \text{a.s..}
\end{aligned}
$$

Hence

$$|Y_h(\tau_M)| = \exp\left(\frac{1}{2}\int_\theta^{\tau_M}y(s)^{2(\gamma-h)}\sigma(s)^2ds\right) \quad \text{a.s.,}$$

and

$$\int_\theta^{\tau_M}y(s)^{2(\gamma-h)}\sigma(s)^2ds = 2\log|Y_h(\tau_M)| \quad \text{a.s..}$$

Hence (4.16) follows from (4.26). $\square$

Proof of Corollary 4.5 follows immediately from Theorem 4.4. $\square$

4.3.3   Estimation based on the representation theorem

Up to the end of this section, we assume that $\sigma(t) \equiv \sigma$ is an unknown positive constant.

We present below estimates of (σ, γ) based on available samples $\{y(t_k)\}$, where $t_k \in [\theta, \tau \wedge T]$ are such that

$$t_{k+1} = t_k + \delta, \quad k = m_0, m_0 + 1, ..., m - 1, \quad \delta = \frac{\tau \wedge T - \theta}{m - m_0},$$

$$t_{m_0} = \theta, \quad t_m = \tau \wedge T.$$

In this setting, $y(t_k) > 0$ for $k = m_0, ..., m$.

For $h \in [0, 1]$, let

$$\eta_{h,k} = \frac{y(t_k) - y(t_{k-1})}{y(t_{k-1})^h}.$$

Estimation of σ under the assumption that γ is known

Let us first suggest an estimate for σ under the assumption that γ is known.

Corollary 4.6 *For any $h \in [0,1]$, the value σ can be estimated as $\widetilde{\sigma}_{\gamma,h}$, where*

$$\widetilde{\sigma}^2_{\gamma,h} = \left(\delta \sum_{k=m_0+1}^{m} y(t_k)^{2(\gamma-h)} \right)^{-1} \sum_{k=m_0+1}^{m} \log(1 + \eta^2_{h,k}). \qquad (4.20)$$

Estimation of γ with excluded σ

It appears that Theorem 4.4 implies some useful properties of the process $Y_h(t)$ allowing us to estimate γ in a setting with unknown constant σ.

Proposition 4.7 *For any $h_1, h_2 \in [0,1]$,*

$$\frac{\int_\theta^\tau y(s)^{2(\gamma-h_1)}ds}{\int_\theta^\tau y(s)^{2(\gamma-h_2)}ds} = \frac{\log|Y_{h_1}(\tau)|}{\log|Y_{h_2}(\tau)|} \quad a.s.. \qquad (4.21)$$

Since calculation of $Y_{h_1}(\tau)$ and $Y_{h_2}(\tau)$ does not require knowing the values of f, γ, and σ, property (4.21) allows us to calculate γ as shown below.

Corollary 4.8 *An estimate $\hat{\gamma}$ of γ can be found as a solution of the equation*

$$\frac{\sum_{k=m_0+1}^m y(t_k)^{2(\gamma-h_1)}}{\sum_{k=m_0+1}^m y(t_k)^{2(\gamma-h_2)}} = \frac{\sum_{k=m_0+1}^m \log(1 + \eta^2_{h_1,k})}{\sum_{k=m_0+1}^m \log(1 + \eta^2_{h_2,k})}, \qquad (4.22)$$

for any pair of preselected h_1 and h_2.

It can be noted that σ remains unused and excluded from the analysis for the method described in Proposition 4.7 and Corollary 4.8; respectively, this method does not lead to an estimate of σ.

Estimation of the pair (σ, γ)

Proposition 4.9 *The process*

$$\frac{1}{t \wedge \tau - \theta} \log|Y_\gamma(t \wedge \tau)| \qquad (4.23)$$

is a.s. constant in $t \in [\theta, T]$.

Let

$$v_{h,k} = \log(1 + \eta^2_{h,j}), \quad k = m_o + 1, ..., m,$$

and

$$\overline{v}_h = \frac{1}{m - m_o} \sum_{j=m_o+1}^{m} v_{h,k}.$$

Corollary 4.10 *An estimate of γ can be found as the solution of the optimization problem*

$$\text{Minimize} \quad \sum_{k=m_0+1}^{m} \left(v_{h,k} - \bar{v}_h\right)^2 \quad over \quad h \in [0,1]. \tag{4.24}$$

In this case, σ can be estimated as

$$\hat{\sigma} = \sqrt{\bar{v}_{\hat{\gamma}}/\delta}, \tag{4.25}$$

where $\hat{\gamma}$ is the estimate of γ obtained as a solution of (4.24).

Remark 4.11 *Corollary 4.10 allows the following modification of estimation of γ for the case where σ is a known constant: an estimate $\hat{\gamma}$ of γ can be found as the solution of the optimization problem*

$$\text{Minimize} \quad \sum_{k=m_0}^{m} \left(v_{h,k}/\delta - \sigma^2\right)^2 \quad over \quad h \in [0,1].$$

Some remarks

Similar to the estimates described in Section 4.1, the estimates for the volatility coefficients and for the power index described here are again pathwise estimates that do not rely on the distribution of the underlying process and on a particular choice of the drift. An attractive feature of these pathwise estimates is that they do not require neither estimation of the drift nor estimation of the distributions of the underlying process. In particular, one does not need to know the shape of the likelihood function required for the maximum likelihood method. Respectively, this method allows consideration of models with a large number of parameters for the drift. This allows covererage of cases where the maximum likelihood method and the method of moments are not applicable due to the high dimensions. This can be especially beneficial for financial applications where the trend for the prices is usually difficult to estimate since it is overshadowed by a relatively large volatility. On the other hand, the method does not lead to an estimation of the drift, since the drift is excluded from the analysis.

Proofs

For $M \in (0, y(\theta))$, let $\tau_M = \tau \wedge \sup\{s \in [\theta, T] : \inf_{q \in [\theta, s]} y(q) \geq M\}$. Clearly,

$$\tau_M \to \tau \quad as \quad M \to 0 \quad \text{a.s.} \tag{4.26}$$

Proof of Corollary 4.6. Let $t_m = \tau_M$, $t_{m_0} = \theta$, and let $t_k = t_{m_0} + (k - m_0)\delta$ if $m_0 \leq k \leq m$. Let $\eta_{h,k}$ be defined as above. The Euler–Maruyama time discretization of (4.17) leads to the stochastic difference equation

$$y_h(t_k) = y_h(t_{k-1}) + iy_h(t_{k-1})\eta_{h,k}, \quad k \geq m_0 + 1,$$
$$y_h(t_{m_0}) = 1.$$

(See, e.g., [78], Ch. 9.) This equation can be rewritten as

$$\mathcal{Y}_h(t_k) = \mathcal{Y}_h(t_{k-1})(1 + i\eta_{h,k}), \quad k \geq m_0 + 1,$$
$$\mathcal{Y}_h(t_{m_0}) = 1.$$

Hence

$$\mathcal{Y}_h(t_m) = \prod_{k=m_0+1}^{m} (1 + i\eta_{h,k}).$$

Clearly,

$$|\mathcal{Y}_h(t_m)| = \prod_{k=m_0+1}^{m} |1 + i\eta_{h,k}| = \prod_{k=m_0+1}^{m} (1 + \eta_{h,k}^2)^{1/2},$$

and

$$\log |\mathcal{Y}_h(t_m)| = \sum_{k=m_0+1}^{m} \log[(1 + \eta_{h,k}^2)^{1/2}] = \frac{1}{2} \sum_{k=m_0+1}^{m} \log(1 + \eta_{h,k}^2). \quad (4.27)$$

Then (4.18) leads to estimate (4.20). $\square$

Proof of Proposition 4.7 and Proposition 4.9 follows immediately from Theorem 4.4 and Proposition 4.5(b). $\square$

Proof of Corollary 4.8 follows from the natural discretization of integration and (4.27). $\square$

Proof of Corollary 4.10. It follows from (4.27) that the sequence $\{\log |\mathcal{Y}_h(t_k)|\}$ represents the discretization of the continuous time process $\log |Y_h(t \wedge \tau)|$ at points $t = t_k$; this process is linear in time for $h = \gamma$ and $2 \log |Y_\gamma(t \wedge \tau)| \equiv (t \wedge \tau - \theta)\sigma^2$. Hence

$$2 \log |Y_\gamma(t_{k+1})| - 2 \log |Y_\gamma(t_k)| = \delta\sigma^2.$$

On the other hand, (4.27) implies that

$$2 \log |\mathcal{Y}_h(t_{k+1})| - 2 \log |\mathcal{Y}_h(t_k)| = v_{h,k}, \quad h \in [0, 1].$$

This leads to an optimization problem

$$\text{Minimize} \quad \sum_{k=m_0+1}^{m} (v_{h,k}/\delta - c)^2 \quad \text{over} \quad h \in [0, 1], \quad c > 0.$$

By the properties of quadratic optimization, this problem can be replaced by the problem

$$\text{Minimize} \quad \sum_{k=m_0+1}^{m} (v_{h,k}/\delta - \bar{v}_h)^2 \quad \text{over} \quad h \in (0, 1]. \quad (4.28)$$

Then the proof follows. $\square$

Proof of Remark 4.11 repeats the previous proof without optimization over c. $\square$

4.3.4 Numerical experiments

To illustrate numerical implementation of the algorithms described above, we applied these algorithms for discretized Monte Carlo simulations of some generalized version of the Cox–Ingersoll–Ross process (4.14). We consider a toy example of a process with a large number of parameters. Presumably, estimation of all these parameters is not feasible due the high dimensions for the maximum likelihood method and for the method of moments.

We consider a process evolving as follows:

$$dy(t) = H\left(y(t), y(\max(t - \lambda, 0))\right) dt + \sigma y(t)^\gamma dw(t), \qquad t > 0, \qquad (4.29)$$

where

$$H(x, y) = \sum_{k=1}^{N} \left[F_k(x) + G_k(y) \right],$$

$$F_k(x) = a_k[b_k - x^{\nu_k + 1/2}] + c_k \cos(d_k x + e_k),$$

$$G_k(x) = 0.1 \, \hat{a}_k[\hat{b}_k - x^{\hat{\nu}_k + 1/2}].$$

The parameters $N, a_k, b_k, \nu_k, c_k, d_k, e_k, \hat{a}_k, \hat{b}_k, \hat{e}_k, \nu_k, \hat{\nu}_k, \lambda$ are randomly selected in each experiment. In particular, the integers N are selected randomly at the set $\{1, 2, 3, 4, 5\}$ with equal probability. The delay parameter λ has a uniform distribution on the interval $[0, 0.2]$. The parameters $a_k, b_k, \nu_k, c_k, d_k, e_k, \hat{a}_k, \hat{b}_k, \hat{e}_k, \nu_k, \hat{\nu}_k$ are uniformly distributed on the interval $[0, 1]$.

For the Monte Carlo simulation, we considered a corresponding discrete-time process $\{y(t_k)\}$ evolving as

$$y(t_{k+1}) = y(t_k) + H(y(t_k), y(t_{\max(k-\ell, 0)}))\delta + \sigma y(t_k)^\gamma \delta^{1/2} \xi_{k+1},$$

$$k = 0, ..., n, \qquad (4.30)$$

with mutually independent random variables ξ_k from the standard normal distribution $N(0, 1)$. Here $\delta = t_{k+1} - t_k = 1/n$ which corresponds to $[\theta, T] = [0, 1]$ for the continuous time underlying model. The delay ℓ is the integer part of $\lambda(T - \theta)/(n + 1)$.

We considered $n \in \{52, 250, 10000, 20000\}$. For the financial applications, the choice of $n = 52$ corresponds to weekly sampling; the choice of $n = 250$ corresponds to daily sampling.

In the Monte Carlo simulation trials, we considered random $y(t_{m_0})$ uniformly distributed on $[0.1, 10]$ and truncated paths $y(t_k)|_{m_0 \le k \le m}$, with the Markov stopping time $m = n \wedge \inf\{k : y(t_k) \le 0.001 y(t_{m_0})\}$. In this case, $y(t_k) > 0$ for $k = m_0, ..., m$. To exclude the possibility that $y(t_m) \le 0$ (which may happen for our discrete time process since the values of ξ_k are unbounded), we replace $y(t_m)$ defined by (4.30) by $y(t_m) = y(t_{m-1}) > 0$ every time that $m < n$ occurs. It can be noted that, for our choice of parameters, the occurrences of the event $m < n$ were very rare and have no impact on the statistics.

We used 10,000 Monte Carlo trials for each trial (i.e., for each entry in each of the Tables 4.14 through 4.16). We found that enlarging the sample does not improve the results. Actually, the experiments with 5,000 trials or even 1,000 Monte Carlo trials produced the same results.

The parameters of the errors obtained in these experiments are quite robust with respect to the change of other parameters as well.

We denote by $\mathbb{E}$ the sample mean of a corresponding value over all Monte Carlo simulation trials. For the estimates $(\hat{\sigma}, \hat{\gamma})$ of (σ, γ), we evaluated the root mean-squared errors (RMSE) $\sqrt{\mathbb{E}\,|\hat{\sigma} - \sigma|^2}$ and $\sqrt{\mathbb{E}\,|\hat{\gamma} - \gamma|^2}$, the mean errors $\mathbb{E}|\hat{\sigma} - \sigma|$ and $\mathbb{E}\,|\hat{\gamma} - \gamma|$, and the biases $\mathbb{E}(\hat{\sigma} - \sigma)$ and $\mathbb{E}\,(\hat{\gamma} - \gamma)$.

In the experiment described below, we used $\sigma = 0.3$, $\gamma = 1/2$, and $\gamma = 0.6$.

Estimation of σ using Corollary 4.6

The numerical implementation of Corollary 4.6 requires use of the value γ. In other words, one has to use a certain hypothesis about the value of γ, for instance, based on estimation of γ that was done separately. This setting leads to an error caused by miscalculation of γ.

To illustrate the dependence of the error for the estimate of σ from the error in the hypothesis on γ, we considered estimates under a fixed hypothesis that $\gamma = 1/2$ for inputs simulated with different "true" $\gamma \in \{0.4, 0.5, 0.6, 0.7\}$.

Table 4.14 shows the parameters of the errors in the experiments described above for estimate (4.20) for σ applied with $h = \gamma = 1/2$ to processes (4.29) simulated with different "true" γ, for $\delta = 1/52$ and $\delta = 1/250$. According to this table, estimate (4.20) is robust with respect to small errors for γ.

Estimation of γ with unknown σ using (4.22) and (4.24)

In these experiments, we used a simulated process with $\gamma = 0.6$ and estimates (4.22) and (4.24).

For solution of equation (4.22) and optimization problem (4.24), we used a simple search over a finite set $\{h_k\}_{k=1}^{N} = \{k/N\}_{k=1}^{N}$. We used $N = 300$ for (4.22) and $N = 30$ for (4.24). Further increasing of N does not improve the results but slows down calculation.

It appears that estimation of γ is more numerically challenging than estimation of σ using (4.20) with known γ. In our experiments, we observed that the dependence of the value of the criterion function in (4.24) depends on h smoothly and the dependence on h for each particular Monte Carlo trial is represented by a U-shaped smooth convex function. However, the minimum point of this function deviates significantly for different Monte Carlo trials, especially in the case of low-frequency sampling. It requires high-frequency sampling to reduce the error $\hat{\gamma} - \gamma$. Table 4.15 shows the parameters of the error $\hat{\gamma} - \gamma$. We found that these parameters are quite robust with respect to the change of other parameters of the simulated process.

(a) $\delta = 1/52$

	$\sqrt{\mathbb{E}\,|\hat{\gamma} - \gamma|^2}$	$\mathbb{E}\,|\hat{\gamma} - \gamma|$	$\mathbb{E}\,(\hat{\gamma} - \gamma)$
$\gamma = 0.5$	0.0312	0.0248	0.0034
$\gamma = 0.4$	0.0458	0.0365	0.0281
$\gamma = 0.6$	0.0358	0.0290	-0.0183
$\gamma = 0.7$	0.0495	0.0413	-0.0370

(b) $\delta = 1/250$

	$\sqrt{\mathbb{E}\,|\hat{\gamma} - \gamma|^2}$	$\mathbb{E}\,|\hat{\gamma} - \gamma|$	$\mathbb{E}\,(\hat{\gamma} - \gamma)$
$\gamma = 0.5$	0.0136	0.0109	0.0006
$\gamma = 0.4$	0.0328	0.0272	0.0259
$\gamma = 0.6$	0.0269	0.0227	-0.0215
$\gamma = 0.7$	0.0468	0.0416	-0.0414

Table 4.14: Parameters of the error $\hat{\sigma} - \sigma$ for $\hat{\sigma}$ obtained from estimates (4.20) applied with $\gamma = h = 1/2$ for $\delta = 1/52$ and $\delta = 1/250$. In the first column, γ is the "true" power used for simulation; a larger mismatch between γ and $1/2$ leads to a larger estimation error.

Estimation of σ using (4.24)

The solution of optimization problem (4.24) gives an estimate of σ, in addition to an estimate of γ, in the setting with unknown σ, via (4.25). This gives a method for estimation of σ that can be an alternative to estimator (4.20). Table 4.16 shows the parameters of the error $\hat{\sigma} - \sigma$. It appears that the RMSE is larger than for estimators (4.20) applied with a correct $\gamma = 1/2$ and has the same order as the RMSE for this estimators applied with "miscalculated" γ.

Comparison with the performance of other methods

In [112] and [126], results were reported for testing of a variety of estimators based on the maximum likelihood method or the method of moments for special cases of (4.13). These works considered simulated processes with a preselected structure for the drift term with a low dimension of the vector of parameters. Due to numerical challenges for the methods used, the number of Monte Carlo trials was relatively low in these works (100 trials in [112] and 1,000 trials in [126]). In [112], model (4.15) was considered with one fixed set of parameters (a, b) for the drift. In [126], model (4.14) was considered for a variety of the parameters (a, b) for the drift. In [112], estimation of (σ, γ) and estimation of the drift parameters were considered.

δ	$\sqrt{\mathbb{E}\,\lvert\hat{\gamma}-\gamma\rvert^2}$ for (4.22)	$\mathbb{E}\lvert\hat{\gamma}-\gamma\rvert$ for (4.22)	$\mathbb{E}(\hat{\gamma}-\gamma)$ for (4.22)	$\sqrt{\mathbb{E}\,\lvert\hat{\gamma}-\gamma\rvert^2}$ for (4.24)	$\mathbb{E}\lvert\hat{\gamma}-\gamma\rvert$ for (4.24)	$\mathbb{E}(\hat{\gamma}-\gamma)$ for (4.24)
$1/250$	0.2078	0.1736	0.1078	0.2304	0.1946	0.1166
$1/10,000$	0.0309	0.0182	0.0039	0.0356	0.0221	0.0042
$1/20,000$	0.0222	0.0109	0.0020	0.0483	0.0294	0.0004

Table 4.15: Parameters of the error $\hat{\gamma}-\gamma$ for the solution of (4.22) and (4.24) with an unknown σ.

	$\sqrt{\mathbb{E}\,\lvert\hat{\sigma}-\sigma\rvert^2}$	$\mathbb{E}\lvert\hat{\sigma}-\sigma\rvert$	$\mathbb{E}(\hat{\sigma}-\sigma)$
$\delta = 1/250$	0.0515	0.0264	0.0092
$\delta = 1/10,000$	0.0063	0.0038	0.0001
$\delta = 1/20,000$	0.0168	0.0108	0.00003

Table 4.16: Parameters of the error $\hat{\sigma}-\sigma$ for $\hat{\sigma}$ obtained from (4.25) and (4.24) with an unknown γ.

In [126], estimation of σ and estimation of the drift parameters with fixed $\gamma = 1/2$ were considered.

The results for σ in Table 5 from [126] reported for $\delta = 1/500$ depends significantly on the choice of the drift parameters (a, b) in (4.14) (in our notations). The minimal RMSE for estimates of σ among all pairs (a, b) is of the same order as the RMSE reported in our Table 4.14(a) for $\delta = 1/250$ for the case of known h; for other choices of the drift, the RMSE in Table 5 from [126] is much larger. Remember that the RMSE is smaller for smaller δ.

The RMSE for σ reported in Table II.1 from [112] for $\delta = 1/500$ (in our notations) is approximately the same as in Table 4.16 for $\delta = 1/250$. The RMSE for γ with $\delta = 1/500$ is three times smaller in Table II.1 from [112] for some estimators than in Table 4.15 with $\delta = 1/250$. However, it may happen that the performance of the estimators in Table II.1 from [112] is not robust with respect to different choices of the drift parameters, similar to the case presented in Table 5 from [126] for $\gamma = 1/2$. On the other hand, the method described above allows inclusion of a high variety of drift models with almost unlimited dimensions, and, as we found in some unreported experiments, the choice of particular drifts does not have an impact on the performance of the estimator.

On analysis of the historical data

For historical financial data, it is not possible to be sure that the inputs obey a CIR-type evolution law. Therefore, any results of the inference for the model parameters depends on the assumptions about the model. Assuming that the prices evolve as a CIR model with unknown (σ, γ), it is possible, using the methods described above, to identify the most fitting (σ, γ). There are other possible methods. For example, estimate (4.20) is supposed to give the same value σ for all choices of $h \in [0, 1]$. Therefore, it is natural to apply this estimate for calculation of $\sigma = \sigma(\gamma, h)$ with different pairs (γ, h) and select $\hat{\gamma}$ to be the estimate of γ if

$$\sup_{h_1, h_2} |\sigma(\hat{\gamma}, h_1) - \sigma(\hat{\gamma}, h_2)| \leq \sup_{h_1, h_2} |\sigma(\gamma, h_1) - \sigma(\gamma, h_2)| \quad \forall \gamma \in [0, 1]. \quad (4.31)$$

Some examples of estimates of $\sigma(\gamma, h)$ of historical data were obtained in [122]. In particular, estimate (4.20) applied in [122] to Amazon stock prices, with $\hat{\gamma} = 0.095$ produces values $\sigma(\hat{\gamma}, h) = 0.609$ for $h = 0.01$, $\sigma(\hat{\gamma}, h) = 0.607$ for $h = 0.1$, $\sigma(\hat{\gamma}, h) = 0.604$ for $h = 0.25$, $\sigma(\hat{\gamma}, h) = 0.602$ for $h = 0.5$, $\sigma(\hat{\gamma}, h) = 0.603$ for $h = 0.75$, and $\sigma(\hat{\gamma}, h) = 0.606$ for $h = 1$, according to the last table in [122]. The corresponding values $\sigma(\gamma, h)$ depend more significantly on h for other choices of γ. This allows acceptance of $\sigma = 0.604$ and $\gamma = 0.95$ for the underlying path of prices.

4.3.5 On the consistency of the method

Let us briefly discuss the consistency of the method. We restrict our consideration to the simulated data based on model (4.13), and we consider consistency of the method as convergence of the estimates to the true values as the sampling frequency is increasing, i.e., $\delta \to 0$. Equations (4.19), (4.21), (4.23) in the continuous time used for our method are exact and hold almost surely. Therefore, the only source of the error is the time discretization error. This error is inevitable since the method requires pathwise evaluation of stochastic integrals.

There are two options. First, one can consider Euler-Maruyama time discretization for the pair (y, Y_h) such as described for the numerical experiments described above. In this case, f and the sampling frequency δ have to be such that a satisfactory approximation is achieved. In particular, by Theorem 9.6.2 from [78], p. 324, these conditions are satisfied for CIR models as well as for the case where $f(y(\cdot), t) = f(y(t), t)$. Some analysis of conditions for the convergence in more general cases can be found in [72]. The numerical experiments described above demonstrate that the required convergence takes place for equations will delay in model (4.29).

Another option is to consider convergence of the method for $\delta \to 0$ given that $\mathcal{Y}(t_k)$ are constructed with the "true" entries $y(t)$. We presume here that it is possible to produce an arbitrarily close approximation of a continuous path $y(t)$ via Monte Carlo simulation with increasing of the simulation frequency. Let $t_\delta(t)$ be selected as t_k if $|t - t_k| \leq |t - t_p|$ for all p (for certainty, we assume that $t_\delta(t) = t_k$ if $|t - t_k| = |t - t_{k+1}|$). Clearly, $\mathbf{E} \sup_{t \in [0, \tau]} |y(t_\delta(t)) - y(t)|^2 \to 0$ as $\delta \to 0$. Hence $\delta \sum_{k=m_0+1}^{m} y(t_k)^{2(\gamma-h)} \to \int_\theta^\tau y(s)^{2(\gamma-h)} ds$ in probability as $\delta \to 0$. Further, it can be shown that $\log |\mathcal{Y}_h(t_\delta(t))| \to \log |Y_h(t)|$ in probability as $\delta \to 0$. This leads to convergence of estimates to their true values in probability.

4.3.6 Some properties of the estimates

The estimates listed in Section 4.3.3 use neither f nor the probability distribution of the process y. In particular, they are invariant with respect to the choice of an equivalent probability measure. This is an attractive feature that allows us to consider models with a large number of parameters for the drift.

Furthermore, it appears that estimation of the power index γ with unknown σ is numerically challenging and requires high-frequency sampling to reduce the error. Perhaps, this can be improved using other modifications of (4.24) and other estimates for the degree of nonlinearity for the implementation of Proposition 4.9. In particular, the standard criterions of linearity for the first-order regressions could be used, and L_2-type criterions could be replaced by L_p-type criterions with $p \neq 2$. So far, we were unable to find a way to reduce the error for lower sampling frequency. We leave it for future research.

The approach suggested in this section does not cover the estimation of the drift f which is a more challenging problem. However, the estimates for (σ, γ) suggested above can be used to simplify statistical inference for f by reduction of the dimension of the numerical problems arising in the maximum likelihood method, methods of moments, or least squares estimators, for (f, σ, γ). This can be illustrated as the following.

Assume that $\gamma = 1/2$ is given and that evolution of $y(t)$ is described by Cox–Ingersoll–Ross equation (4.14). It is known that

$$\mathbf{E}y(T) = b(1 - e^{-aT}) + e^{-aT}y(0),$$
$$\operatorname{Var} y(T) = \frac{\sigma^2}{2a}b(1 - e^{-aT}) + e^{-aT}\frac{\sigma^2}{a^2}(1 - e^{-aT})y(0).$$

(See, e.g.[60].) This system can be solved with respect to (a, b) given that $\mathbf{E}y(T)$ and $\operatorname{Var} y(T)$ are estimated by their sampling values, and σ is estimated as suggested above.

The approach suggested in this section focuses on the case where σ is constant. However, some results can be extended to the case of time depending and random $\sigma = \sigma(t)$. For example, the proofs given above imply that $\int_\theta^{t_m} \sigma(s)^2 ds$ can be estimated as

$$\tilde{\sigma}^2_{\gamma,\gamma} = \sum_{k=m_0+1}^{m} \log(1 + \eta^2_{\gamma,k}).$$

The experiments described above for simulated data demonstrated feasibility of pathwise inference for (σ, γ).

4.4 Estimation of the appreciation rates

Consider a diffusion model of a market consisting of a locally risk-free bank account or bond with price $B(t)$, $t \geq 0$, and n risky stocks with prices $S_i(t)$, $t \geq 0$, $i = 1, 2, \ldots, n$, where $n < +\infty$ is given. The prices of the stocks evolve according to the following equations:

$$dS_i(t) = S_i(t) \left(a_i(t)dt + \sum_{j=1}^{n} \sigma_{ij}(t)dw_j(t) \right), \quad t > 0, \tag{4.32}$$

where $w_i(t)$ are standard independent Wiener processes, $a_i(t)$ are appreciation rates, and $\sigma_{ij}(t)$ are volatility coefficients. The initial price $S_i(0) > 0$ is a given non-random constant. The price of the bond evolves according to the following equation

$$B(t) = B(0) \exp \left(\int_0^t r(t)dt \right), \tag{4.33}$$

where $B(0)$ is a given constant which we take to be 1 without loss of generality, and $r(t)$ is the random process of the risk-free interest rate.

Let

$$w(t) = (w_1(t), \ldots, w_n(t))^\top, \quad S(t) = (S_1(t), \ldots, S_n(t))^\top,$$
$$a(t) = (a_1(t), \ldots, a_n(t))^\top,$$

and $\sigma(t) = \{\sigma_{ij}(t)\}_{i,j=1}^n$.

Let $\mathbf{1} \triangleq (1, \ldots, 1)^\top \in \mathbf{R}^n$, and $\widetilde{a}(t) \triangleq a(t) - r(t)\mathbf{1}$.

Let $\{\mathcal{F}_t^{S,r}\}_{0 \leq t \leq T}$ be the filtration generated by the process $(r(t), S(t))$ completed with the null sets of $\mathcal{F}$.

Set $\widetilde{S}(t) \triangleq \exp \left(- \int_0^t r(s)ds \right) S(t)$.

For an Euclidean space E, we denote by $\mathrm{B}([0,T]; E)$ the set of bounded measurable functions $f(t) : [0,T] \to E$. We denote by I_n the identity matrix in $\mathbf{R}^{n \times n}$.

We say that $A < B$ for symmetric matrices if the matrix $B - A$ is definitely positive.

The model for r, σ, and a

To describe the distribution of $\widetilde{a}(t)$, we will use the model introduced in [80], p.84, generalized for our case of random r, non-constant coefficients for the equation for $\widetilde{a}$, and correlated r, $\widetilde{a}$, and w. We assume that we are given measurable deterministic processes $\alpha(t)$, $\beta(t)$, $b(t)$, and $\delta(t)$ such that

$$d\widetilde{a}(t) = \alpha(t)[\delta(t) - \widetilde{a}(t)]dt + b(t)d\widetilde{R}(t) + \beta(t)dW(t), \tag{4.34}$$

where $\alpha(t) \in \mathbf{R}^{n \times n}$, $\beta(t) \in \mathbf{R}^{n \times n}$, $b(t) \in \mathbf{R}^{n \times n}$, $\delta(t) \in \mathbf{R}^n$, and where W is an n-dimensional Wiener process in $(\Omega, \mathcal{F}, P)$. We assume that $\alpha(t)$, $\beta(t)$, $b(t)$, and

$\delta(t)$ are continuous in t and such that the matrix $\beta(t)$ is invertible and $|\beta(t)^{-1}| \leq c$, where $c > 0$ is a constant. Further, we assume that $\widetilde{a}(0)$ follows an n-dimensional normal distribution with mean vector m_0 and covariance matrix γ_0. The vector m_0 and the matrix γ_0 are assumed to be known. We note that this setting covers the case when $\widetilde{a}$ is an n-dimensional Ornstein-Uhlenbeck process with mean-reverting drift.

Clearly, equation (4.34) can be rewritten as

$$d\widetilde{a}(t) = \Big(\alpha(t)\delta(t) + [b(t) - \alpha(t)]\widetilde{a}(t) \Big) dt + b(t)\sigma(t)dw(t) + \beta(t)dW(t). \quad (4.35)$$

We assume that the process $\sigma(t)$ is continuous in t, non-random and such that $\sigma(t)\sigma(t)^\top \geq c_\sigma I_n$, where $c_\sigma > 0$ is a constant.

Further, we assume that $r(\cdot) = \phi_r(\widetilde{R}(\cdot), \Theta)$, where Θ is a random element in a metric space $\mathcal{X}_r$, and where $\phi_r : C([0,T]; \mathbf{R}^n) \times \mathcal{X}_r \to \mathrm{B}([0,T]; \mathbf{R})$ is a measurable function, and Θ does not depend on $(w(\cdot), W(\cdot), \widetilde{a}(0))$. In addition, we assume that the process $r(t)$ is adapted to the filtration generated by $(\widetilde{R}(t), \Theta)$. Note that a closed system for the pair $(\widetilde{a}(t), \widetilde{R}(t))$ does not include $r(\cdot)$, and $(\widetilde{a}(\cdot), \widetilde{R}(\cdot))$ does not depend on Θ. Therefore, the market model is well defined. The assumptions for measurability of r do not look very natural. However, they cover generic models when r is independent on $\widetilde{R}$ or non-random, and we can still consider some models with correlated r and $\widetilde{R}$.

Under these assumptions, the solution of (4.32) is well defined, but the market is incomplete.

Let $\widetilde{\phi}_m(t, s)$, $m = 0, 1$, be the solution of the matrix equation

$$\begin{cases} \frac{d\widetilde{\phi}_m}{dt}(t, s) = [m \cdot b(t) - \alpha(t)]\widetilde{\phi}_m(t, s), \\ \widetilde{\phi}_m(s, s) = I_n. \end{cases}$$

Let

$$\widetilde{K}_m(t) \triangleq \int_0^t \widetilde{\phi}_m(t, s)b(s)\sigma(s)\sigma(s)^\top b(s)^\top \widetilde{\phi}_m(t, s)^\top ds, \quad m = 0, 1. \quad (4.36)$$

We have that

$$\widetilde{a}(t) = \widetilde{\phi}_1(t, 0)\widetilde{a}(0) + \int_0^t \widetilde{\phi}_1(t, s)[\alpha(s)\delta(s)ds + b(s)\sigma(s)dw(s) + \beta(s)dW(s)].$$

It follows that $\widetilde{K}_1(t)$ is the covariance matrix for $\widetilde{a}(t)$ calculated with $\beta(t) \equiv 0$ and $\widetilde{a}(0) = 0$. By the linearity of (4.35), it follows that $\widetilde{K}_1(t)$ is the conditional covariance for $\widetilde{a}(t)$ given $(W(\cdot)|_{[0,t]}, \widetilde{a}(0))$ or $(W(\cdot)|_{[0,T]}, \widetilde{a}(0))$.

Note that $\widetilde{K}_m(t)$ can be found as solutions of linear equations that one can easily derive from (4.35) and (4.39) (see, e.g., [15], Chapter 8).

We assume that b is "small." More precisely, we assume that there exists $\varepsilon > 0$ such that

$$T\widetilde{K}_m(t) + \varepsilon I_n < \sigma(t)\sigma(t)^\top \quad \forall t \in [0, T], \quad m = 0, 1. \quad (4.37)$$

Solution via conditional expectation

Let
$$\hat{a}(t) \triangleq \mathbf{E}\{\tilde{a}(t) \mid \mathcal{F}_t^{S,r}\}.$$

Set $\tilde{\alpha}(t) \triangleq \alpha(t) - b(t)$ and $m_0 \triangleq \mathbf{E}\tilde{a}(0)$.

Let $\gamma(t) \in \mathbf{R}^{n \times n}$ be the unique solution (in the class of symmetric non-negative definite matrices) of the deterministic Riccati equation

$$\frac{d\gamma}{dt}(t) = -[b(t)\sigma(t)^\top + \gamma(t)]Q(t)[b(t)\sigma(t)^\top + \gamma(t)]^\top$$
$$- \tilde{\alpha}(t)\gamma(t) - \gamma(t)\tilde{\alpha}(t)^\top + \beta(t)\beta(t)^\top, \qquad (4.38)$$

$$\gamma(0) = \gamma_0.$$

Here $\gamma_0 \triangleq \mathbf{E}[\tilde{a}(0) - m_0][\tilde{a}(0) - m_0]^\top$. In fact, $\gamma(t) = \mathbf{E}\left\{[\tilde{a}(t) - \hat{a}(t)][\tilde{a}(t) - \hat{a}(t)]^\top \mid \mathcal{F}_t^{S,r}\right\}$.

Let $A(t) \triangleq -\tilde{\alpha}(t) - \gamma(t)Q(t)$, and let $\phi(t)$ be the solution of the matrix equation

$$\begin{cases} \frac{d\phi}{dt}(t) = A(t)\phi(t), \\ \phi(0) = I_n, \end{cases}$$

where I_n is the unit matrix in $\mathbf{R}^{n \times n}$.

Let $A(t)$ and $\gamma(t)$ be the matrices defined above, let $\gamma(t)$ be the solution of (4.38), and

$$x = (x_1, \ldots, x_{n+1})^\top = \begin{pmatrix} \hat{x} \\ x_{n+1} \end{pmatrix}, \qquad \hat{x} = (x_1, \ldots, x_n)^\top.$$

By Theorem 10.3 from [85], p.396, the equation for $\hat{a}(t)$ is

$$d\hat{a}(t) = [A(t)\hat{a}(t) - b(t)\sigma(t)^\top Q(t)\hat{a}(t) + \alpha(t)\delta(t)]dt$$
$$+ [b(t)\sigma(t)^\top + \gamma(t)]Q(t)d\tilde{R}(t), \qquad (4.39)$$

$$\hat{a}(0) = m_0.$$

The case of a single stock and constant $\tilde{a}$

As an example, consider a special case where $n = 1$ and where $\tilde{a} = \tilde{a}(t)$ is constant in time almost surely and has Gaussian distribution such that $\operatorname{Var} \tilde{a} = v_0^2$, and $\mathbf{E}\tilde{a} = 0$. Further, let $\sigma(t) \equiv \sigma$ be a non-random constant. Then the Kalman–Bucy filter gives

$$\begin{cases} d\hat{a}(t) = \frac{v(t)}{\sigma^2}\left[\frac{d\tilde{S}(t)}{\tilde{S}(t)} - \hat{a}(t)dt\right], \\ dv(t) = -\frac{v(t)^2}{\sigma^2}dt, \\ \hat{a}(0) = a_0, \quad v(0) = v_0. \end{cases}$$

Here $v(t) = \mathbf{E}\{(\tilde{a} - \hat{a})^2 \mid \mathcal{F}_t\}$.

4.5 Bibliographic notes and literature review

Section 4.1

The method described in Section 4.1 was suggested in [47].

A rigorous analysis of convergency for the time discretization requires significant analytical efforts; see e.g. [78, 72], where a review of the recent literature can be found.

Section 4.2

The commonly recognized explanation for the dependence of volatility on different timescales is the presence market micro-structure noise [4, 7, 14]. This term refers to imperfections in the trading process of financial assets which causes the observed prices to differ from the underlying 'true' price. Another approach is to find some optimal sampling frequencies that can ensure the most accurate volatility estimation for the underlying assets; see, e.g., [9, 55]. One more approach is to introduce bias-correction in measuring the volatility, such as the two timescales estimator [2]. For this method, the highest frequency sample is used and the market micro-structure noise is subtracted from a sub-sampled estimator that is calculated by using a "sparse" sampling frequency.

Some of the properties of high-frequency volatility estimates have been discussed in [111]. The choices of the frequency for the data is discussed in [3].

Section 3.2 is based on the results [88].

Section 4.3

Some applications of continuous time stochastic processes with a power type dependence are described in [62, 58, 32, 60, 10, 56, 64, 84]; see also the bibliographies therein. Estimation of the parameters for these models was widely studied; see e.g., [58, 60, 10, 75, 112, 39, 1, 56, 65, 84]. The approach of these works uses estimation of the parameters for the drift term; the method suggested in Section 4.3 bypasses this task.

The assumptions on the process y allows us to use it for a variety of financial models. In particular, the assumptions on the drift coefficient f cover a path depending evolution such as described by equations with delay; see some examples in [88, 48, 108].

The assumptions on the diffusion coefficient allow us to cover many important financial models. In particular, the so-called Cox–Ingersoll–Ross process (4.14) is used for the interest rate models and the volatility of stock prices (see e.g., [62]). An extension of this model (4.15) is called a *Chan–Karolyi–Longstaff–Sanders (CKLS) model* in the econometric literature; see e.g., [68]. This model was introduced in [33]. This equation with $\gamma = 2/3$ is used for volatility modeling; see, e.g., [32, 84]. The method described in Section 4.3 was suggested in [49].

Section 4.4

Section 4.4 is based on the results [41] and the method suggested in [80]. Some references for the estimation of the appreciation rate can be found in [41] and [46], Chapter 9, p.128.

5

Some background on bond pricing

Bonds are sold at an initial time for a certain price, and gives the right to obtain a certain amount of cash (higher than this initial price) in a fixed time (we restrict our consideration to zero coupon bonds only). Therefore, the owner can have a fixed income. Typically, there are many different bonds on the market with different times for maturity, and they are actively traded, so the analysis of bonds is very important for applications.

For the bond-and-stock market models introduced above, we refer to bonds as a risk-free investment similar to a cash account. For instance, this is typical for the Black and Scholes market model, where the bank interest rate is supposed to be constant. In reality, the bank interest rate is fluctuating, and its future evolution is unknown. Investments in bonds are such that money is trapped for some time period with a fixed interest rate. Therefore, the investment in bonds may be more or less profitable than the investment in a cash account. Thus, there is risk and uncertainty for the bond market, which requires stochastic analysis, similarly to the stock market.

5.1 Zero-coupon bonds

Let $C(t)$ represent the value at time t of \$1 invested at time $t = 0$ at the cash account earning the interest r. The equation for $C(t)$ is

$$dC(t) = r(t)C(t)dt, \quad C(0) = 1,$$

i.e.

$$C(t) = C(0) \exp \int_0^t r(u)du.$$

This means that the investment gain made on a unit of the cash account from t to $t + dt$ is equal to $r(t)C(t)dt$.

We say that $r(t)$ is an instantaneous risk-free rate of interest at t, or the *short rate*. If $r(t) \equiv r$ is constant, then

$$C(t) = e^{rt}C(0).$$

We assume that $r(t)$ is a random process such that it is currently observable but

113

its future values are unknown. In this random setting, the value $C(T)$ is random and unknown at time $t < T$, and this investment is not exactly risk-free.

In this chapter, we consider a cash account as the numérare; all other investments will be compared with this investment.

Let us consider market models where it is possible to invest in a cash account and invest in bonds. A bond is a tradable asset that gives the right to receive \$1 at time T (maturity time). For simplicity, we assume that it is a zero coupon bond, i.e., there are no other payments until T. We assume also that there is no risk of default, i.e., the payment will be made in full.

General requirements for bond market models

The main features of models for a bond market that generate requirements for the pricing rules are the following:

(i) The process $r(t)$ of the bank interest rate is assumed to be random;

(ii) The number of securities can be large; in particular, this number can be larger than the number of driving Wiener processes in the case of the diffusion model.

The last feature (ii) has explicit economic sense: there are many different bonds (since bonds with different maturities represent different assets) but their evolution depends on a few factors only, and the main factors are the ones that describe the evolution of $r(t)$.

The multi-stock market model can be used as a model for a market with many different bonds (or fixed income securities). Assume that we are using a multi-stock market model described above as the model for bonds (i.e., $S_i(t)$ are the bond prices). Feature (ii) can be expressed as the condition that $\sigma_{ij}(t) \equiv 0$ for all $j > n$, $= 1, \ldots, N$, where n is the number of the driving Wiener processes, N is the number of the bonds, $N >> n$. It follows that the matrix σ is degenerate. This is a very essential feature of the bond market. To ensure that the process $\theta(t)$ is finite and the model is arbitrage free, some special conditions on $\widetilde{a}$ must be imposed such that equation (2.19) is solvable with respect to θ. To satisfy these restrictions, the bond market models deals with $\widetilde{a}$ being linear functions of σ.

Models for bond prices are widely studied in the literature; see a review in [81].

Bond price in the case of non-random $r(t)$

We denote by $P(t, T)$ the zero coupon bond price at time t (the fair price). We assume a market model where bonds are tradable assets, and where there is an alternative investment in a cash account. We assume that the fair price of a bond is a price that does not allow arbitrage.

Obviously,

$$P(T, T) = 1.$$

Lemma 5.1 *Consider the case where $r(t)$ is non-random and time variable. In this case,*

$$P(t,T) = \exp\left(-\int_t^T r(u)du\right). \tag{5.1}$$

This is the only price of the bond that does not allow arbitrage for the seller and for the buyer. With this price, investment in the bond gives the same profit as the investment in a cash account.

The formula in Lemma 5.1 cannot be used directly for the case where $r(s)$ is a random process, since it requires future values of r. In fact, a model for bond prices would suggest using a modification of this formula that includes conditional expectation, i.e.,

$$P(t,T) = \mathbf{E}_Q\left\{\exp\left(-\int_t^T r(s)ds\right) \,\Big|\, \mathcal{F}_t\right\}, \tag{5.2}$$

where $\mathbf{E}_Q$ is the expectation generated by a probability measure Q, and $\mathcal{F}_t$ is the filtration generated by all observable data. The measure Q has to be chosen to satisfy the given requirements; typically, $Q = Q_t$ depends on time. The choice of this measure may be affected by risk and the risk premium associated with particular bonds. (For instance, some bonds are considered more risky than others; to ensure liquidity, they are offered at some lower price, so the possible reward for an investor may be higher.)

5.2 One-factor model

Let a historical probability measure **P** be given. Let $\mathcal{F}_t$ be the filtration generated by the random process $r(t)$.

The main feature of models for a bond market is that the number of securities is larger than the number of random factors. It has explicit economical sense: there are many different bonds (since bonds with different maturities represent different assets) but their price evolution depends on few factors only, and the main factors are the ones that describe the evolution of $r(t)$.

In this chapter, we assume that there is just one random factor: more precisely, we assume that there is a standard one-dimensional Brownian motion process $w_P(t)$ under this measure (i.e., a Wiener process), such that the randomness of $r(t)$ is generated by this Brownian motion. This means that $r(t)$ is adapted to the filtration generated by $w_P(t)$ and this filtration is the same as $\mathcal{F}_t$. We say in this case that $w_P(t)$ is the driving Brownian motion.

Similar to the discounted wealth, it will be convenient to use the discounted bond prices

$$\widetilde{P}(t,T) = P(t,T)\exp\left(-\int_0^t r(u)du\right). \tag{5.3}$$

We will consider the so-called *one-factor model*; there is a measure Q such that

I.   The discounted bond price is a martingale under Q;

II. There is a standard one-dimensional Brownian motion process $w_Q(t)$ under this measure (i.e., a Wiener process) that generates the same filtration $\mathcal{F}_t$ (i.e., the same flow of events as $r(t)$ and $w_P(t)$). This means that the randomness of $r(t)$ is generated by this Brownian motion.

In particular, it can be shown that the market model with these bonds as tradable assets is arbitrage free.

Note that we do not require yet that this measure be unique.

We will be using a modification of formula (5.1) for random $r(t)$: the bond price has to be calculated using conditional expectation with respect to the measure Q. More precisely, we assume that the price of the bond at time t is defined by formula (5.2), where $\mathbf{E}_Q$ is the expectation generated by a probability measure Q.

Since $\mathcal{F}_t$ is the filtration generated by observations of $r(t)$, it can be rewritten as

$$P(t,T) = \mathbf{E}_Q\left\{\exp\left(-\int_t^T r(s)ds\right)\,\middle|\,r(\tau),\ \tau \le t\right). \tag{5.4}$$

We consider a set of bonds with different maturity times $T = T_1, T_2, ...,$ with the common driving Brownian motion.

5.2.1   Dynamics of discounted bond prices

Theorem 5.2 *For every choice of the measure Q, the discounted prices for the bonds are martingales under the measure Q*

Proof: By the definitions,

$$\widetilde{P}(T,T) = P(T,T)\exp\left(-\int_0^T r(s)ds\right) = \exp\left(-\int_0^T r(s)ds\right)$$

since $P(T,T) = 1$. Hence

$$\widetilde{P}(t,T) = \exp\left(-\int_0^t r(u)du\right)\mathbf{E}_Q\left\{\exp\left(-\int_t^T r(s)ds\right)\,\middle|\,\mathcal{F}_t\right\}$$

$$= \mathbf{E}_Q\left\{\exp\left(-\int_0^T r(s)ds\right)\,\middle|\,\mathcal{F}_t\right\} = \mathbf{E}_Q\{\widetilde{P}(T,T)\,|\,\mathcal{F}_t\}.$$

This completes the proof. $\square$

Lemma 5.3 *For every* $T = T_k$, *there exists an* $\mathcal{F}_t$*-adapted process* $\sigma(t, T)$ *such that*

$$\widetilde{P}(t, T) = \mathbf{E}_Q \widetilde{P}(t, T) + \int_0^t \sigma(s, T) \widetilde{P}(s, T) dw_Q(s).$$

Proof. By the Martingale representation theorem (Clark theorem), for every $T = T_k$, there exists an $\mathcal{F}_t$-adapted process $f(t, T)$ such that

$$\widetilde{P}(t, T) = \mathbf{E}_Q \widetilde{P}(t, T) + \int_0^t f(s, T) dw_Q(s), \quad t \leq T.$$

(Remember that $\mathcal{F}_t$ can be considered the filtration generated by $w_Q(t)$). The desired process can be obtained as

$$\sigma(t, T) = \frac{f(t, T)}{\widetilde{P}(t, T)}.$$

This completes the proof. $\square$

From the last lemma,

$$d_t \widetilde{P}(t, T) = \sigma(t, T) \widetilde{P}(t, T) dw_Q(t).$$

(We also note that $\mathbf{E}_Q \widetilde{P}(t, T) = \mathbf{E}_Q \widetilde{P}(T, T)$ does not depend on t.)

Lemma 5.4

$$d_t P(t, T) = P(t, T)[r(t)dt + \sigma(t, T) dw_Q(t)].$$

Proof: It is sufficient to find the differential of the process $P(t, T) = \widetilde{P}(t, T) \exp\left(\int_0^t r(u) du\right)$ using the Ito formula for the product of two processes.

5.2.2 Dynamics of the bond prices under the original measure

We assume now that Q is equivalent to the original measure $\mathbf{P}$. In this case, it can be shown that there exists a random process $q(t)$ that is adapted to $\mathcal{F}_t$ and is such that

$$w_Q(t) = w_P(t) + \int_0^t q(s) ds,$$

i.e.,

$$dw_Q(t) = dw_P(t) + q(t) dt.$$

(For the special case where $q = (\mu - r)/\sigma$, it is the Cameron–Martin–Girsanov Theorem.)

In this case, we have that

$$\begin{aligned}
d_t P(t, T) &= P(t, T)[r(t)dt + \sigma(t, T) dw_Q(t)] \\
&= P(t, T)[r(t)dt + \sigma(t, T)q(t)dt + \sigma(t, T) dw_P(t)].
\end{aligned}$$

It can be rewritten as the following.

Theorem 5.5 *The equation for bond prices is*

$$d_t P(t,T)$$
$$= P(t,T)[r(t)dt + \sigma(t,T)q(t)dt + \sigma(t,T)dw_P(t)]. \tag{5.5}$$

The process $q(t)$ is said to be *the market price of risk*; it has the same role as $(\mu-r)/\sigma$ for the stock price, in the notations of Chapter 9. Similarly, it can be shown that the equation for the discounted bond prices is

$$d_t \widetilde{P}(t,T) = \widetilde{P}(t,T)[\sigma(t,T)q(t)dt + \sigma(t,T)dw_P(t)]. \tag{5.6}$$

Representation of E_Q via historical measure P

Theorem 5.6 *Let F be a random variable, then*

$$\mathbf{E}_Q(F|\mathcal{F}_t) = \mathbf{E}(\eta_{t,T}F|\mathcal{F}_t),$$

where

$$\eta_{t,T} = \exp\left(-\frac{1}{2}\int_t^T q(s)^2 ds - \int_t^T q(s)dw(s)\right).$$

It can be observed that $\eta_{t,T} = A(T)/A(t)$, where $A(t) = \eta_{0,t}$ is called the *state price deflator*. Therefore, the pricing formula can be rewritten as

$$\mathbf{E}_Q(F|\mathcal{F}_t) = \frac{\mathbf{E}(A_T F|\mathcal{F}_t)}{A_t}.$$

This theorem is given for reference only; we will not be using it for calculations. In our examples, we will assume for simplicity that the distribution under Q is known.

On the selection of $q(t)$ and Q

Note that every process $q(t)$ generates its own unique measure Q, and, respectively, every measure Q generates its own $q(t)$. The process $q(t)$ must be the same for all maturities T.

To use this pricing model in practice, one has to estimate $q(t)$ as a parameter of equation (5.5) using statistical methods applied to historical bond prices matched to this equation. It is a problem from *financial econometrics*.

In theory, $q(t)$ is uniquely defined by historical bond prices; therefore, the measure Q is also uniquely defined.

The process $q(t)$ defines the reward for the risk associated with the investment in the entire bond market. The choice of $\sigma(t,T)$ is an analog of the volatility for the stock prices; it defines the risk associated with particular bonds with particular maturity T. In other words, the bonds with larger $\sigma(t,T)$ are considered to be more risky than others; however, they are still liquid since they promise larger expected profit.

5.2.3 An example: The Cox–Ross–Ingresoll model

The Cox-Ross-Ingresoll model:

$$dr(t) = \alpha(\mu - r(t))dt + \sigma\sqrt{r(t)}dw_Q(t),$$

where $w_Q(t)$ is a standard Brownian motion under the risk-neutral measure Q. A benefit of using this model is that $r(t) > 0$.

An example: Conditionally log-normal model

Assume that, under the measure Q, for every $t < T$, the random value $\int_t^T r(s)ds$ has a conditionally normal distribution $N(\mu, s^2)$ that is conditional given the history $r(\tau)$, $\tau \le t$, and such that

$$\mu = f(t, r(t)), \quad s = g(t, r(t)).$$

The functions $f(t, x)$ and $g(t, x)$ are known and deterministic.

Let us find the bond price $P(t, T)$ suggested by this model.

We have that

$$P(t, T) = \mathbf{E}_Q\left(e^{-Z} \middle| r(\tau), \ \tau \le t\right),$$

where $Z \sim N(\mu, s^2)$ conditionally given $\{r(\tau), \ \tau \le t\}$. Hence $-Z \sim N(-\mu, s^2)$ and $Y = e^{-Z} \sim LN(-\mu, s^2)$ conditionally given $\{r(\tau), \ \tau \le t\}$. We obtain that

$$P(t, T) = e^{-\mu + s^2/2} = e^{-f(t, r(t)) + g(t, r(t))^2/2}. \tag{5.7}$$

This model will be considered in next chapters in detail.

5.3 Vasicek model

Let us consider in detail a particular model of the term structure of interest rates, the so-called Vasicek model. For this model, under the probability measure Q, the dynamic of $r(t)$ is described as an Ornstein–Uhlenbek proceess

$$dr(t) = \alpha(\mu - r)dt + \sigma dw_Q(t),$$

where $w_Q(t)$ is the standard Brownian motion under the risk-neutral measure Q, and where $\alpha > 0$, $\mu \ge 0, \sigma > 0$ are some given numbers.

Theorem 5.7 *For the Vasicek model, $r(s)$ has conditionally normal distribution given $r(s)$, and*

$$r(s) = \mu + (r(t) - \mu)e^{-\alpha(s-t)} + e^{-\alpha s}\sigma \int_t^s e^{\alpha q}dw_Q(q)$$

for any $s \ge t \ge 0$.

Proof: (i) For simplicity, assume first that $\mu = 0$, i.e., that

$$dr(t) = -ar(t)dt + \sigma dw_Q(t).$$

Let us show that

$$r(s) = r(t)e^{-\alpha(s-t)} + e^{-\alpha s}\sigma \int_t^s e^{\alpha q} dw_Q(q)$$

for any $s \geq t \geq 0$.

The integral equation gives $r(s) = r(t)$ for $s = t$. In addition, we find that the differential of the integral expression for $r(s)$ is

$$d_s r(s) = d_s \left[r(t)e^{-\alpha(s-t)} + e^{-\alpha s}\sigma \int_t^s e^{\alpha q} dw_Q(q) \right]$$

$$= -\alpha r(t)e^{-\alpha(s-t)} ds - \alpha e^{-\alpha s} ds \int_t^s e^{\alpha q}\sigma dw_Q(q) + e^{-\alpha s}e^{\alpha s}\sigma dw_Q(s)$$

$$= -\alpha r(s)ds + \sigma dw_Q(s).$$

Hence, the equation is satisfied.

(ii) Assume now that $\mu \geq 0$. Again, the integral equation gives $r(s) = r(t)$ for $s = t$. In addition, we find that the differential of the integral expression for $r(s)$ is

$$d_s r(s) = d_s \left[\mu + (r(t) - \mu)e^{-\alpha(s-t)} + e^{-\alpha s}\sigma \int_t^s e^{\alpha q} dw_Q(q) \right]$$

$$= -\alpha(r(t) - \mu)e^{-\alpha(s-t)} ds - \alpha e^{-\alpha s} ds\sigma \int_t^s e^{\alpha q}\sigma dw_Q(q) + e^{-\alpha s}e^{\alpha s} dw_Q(s)$$

$$= \alpha(\mu - r(s))ds + \sigma dw_Q(s).$$

Hence, the equation is satisfied.

In this equation, the stochastic integral is independent from $r(s)$ and has normal distribution. It follows that $r(s)$ has conditionally normal distribution given $r(s)$.

This completes the proof. $\square$

Pricing under Vacisek model

The case where $\mu = 0$.

Theorem 5.8 *If $\mu = 0$, then*

$$f(t, r(t)) = \mathbf{E}_Q \left(\int_t^T r(q)dq \,\Big|\, r(\tau),\ \tau \leq t \right) = r(t)b(\tau),$$

$$\tau = T - t, \quad b(\tau) = \frac{1 - e^{-\alpha\tau}}{\alpha},$$

and

$$\frac{1}{2}g(t, r(t))^2 = \frac{1}{2}\mathrm{Var}_Q \left(\int_t^T r(q)dq \,\Big|\, r(\tau),\ \tau \leq t \right)$$

$$= -(b(\tau) - \tau)\frac{\sigma^2}{2\alpha^2} - \frac{\sigma^2}{4\alpha}b(\tau)^2.$$

Proof (for f only): We have that

$$r(s) = r(t)e^{-\alpha(s-t)} + e^{-\alpha s}\sigma \int_t^s e^{\alpha q}dw_Q(q).$$

Taking the conditional expectation $\mathbf{E}_Q(\cdot \mid r(\tau), \ \tau \le t)$ for both parts gives

$$\mathbf{E}_Q(r(s) \mid r(\tau), \ \tau \le t) = r(t)e^{-\alpha(s-t)}.$$

Hence

$$f(t, r(t)) = \mathbf{E}_Q\left(\int_t^T r(s)ds \,\Big|\, r(\tau), \ \tau \le t\right) = \int_t^T r(t)e^{-\alpha(s-t)})ds$$

$$= r(t)\int_t^T e^{-\alpha(s-t)}ds = r(t)\int_0^{T-t} e^{-\alpha y}dy$$

$$= r(t)\frac{e^{-\alpha(T-t)} - 1}{-\alpha} = r(t)\frac{1 - e^{-\alpha(T-t)}}{\alpha} = r(t)b(\tau). \quad \square$$

Theorem 5.9 *If $\mu = 0$, then*

$$P(t, T) = e^{-f(t,r(t)) + g(t,r(t))^2/2}$$

$$= \exp\left(-r(t)b(\tau) - (b(\tau) - \tau)\frac{\sigma^2}{2\alpha^2} - \frac{\sigma^2}{4\alpha}b(\tau)^2\right).$$

Proof follows from Theorem 5.8 and equation (5.7).

Example 5.10 *Let us calculate the bond price for Vacisek model with*

$$t = 0, \quad T = 1, \quad r(0) = 0.07, \quad \alpha = 0.5, \quad \mu = 0, \quad \sigma = 0.2.$$

Substituting, we obtain $b(T) = 0.7869$, $f(0, r(0)) = r(0)b(T) = 0.07 * (1 - exp(-0.5 * (1 - 0)))/0.5 = 0.0551$. Further, we obtain $g(0, r(0))^2/2 = 0.047$ and $P(0, T) = 0.9508$.

Theorem 5.11 *If $\mu \ge 0$, then*

$$f(t, r(t)) = \mathbf{E}_Q\left(\int_t^T r(q)dq \,\Big|\, r(\tau), \ \tau \le t\right)$$

$$= (T - t)\mu + (r(t) - \mu)\frac{1 - e^{-\alpha(T-t)}}{\alpha}$$

$$= b(\tau)r(t) + \tau\mu - b(\tau)\mu,$$

in the notations of Theorem 5.8, and

$$P(t, T) = e^{-f(t,r(t)) + g(t,r(t))^2/2}$$

$$= \exp\left(-r(t)b(\tau) + (b(\tau) - \tau)\left(\mu - \frac{\sigma^2}{2\alpha^2}\right) - \frac{\sigma^2}{4\alpha}b(\tau)^2\right).$$

Proof (for f only): We find that

$$r(s) = \mu + (r(t) - \mu)e^{-\alpha(s-t)} + e^{-\alpha s}\sigma \int_t^s e^{\alpha q}dw_Q(q).$$

Taking the conditional expectation $\mathbf{E}_Q(\cdot \mid r(\tau), \tau \leq t)$ for both parts gives

$$\mathbf{E}_Q(r(s) \mid r(\tau), \tau \leq t) = \mu + (r(t) - \mu)e^{-\alpha(s-t)}.$$

Hence

$$f(t, r(t)) = \mathbf{E}_Q\left(\int_t^T r(s)ds \,\middle|\, r(\tau), \tau \leq t\right) = \int_t^T (\mu + (r(t) - \mu)e^{-\alpha(s-t)})ds$$

$$= (T - t)\mu + (r(t) - \mu)\int_t^T e^{-\alpha(s-t)}ds$$

$$= (T - t)\mu + (r(t) - \mu)\int_0^{T-t} e^{-\alpha y}dy$$

$$= (T - t)\mu + (r(t) - \mu)\frac{e^{-\alpha(T-t)} - 1}{-\alpha}$$

$$= (T - t)\mu + (r(t) - \mu)\frac{1 - e^{-\alpha(T-t)}}{\alpha}.$$

Example 5.12 *Let us calculate the bond price for*

$$t = 0, \quad T = 1, \quad r(0) = 0.05, \quad \alpha = 0.5, \quad \mu = 0.5, \quad \sigma = 0.2.$$

Substituting, we obtain $\tau = T = 1$, $b(T) = 0.7869$ *again, and* $P(0, T) = 0.8683$.

Note that since $r(t)$ is Gaussian in the Vacisek model, the case of $r(t) < 0$ is not excluded.

5.4   An example of a multi-bond market model

Let us describe a possible model of a market with N zero coupon bonds with bond prices $P(t, T_k)$, where $t \in [0, T_k]$, and where $\{T_k\}_{k=1}^N$ is a given set of maturing times, $T_k \in (0, T]$, $P(T_k, T_k) = 1$.

We consider the case where there is a driving n-dimensional Wiener process $w(t)$. Let $\mathcal{F}_t$ be a filtration generated by this Wiener process. We assume that the process $r(t)$ is adapted to $\mathcal{F}_t$. (To cover some special models, we do not assume that $r(t) \geq 0$.) In addition, we assume that we are given an $\mathcal{F}_t$-adapted and bounded process $q(t)$ that takes values in $\mathbf{R}^n$.

Set the bond prices as

$$P(t, T_k) = \mathbf{E}\left\{\exp\left(-\int_t^{T_k} r(s)ds + \int_t^{T_k} q(s)^\top dw(s) - \frac{1}{2}\int_t^{T_k} |q(s)|^2 ds\right) \middle| \mathcal{F}_t\right\}. \quad (5.8)$$

In this model, different bonds are defined by their maturity times.

Clearly, the processes $P(t, T_k)$ are adapted to $\mathcal{F}_t$, and $P(T_k, T_k) = 1$, $P(t, T_k) \geq 0$. If $r(t) \geq 0$, then $P_k(0) \in [0, 1]$, and $\widetilde{P}(t, T_k) \triangleq P(t, T_k) \exp\left(-\int_0^t r(s)ds\right) \in [0, 1]$ a.s. In addition, it can be seen that (5.2) holds for the measure $\mathbf{Q}_t = \mathbf{Q}_{t,k}$ such that $\mathbf{Q}_{t,k}/d\mathbf{P} = Z_k(t)$, where

$$Z_k(t) \triangleq \exp\left(\int_t^{T_k} q(s)^\top dw(s) - \frac{1}{2}\int_t^{T_k} |q(s)|^2 ds\right).$$

Theorem 5.13 *Pricing rule (5.8) ensures that, for any k, there exists an $\mathcal{F}_t$-adapted process $\sigma_k(t)$ with values in $\mathbf{R}^n$ such that*

$$d_t P(t, T_k) = P(t, T_k)\left(\left[r(t) - \sigma_k(t)^\top q(t)\right] dt + \sigma_k(t)^\top dw(t)\right), \quad t < T_k. \quad (5.9)$$

Proof. Let k be fixed. We have that

$$P(t, T_k) = y(t)z(t) \exp\left(\int_0^t r(s)ds\right),$$

where

$$y(t) \triangleq \mathbf{E}\left\{\exp\left(-\int_0^{T_k} r(s)ds + \int_0^{T_k} q(s)^\top dw(s) - \frac{1}{2}\int_0^{T_k} |q(s)|^2 ds\right)\bigg| \mathcal{F}_t\right\},$$

$$z(t) \triangleq \exp\left(-\int_0^t q(s)^\top dw(s) + \frac{1}{2}\int_0^t |q(s)|^2 ds\right).$$

It follows from the Clark Theorem (1.72) that there exists a square integrable n-dimensional $\mathcal{F}_t$-adapted process $\hat{y}(t) = \hat{y}(t, T_k)$ with values in $\mathbf{R}^n$ such that

$$y(T) = \mathbf{E}y(T) + \int_0^T \hat{y}(t)^\top dw(t).$$

Note that $y(t) > 0$. Set $\delta_k(t) = \hat{y}(t)/y(t)$. Then

$$y(T) = \mathbf{E}y(T) + \int_0^T y(t)\delta_k(t)^\top dw(t).$$

By Ito's formula, it follows that

$$dz(t) = z(t)\left(|q(t)|^2 dt - q(t)^\top dw(t)\right).$$

Set $\sigma_k(t) \triangleq \delta_k(t) - q(t)$. Finally, the Ito formula applied to (5.10) implies that (5.9) holds. This completes the proof. $\square$

It follows from (5.9) that this bond market is a special case of the multi-stock market described above, where $S_k(t) = P(t, T_k)$, $k = 1, \ldots, N$, where $\widetilde{a}(t) \equiv (\widetilde{a}_1(t), \ldots, \widetilde{a}_N(t))^\top \in \mathbf{R}^N$, and where $\sigma(t)$ is a matrix process with values in $\mathbf{R}^{N \times n}$

such that its kth row is zero for $t > T_k$ and it is equal to $\sigma_k(t)^\top$ for $t \leq T_k$. The process $\widetilde{a}(t)$ is such that

$$\widetilde{a}_k(t) = -\sigma_k(t)^\top q(t), \quad t \leq T_k,$$
$$\widetilde{a}_k(t) = 0, \quad t > T_k.$$

Then the corresponding market price of risk process $\theta(t)$ is $\theta(t) \equiv -q(t)$. This process $\theta(t)$ is bounded if $q(t)$ is bounded, since $\theta(t) \equiv -q(t)$. Note that the case $N \gg n$ is allowed, and the bond market is still arbitrage free.

   To derive an explicit equation for $P(t, T_k)$ and $\sigma_k(t)$, we need to specify a model for the evolution of the process $(r(t), q(t))$. The choice of this model defines the model for the bond prices. For instance, let $n = 1$, let the process q be constant, and let $r(t)$ be an Ornstein–Uhlenbek process described in Example 1.68. Then this case corresponds to the so-called Vasicek model (see, e.g., [81], p.127). In that case, $P(t, T_k)$ can be found explicitly from (5.8).

   A more comprehensive review of bond pricing can be found in, e.g., [81].

6

Implied volatility and other implied market parameters

This chapter studies inference of implied parameters of different financial models from observed prices. In particular, it considers inference of a pair $(\sigma_{imp}(t), \rho_{imp}(t))$ of two *unconditionally* implied parameters, where $\sigma_{imp}(t)$ is the unconditionally implied volatility, and $\rho_{imp}(t)$ is the unconditionally implied value of $\rho(t)$. This pair can be found in a system of two equations with option prices for different strike prices. In addition this chapter considers the implied zero-coupon bond price and implied market price of risk.

6.1 Risk-neutral pricing in a Black–Scoles setting

Background

Let us consider the diffusion model of a securities market consisting of a risk-free bank account with the price of a unit $B(t)$, $t \geq 0$, and a risky stock with price $S(t)$, $t \geq 0$. The prices of the stocks evolve as

$$dS(t) = S(t)\left(a(t)dt + \sigma(t)dw(t)\right), \quad t > 0, \tag{6.1}$$

where $w(t)$ is a Wiener process, and $a(t)$ is a random appreciation rate, and $\sigma(t)$ is a random volatility coefficient. The initial price $S(0) > 0$ is a given deterministic constant. The price of the unit in the bank account evolves as

$$B(t) = \exp\left(\int_0^t r(s)ds\right)B(0), \tag{6.2}$$

where $r(t) \geq 0$ is a random process and $B(0)$ is given.

We assume that $w(\cdot)$ is a standard Wiener process on a given standard probability space $(\Omega, \mathcal{F}, \mathbf{P})$, where Ω is a set of elementary events, $\mathcal{F}$ is a complete σ-algebra of events, and $\mathbf{P}$ is a probability measure.

Let $\mathcal{F}_t$ be a filtration generated by the currently observable data. We assume that the process $(S(t), \sigma(t), r(t))$ is $\mathcal{F}_t$-adapted and that $\mathcal{F}_t$ is independent from $\{w(t_2) - w(t_1)\}_{t_2 \geq t_2 \geq t}$. In particular, this means that the process $(S(t), \sigma(t), r(t))$ is currently observable and $\sigma(t)$ is independent from $\{w(t_2) - w(t_1)\}_{t_2 \geq t_2 \geq t}$. We assume that $\mathcal{F}_0$ is trivial, i.e., it is the $\mathbf{P}$-augmentation of the set $\{\emptyset, \Omega\}$.

We assume that $a(t)$ is independent from $\{w(t_2) - w(t_1)\}_{t_2 \geq t_1 \geq t}$. For simplicity, we assume that $a(t)$ is a bounded process.

Set

$$\tilde{S}(t) \triangleq S(t) \exp\left(-\int_0^t r(s)ds\right).$$

We assume that, for any (a, σ, r) under consideration, there exists a risk-neutral measure Q such that the process $\tilde{S}(t)$ is a martingale under Q, i.e., $\mathbf{E}_Q\{\tilde{S}(T) | \mathcal{F}_t\} = \tilde{S}(t)$, where $\mathbf{E}_Q$ is the corresponding expectation.

Black–Scholes prices

Let terminal time $T > 0$ and strike price $K > 0$ be fixed. Let $H_{BS,c}(t, x, \sigma, r, K)$ and $H_{BS,p}(t, x, \sigma, r, K)$ denote prices (6.5) for the vanilla put and call options, with the payoff functions $F(S(T), K) = (S(T) - K)^+$ and $F(S(T), K) = (K - S(T))^+$ respectively, under the assumption that $S(t) = x$, $(\sigma(s), r(s)) = (\sigma, r)$ $(\forall s > t)$, where $\sigma \in (0, +\infty)$ is non-random. In other words, these are the Black–Scholes prices, and the celebrated Black–Scholes formula for their explicit values can be rewritten as

$$H_{BS,c}(t, x, \sigma, r, K) \tag{6.3}$$
$$= x\Phi(d_+(t, x, \sigma, r, K)) - Ke^{-r(T-t)}\Phi(d_-(t, x, \sigma, r, K)),$$
$$H_{BS,p}(t, x, \sigma, r, K) = H_{BS,c}(t, x, \sigma, r, K) - x + Ke^{-r(T-t)}, \tag{6.4}$$

where $\Phi(x) \triangleq \frac{1}{\sqrt{2\pi}} \int_{-\infty}^x e^{-\frac{s^2}{2}} ds$, and where

$$d_+(t, x, \sigma, r, K) \triangleq \frac{\ln(x/\tilde{K}(t))}{\sigma\sqrt{(T-t)}} + \frac{\sigma\sqrt{(T-t)}}{2},$$
$$d_-(t, x, \sigma, r, K) \triangleq d_+(t, x, \sigma, r, K) - \sigma\sqrt{(T-t)},$$
$$\tilde{K}(t) \triangleq Ke^{-r(T-t)}.$$

Let $H_{BS,s}(t, x, \sigma, r, K)$ denote the price for the share-or-nothing call options with the payoff function $F(S(T), K) = S(T)\mathbb{I}_{\{S(T)>K\}}$ under the assumption that $S(t) = x$, $(\sigma(s), r(s)) = (\sigma, r)$ $(\forall s > t)$, where $v \in (0, +\infty)$ is non-random. The analog of the Black–Scholes formula for this case is known:

$$H_{BS,s}(t, x, \sigma, r, K) = x\Phi(d_+(t, x, \sigma, r, K)).$$

For brevity, we denote by H_{BS} the corresponding Black–Scholes prices of different options, i.e., $H_{BS} = H_{BS,c}$ $H_{BS} = H_{BS,p}$, or $H_{BS} = H_{BS,s}$, for vanilla call, vanilla put, and share-or-nothing call, respectively. In addition, we will denote by $F(x, K)$ the functions $(x - K)^+$, $(K - x)^+$, and $x\mathbb{I}_{\{x>K\}}$ corresponding to these options.

Pricing rules

The basic pricing rule for models with random volatility is risk-neutral valuation, when the option price is given as the expected value of its future payoff with respect to a risk-neutral measure discounted back to the present time t (see, e.g., [37]). This method has been developed pricing rules based on optimal choice of the risk-neutral measures such as local risk minimization, mean variance hedging, q-optimal measures, and minimal entropy measures. These pricing rules are applicable for the most complicated models with an Ito equation for volatility, and they lead to the following pricing rule for the option with payoff $F(S(T), K)$: given $a(\cdot), \sigma(\cdot), r(\cdot)$, the option price is

$$P_{RN}(t, \sigma(\cdot), r(\cdot)) \triangleq \mathbf{E}_Q\{e^{-\int_t^T r(s)ds} F(S(T), K) \,|\, \mathcal{F}_t\}, \tag{6.5}$$

where Q is some risk-neutral measure such that the process

$$\tilde{S}(t) \triangleq S(t) \exp\left(-\int_0^t r(s)ds\right)$$

is a martingale under Q, i.e., $\mathbf{E}_Q\{\tilde{S}(T) \,|\, \mathcal{F}_t\} = \tilde{S}(t)$, where $\mathbf{E}_Q$ is the corresponding expectation.

Usually, Q is uniquely defined by $(a(\cdot), \sigma(\cdot), r(\cdot))$, provided that a certain optimality criterion for Q is used.

We assume that we have chosen one of the pricing methods (for example, implied by a preselected optimality criterion). Under this assumption, the risk-neutral measure Q is uniquely defined by $(a(\cdot), \sigma(\cdot), r(\cdot))$.

Let

$$v(t) \triangleq \frac{1}{T-t} \int_t^T \sigma(s)^2 ds, \quad \rho(t) \triangleq \frac{1}{T-t} \int_t^T r(s)ds. \tag{6.6}$$

The following lemma is a generalization on the case of random r of a lemma from [67], p.245.

Lemma 6.1 *Let $t \in [0, T)$ be fixed. Let $v(t)$ and $\rho(t)$ be $\mathcal{F}_t$-measurable. Then*

$$\mathbf{E}_Q\{e^{-\int_t^T r(s)ds} F(S(T), K) | \mathcal{F}_t\} = H_{BS}(t, S(t), \sqrt{v(t)}, \rho(t), K).$$

(Note that $(\sigma(s), r(s))$ is not necessarily $\mathcal{F}_t$-measurable for $s > t$.)

Clearly, $\frac{1}{T-t}\int_t^T \sigma(s)^2 ds$ and $\frac{1}{T-t}\int_t^T r(s)ds$ are not $\mathcal{F}_t$-measurable in the general case of stochastic $(r(\cdot), \sigma(\cdot))$, and the assumptions of Lemma 6.1 are not satisfied. (However, there are exemptions: for instance, $v(0)$ is non-random if $\sigma(t) = \sigma_1$ for $t \in I$, $\sigma(t) = \sigma_2$ for $t \notin I$, where $I = [t_1, t_1 + d] \subset [0, T]$, and where t_1 is random, and $d > 0$, σ_i are given and non-random.) In fact, we need Lemma 6.1 only for the proof of Corollary 6.4 below.

Proof of Lemma 6.1. It suffices to consider the case when $t = 0$ and $v(0)$ and $\rho(0)$ are non-random. Set $R(t) \triangleq e^{\rho(0)t}$, $\tilde{K} \triangleq R(T)^{-1}K$.

Set $\hat{H}_{BS}(t, x) \triangleq R(t)^{-1} H_{BS}(t, R(t)x, \sqrt{v(0)}, \rho(0), K)$. As is known,

$$\frac{\partial H_{BS}}{\partial t}(t, x, \sigma, r, K) + \frac{\sigma^2 x^2}{2} \frac{\partial^2 H_{BS}}{\partial x^2}(t, x, \sigma, r, K)$$
$$= r \left[H_{BS}(t, x, \sigma, r, K) - x \frac{\partial H_{BS}}{\partial x}(t, x, \sigma, r, K) \right],$$
$$H_{BS}(T, x, \sigma, r, K) = F(x).$$

It follows that

$$\frac{\partial \hat{H}_{BS}}{\partial t}(t, x) + \frac{v(0)x^2}{2} \frac{\partial^2 \hat{H}_{BS}}{\partial x^2}(t, x) = 0, \quad \hat{H}_{BS}(T, x) = F(x, \tilde{K}).$$

Let

$$\tau(t) \triangleq \frac{1}{v(0)} \int_0^t \sigma(s)^2 ds, \quad \tilde{X}(t) \triangleq \tilde{H}_{BS}(\tau(t), \tilde{S}(t)).$$

Similar to the proof of Lemma 5.2 from [46], we obtain

$$d\tilde{X}(t) = \frac{\partial \hat{H}_{BS}}{\partial x}(\tau(t), \tilde{S}(t)) d\tilde{S}(t), \quad \tilde{X}(T) = F(\tilde{S}(T), \tilde{K}).$$

Hence

$$\tilde{X}(0) = \hat{H}_{BS}(0, S(0), \sigma, r, K) = \mathbf{E}_Q \tilde{X}(T) = \mathbf{E}_Q F(\tilde{S}(T), \tilde{K})$$
$$= \mathbf{E}_Q R(T)^{-1} F(S(T), K).$$

This completes the proof. $\square$

Corollary 6.2 *Assume that $H_{BS} = H_{BS,c}$ or $H_{BS} = H_{BS,p}$. Consider a market model with pricing rule (6.5). Let the processes $(\sigma(t), r(t))$, and $w(t)$ be independent under Q. Then*

$$P_{RN}(t) = \mathbf{E}_Q \{ H_{BS}(t, S(t), \sqrt{v(t)}, \rho(t), K) \mid \mathcal{F}_t \},$$

where (v, ρ) are defined by (6.6). (They are not necessarily $\mathcal{F}_t$-measurable for $s > t$.)

Proof. Again, it suffices to consider $t = 0$. Let $F(S(T), K)$ be the payoff corresponding to the choice of H_{BS}. By Lemma 6.1, it follows that

$$P_{RN}(0) = \mathbf{E}_Q e^{-\int_0^T r(s)ds} F(S(T), K)$$
$$= \mathbf{E}_Q \mathbf{E}_Q \{ e^{-\int_0^T r(s)ds} F(S(T), K) \mid \rho(0), v(0) \}$$
$$= \mathbf{E}_Q H_{BS}(0, S(0), \sqrt{v(0)}, \rho(0), K).$$

$\square$

6.2 Implied volatility: The case of constant r

In this section, we assume that $r(t)$ is a constant, $r(t) \equiv r \geq 0$. In this case,

$$B(t) = e^{rt}B(0). \tag{6.7}$$

Definition 6.3 *We say that $\sigma_{imp}(t)$ is the implied volatility at time t, if the current market option price is $H_{BS}(t, S(t), \sigma_{imp}(t), r, K)$.*

For constant r, and $\mathcal{F}_t$-measurable $v(t)$, $t \in [0, T)$, pricing rule (6.5) has the form

$$e^{-r(T-t)}\mathbf{E}_{Q_{\sigma(\cdot)}}\{F(S(T), K)|\mathcal{F}_t\} = H_{BS}(t, S(t), \sqrt{v(t)}, r, K)$$

([67], p.245). This holds even if σ and w are correlated. However, the value $\frac{1}{T-t}\int_t^T \sigma(s)^2 ds$ is not $\mathcal{F}_t$-measurable in the general case of random σ.

Corollary 6.4 *Assume that $H_{BS} = H_{BS,c}$ or $H_{BS} = H_{BS,p}$. Consider a market model with pricing rule (6.5). Let σ be independent from w under Q. Then*

$$P_{RN}(t, \sigma(\cdot)) = \mathbf{E}_Q\{H_{BS}(t, S(t), \sqrt{v(t)}, r, K)\,|\,\mathcal{F}_t\}.$$

By Corollary 6.4, it is natural to accept

$$\hat{\sigma}_1(t) \triangleq \mathbf{E}_Q\left\{\sqrt{v(t)}\,\Big|\,\mathcal{F}_t\right\} = \mathbf{E}_Q\left\{\left[\frac{1}{T-t}\int_t^T \sigma(s)^2 ds\right]^{1/2}\Big|\,\mathcal{F}_t\right\} \tag{6.8}$$

as the forecast (estimate) of $\sqrt{v(t)}$.

Another possible version of the estimate for $\sqrt{v(t)}$ is

$$\hat{\sigma}_2(t) \triangleq \left(\mathbf{E}_Q\left\{v(t)\,\Big|\,\mathcal{F}_t\right\}\right)^{1/2} = \left(\frac{1}{T-t}\mathbf{E}_Q\left\{\int_t^T \sigma(s)^2 ds\,\Big|\,\mathcal{F}_t\right\}\right)^{1/2}. \tag{6.9}$$

This estimate is convenient because the corresponding conditional expectation can be calculated using well-developed ARCH and GARCH models for heteroscedastic time series describing stock prices with random volatility. If $v(t)$ is a martingale under Q, then $\hat{\sigma}_2(t) \equiv \sigma(t)$.

The estimates (6.8) and (6.9) can be generalized as

$$\hat{\sigma}_\nu(t) \triangleq \left(\mathbf{E}_Q\left\{v(t)^{\nu/2}\,\Big|\,\mathcal{F}_t\right\}\right)^{1/\nu}, \quad \nu \geq 1. \tag{6.10}$$

By Jensen's inequality, it follows that $\hat{\sigma}_1(t) \leq \hat{\sigma}_\nu(t)$ with probability 1 for any $\nu \geq 1$.

The shape of the implied volatility for constant r

The following lemma is a corollary from [100], Theorem 4.2.

Lemma 6.5 *Consider a market model with pricing rule (6.5) given some Q with deterministic and constant r and with random (σ, a) such that $\sigma(\cdot)$ is independent from $w(\cdot)$ under Q. Let $P_{RN}(t)$ be the price (6.5) for a vanilla call or put option. Let $H_{BS} = H_{BS,c}$, or $H_{BS} = H_{BS,p}$. Let $t \in [0, T)$ be given. Let $\sigma_{imp}(t) = \sigma_{imp}(t, K, S(t))$ be defined by Definition 6.3, then*

$$
\left. \frac{\partial \sigma_{imp}(t, K, S(t))}{\partial S} \right|_{S(t) = \tilde{K}(t)} = 0, \quad \left. \frac{\partial \sigma_{imp}(t, K, S(t))}{\partial K} \right|_{S(t) = \tilde{K}(t)} = 0, \quad (6.11)
$$

where $\tilde{K}(t) = K \exp(-(T - t)r)$.

Lemma 6.5 gives a reason for why the pricing rule (6.5) generates some volatility smile, or special type of dependence of the implied volatility on K given $S(t)$ with an extremum for the at-the-money option. Notice that these results are valid for all possible measures Q.

6.3   Correction of the volatility smile for constant r

In this section, we consider the case where $r(t) \equiv r > 0$ is constant.

6.3.1   Imperfection of the volatility smile for constant r

Let $P(t)$ be the price of the option at time $t \in [0, T]$ calculated under a given rule.

Let $P_\nu(t)$ be the price calculated under the same rule applied for an auxiliary market model defined at time t such that the process $\{\sigma(s)\}_{s \geq t}$ is replaced by a $\mathcal{F}_t$-adapted process $\overline{\sigma}(t) = \overline{\sigma}_\nu(t)$ such that

$$
\frac{1}{T - t} \int_t^T \overline{\sigma}(s)^2 ds = \hat{\sigma}_\nu(t)^2,
$$

where $\hat{\sigma}_\nu(t)$ is defined by (6.10), i.e., for a market where $v(t)$ is $\mathcal{F}_t$-measurable, or, in other words, it can be forecasted with zero error (for instance, one may take $\overline{\sigma}(s) \equiv \hat{\sigma}_\nu(t)$, $s \in [t, T]$). By Lemma 6.1, $P_\nu(t) = H_{BS}(t, S(t), \hat{\sigma}_\nu(t), r, K)$.

Let us investigate if it is feasible to ensure that the following condition is satisfied for a pricing rule.

Condition 6.6 $P(t) \geq P_\nu(t) = H_{BS}(t, S(t), \hat{\sigma}_\nu(t), r, K)$ *for some $\nu \geq 1$.*

This condition is desirable from a practical point of view, because the additional risk of an error of volatility forecast should lead to increasing of the option price rather than to its decreasing.

Let $\mathcal{A}^{\perp} = \{\sigma(\cdot)\}$ be a set of processes $\sigma(\cdot)$ such that any $\sigma(\cdot) \in \mathcal{A}^{\perp}$ is independent from $w(\cdot)$ under Q.

Theorem 6.7 *Let* $\sigma(\cdot) \in \mathcal{A}^{\perp}$. *If* $S(t) = \tilde{K}(t) = Ke^{-r(T-t)}$, *and* $\frac{1}{T-t} \int_t^T \sigma(s)^2 ds$ *is not* $\mathcal{F}_t$-*measurable, then*

$$P_{RN}(t) < H_{BS}(t, S(t), \hat{\sigma}_\nu(t), r, K) \qquad (6.12)$$

for any $\nu \geq 1$, *where the volatility estimate* $\hat{\sigma}_\nu(t)$ *is defined by (6.10).*

Corollary 6.8 *If* $\sigma(\cdot) \in \mathcal{A}^{\perp}$, $S(t) = \tilde{K}(t)$, *and* $v(t)$ *is not* $\mathcal{F}_t$-*measurable, then the inequality in Condition (6.6) does not hold for any* $\nu \geq 1$, *and this condition is not satisfied for rule (6.5). In this case,*

$$\sigma_{imp}(t) < \hat{\sigma}_\nu(t)$$

for any $\nu \geq 1$, *where* $\sigma_{imp}(t)$ *is the implied volatility defined by Definition 6.3, i.e., the implied volatility is less than the forecasted volatility for at-the-money options.*

6.3.2 A pricing rule correcting the volatility smile

We have assumed that the risk-neutral measure Q is uniquely defined by (σ, a, r, T). Let us assume that (a, r, T) is fixed, then $Q = Q_{\sigma(\cdot)}$ is uniquely defined by $\sigma(\cdot)$. We assume that it is known that the process $\sigma(\cdot)$ is an element of a given set $\mathcal{A}$ of possible volatility processes.

We suggest the following pricing model.

Definition 6.9 *The price* $P_{\max}(t, \mathcal{A})$ *of the option given a class* $\mathcal{A} = \{\sigma(\cdot)\}$ *of possible volatility processes* $\sigma(\cdot)$ *is*

$$P_{\max}(t, \mathcal{A}) \triangleq e^{-r(T-t)} \sup_{\sigma(\cdot) \in \mathcal{A}} \mathbf{E}_{Q_{\sigma(\cdot)}} \{F(S(T), K))|\mathcal{F}_t\}.$$

Let us describe the properties of this pricing rule for some special classes $\mathcal{A}$.

A class of volatilities such that $\sigma_{imp}(t) = \hat{\sigma}_\nu(t)$ **for at-the-money options**

Theorem 6.10 *Let* $\nu \geq 1$ *and* $t \in [0, T)$ *be given. Let* $\mathcal{A}$ *be a class of volatilities such that the following holds:*

(i) *The estimate* $\hat{\sigma}_\nu(t)$ *defined by (6.10) is the same for all* $\sigma(\cdot) \in \mathcal{A}$; *and*

(ii) *There exists a process* $\overline{\sigma}(\cdot) \in \mathcal{A}$ *such that*

$$\hat{\sigma}_\nu(t)^2 = \frac{1}{T-t} \int_t^T \overline{\sigma}(s)^2 ds.$$

Then the price $P_{\max}(t, \mathcal{A})$ is such that condition 6.6 is satisfied with this ν; in particular,

$$P_{\max}(t, \mathcal{A}) \geq H_{BS}(t, S(t), \hat{\sigma}_\nu(t), r, K).$$

Moreover, if $\mathcal{A} \subset \mathcal{A}^\perp$ and $S(t) = \tilde{K}(t)$, then

$$P_{\max}(t, \mathcal{A}) = H_{BS}(t, S(t), \hat{\sigma}_\nu(t), r, K),$$

where $\hat{\sigma}_\nu(t)$ is the volatility forecast defined by (6.10).

In particular, it follows that if $S(t) \sim \tilde{K}(t)$, then $\sigma_{imp}(t)^2 \sim \hat{\sigma}_\nu(t)$, and the price $P_{\max}(t, \mathcal{A}) \sim H_{BS}(t, S(t), \hat{\sigma}_\nu(t), r, K)$, i.e., it is close to the Black–Scholes price.

6.3.3   A class of volatilities in a Markovian setting

We present below an example of a Markovian setting, when maximization over a class of volatilities can be reduced to the solution of some nonlinear parabolic equations.

Instead of functions $F(x, K)$ introduced above for call, put, and share-or-nothing options, we consider in this subsection functions $F(x)$ of a general form.

Let us consider the case when $\sigma(t)^2 = f(\tilde{S}(t), Y(t), t)$, where $f(\cdot) : (0, +\infty) \times \mathbf{R} \times [0, T] \to [0, +\infty)$ is a known function, and where the process $Y(t)$ evolves as

$$dY(t) = g(\tilde{S}(t), Y(t), \eta(t), t)dt + b(\tilde{S}(t), Y(t), \eta(t), t)d\hat{w}(t), \quad t > 0,$$
$$Y(0) = Y_0.$$

Here $\hat{w}(\cdot)$ is a standard Wiener process independent from $w(\cdot)$, and $\eta(t)$ is an n-dimensional random process such that $\eta(t)$ is independent from $\{w(t_2) - w(t_1), \hat{w}(t_2) - \hat{w}(t_1)\}_{t_2 \geq t_1 \geq t}$. The initial value Y_0 is given and deterministic. The functions $g(\cdot) : (0, +\infty) \times \mathbf{R} \times \mathbf{R} \times [0, T] \to \mathbf{R}$ and $b(\cdot) : (0, +\infty) \times \mathbf{R} \times \mathbf{R} \times [0, T] \to \mathbf{R}$ are given.

We assume that $\mathcal{F}_t$ is the filtration generated by $(S(t), \sigma(t)^2, Y(t), \eta(t))$.

In particular, if $g \equiv 0$ and $f(x, y, t) \equiv y$, then estimate (6.9) is such that $\hat{\sigma}_2(t)^2 \equiv \sigma(t)^2$. If, in this case, the process $Y(t)$ is independent from $S(\cdot)$ under Q, then pricing rule (6.5) is such that if $S(t) = \tilde{K}(t)$, then the implied volatility is less than the historical volatility, i.e., $\sigma_{imp}(t)^2 < \sigma(t)^2$.

Let $\mathcal{U}$ be defined as the set of all measurable functions $U : D \to \Delta$, where $\Delta \subset \mathbf{R}$ is a given compact set, $D \overset{\Delta}{=} (0, +\infty) \times \mathbf{R} \times [0, T]$.

Let $\mathcal{A}_\mathcal{U}$ be defined as the set of all processes $\sigma(t)$ such that $\sigma(t)^2 = f(\tilde{S}(t), Y(t), t)$ and $\eta(t) = u(\tilde{S}(t), Y(t), t)$ for some $u(\cdot) \in \mathcal{U}$. Clearly, $\mathcal{A}_\mathcal{U}$ is defined by f, g, b, and Δ.

For simplicity, we assume that there exists a constants $\delta > 0$ such that $f(x, y, t) \geq \delta$ and $\sup_{u \in \Delta} b(x, y, u, t)^2 > 0$ for all x, y, t.

Note that $\mathcal{A}_\mathcal{U}$ covers models when $\sigma(t)^2$ is generated by a mean-reverting process, log-normal process, etc. For instance, the mean-reverting models can be included using $f(x, y, t) = y + \delta$, where $\delta > 0$, and where $Y(t)$ is such that

$dY(t) = Y(t)[\eta_1(t)dt + \eta_2(t)d\hat{w}(t)]$, where $\eta_k(\cdot)$ are some processes. A modification of this example can include a case of the volatility that depends on the stock prices: for instance, one can take $f(x, y, t) = yx^q + \delta$, where $\delta > 0$ and $q \in \mathbf{R}$, with $Y(t)$ that evolves as $dY(t) = Y(t)[\eta_1(t)dt + \eta_2(t)d\hat{w}(t)]$ again.

We restrict our consideration to the framework of the local risk minimization method. In this framework, the risk-neutral measure $Q_{\sigma(\cdot)}$ is such that

$$\frac{dQ_{\sigma(\cdot)}}{d\mathbf{P}} = \exp\left(-\int_0^T a(t)f(S(t), Y(t), t)^{-1/2}dw(t)\right.$$
$$\left. -\frac{1}{2}\int_0^T a(t)^2 f(S(t), Y(t), t)^{-1}dt\right).$$

Let us consider option price defined as

$$P_{\max}(t, \mathcal{A}_u) = e^{r(T-t)} \sup_{\sigma(\cdot)\in\mathcal{A}_u} \mathbf{E}_{Q_{\sigma(\cdot)}}\{F(S(T))|\mathfrak{F}_t\}. \qquad (6.13)$$

For $x, y \in \mathbf{R}$, $u \in \Delta$, $t \in [0, T]$, let

$$\phi(x, y, u, t) \triangleq (f(e^x, y, t), g(e^x, y, u, t), b(e^x, y, u, t)),$$
$$\tilde{F}(x) \triangleq e^{-rT}F(e^{rT}x), \qquad \Phi(x) \triangleq \tilde{F}(e^x).$$

Let $H(x, y, t)$ be a solution of the boundary value problem for the following nonlinear parabolic Bellman equation in D

$$\frac{\partial H}{\partial t}(x, y, t) + \frac{1}{2}f(x, y, t)x^2\frac{\partial^2 H}{\partial x^2}(x, y, t)$$
$$+ \sup_{u\in\Delta}\left\{g(x, y, u, t)\frac{\partial H}{\partial y}(x, y, t) + \frac{1}{2}b(x, y, u, t)^2\frac{\partial^2 H}{\partial y^2}(x, y, t)\right\} = 0,$$
$$H(x, y, T) = \tilde{F}(x). \qquad (6.14)$$

(The equations of these type are also called *Hamilton–Jacobi equations*, or Hamilton–Jacobi–Bellman equations.) It is shown below that (6.14) can be rewritten in more convenient form (6.22). If ϕ is bounded and continuous, then problem (6.14) has a unique viscosity solution.

Theorem 6.11 *Suppose that there exist $m \geq 0$, $C > 0$ such that*

$$|\phi(x, y, u, t) - \phi(x_1, y_1, u, t)| \leq C(|x - x_1| + |y - y_1|),$$
$$|\phi(x, y, u, t)| \leq C(|x| + |y| + 1),$$
$$|\Phi(x)| \leq C(|x| + 1)^m, \qquad \forall x, y, x_1, y_1 \in \mathbf{R}, u \in \Delta, t \in [0, T]. \qquad (6.15)$$

Then the solution H of (6.14) is continuous, and

$$P_{\max}(t, \mathcal{A}_u) = H(\tilde{S}(t), Y(t), t). \qquad (6.16)$$

In addition, suppose that there exist $m \geq 0$, $C_0 > 0$ such that

$$|\Phi(x) - \Phi(x_1)| \leq C_0(1 + R^m)|x - x_1|$$
$$\forall R > 0, \ \forall x, x_1 : \ |x|^2 + |x_1|^2 \leq R^2. \tag{6.17}$$

Then there exists $\hat{C}_0 > 0$ such that

$$|xH_x'(x, y, t) + |H_y'(x, y, t)| \leq \hat{C}_0(1 + |\ln x| + |y|)^{2m}$$

for all t for a.e. $x > 0$, $y \in \mathbf{R}$.

In particular, it follows that H is bounded if $\tilde{F}$ is bounded.

Let

$$G(x, y, u, t) \triangleq (f(x, y, t), g(x, y, u, t), b(x, y, u, t), \tilde{F}(x)).$$

Theorem 6.12 *Let there exist $m \geq 0$, $C > 0$ such that*

$$|G(x, y, u, t)| + |G_t'(x, y, u, t)| + |xG_x'(x, y, u, t)| + |G_y'(x, y, u, t)|$$
$$+|xG_{xy}''(x, y, u, t)| + |x^2 G_{xx}''(x, y, u, t)| + |G_{yy}''(x, y, u, t)|$$
$$\leq C(1 + |\ln x| + |y|)^m \qquad \forall x > 0, \ y \in \mathbf{R}, \ u \in \Delta, \ t \in [0, T]$$

and the corresponding functions and derivatives are continuous. Then problem (6.14) has a unique continuous solution H, and there exists $C_1 > 0$ such that

$$|H(x, y, t)| + |H_t'(x, y, t)| + |xH_x'(x, y, t)| + |H_y'(x, y, t)| + |xH_{xy}''(x, y, t)|$$
$$+|x^2 H_{xx}''(x, y, t)| + |H_{yy}''(x, y, t)| \leq C_1(1 + |\ln x| + |y|)^{3m}$$
$$\forall x > 0, \ y \in \mathbf{R}, \ t \in [0, T].$$

In particular, the corresponding generalized derivatives are locally square integrable and locally bounded in D.

Theorem 6.13 *Let $\tilde{F}(x) = \tilde{F}_1(x) + M_1 x - M_2$, where $\tilde{F}_1(x)$ is such that (6.15) is satisfied with some $m > 0$, $C > 0$ for the corresponding function $\Phi_1(x) = \tilde{F}(e^x)$, and where M_1 and M_2 are constants. Let H_1 be the solution of (6.14) with $\tilde{F}(\cdot) = \tilde{F}_1(\cdot)$. Then $H(x, y, t) \triangleq H_1(x, y, t) + M_1 x + M_2$ is the solution of (6.14), and (6.16) holds for H.*

Theorem 6.14 *Let problem (6.14) have a unique solution such that its generalized derivatives H_t', H_x', H_y', H_{xx}'', H_{yy}'' are locally bounded in D. Then*

(i) *For $X(0) = P_{\max}(0, \mathcal{A}_u) = H(S(0), Y(0), 0)$, there exists a self-financing strategy such that the corresponding discounted wealth is*

$$\tilde{X}(t) = H(\tilde{S}(t), Y(t), t)$$
$$+ \int_0^t \alpha(s)ds - \int_0^t \frac{\partial H}{\partial y}(\tilde{S}(s), Y(s), s)b(\tilde{S}(t), Y(t), \eta(t), t)d\hat{w}(s),$$

where

$$\alpha(t) = \alpha(t, \sigma(\cdot))$$

$$\triangleq \sup_{u \in \Delta} \Big\{ [g(\tilde{S}(t), Y(t), u, t) - g(\tilde{S}(t), Y(t), \eta(t), t)] \frac{\partial H}{\partial y}(\tilde{S}(t), Y(t), t)$$

$$+ \frac{1}{2} [b(\tilde{S}(t), Y(t), u, t)^2 - b(\tilde{S}(t), Y(t), \eta(t), t)^2] \frac{\partial^2 H}{\partial y^2}(\tilde{S}(t), Y(t), t) \Big\},$$

and $\alpha(\cdot) = \alpha(t, \sigma(\cdot))$ *is such that* $\alpha(t) \geq 0$ *a.s. for a.e. t. for all* $\sigma(\cdot) \in \mathcal{A}_u$.

(ii) *The value* $X(0) = H(S(0), Y(0), 0)$ *is the minimal initial wealth such that, for any* $\sigma(\cdot) \in \mathcal{A}_u$, *there exists a self-financing strategy and an* $\mathcal{F}_t$-*adapted square integrable process* $\xi(\cdot) = \xi(\cdot, \sigma(\cdot))$ *such that the corresponding discounted wealth* $\tilde{X}(t) = \tilde{X}(t, \sigma(\cdot))$ *satisfies*

$$\tilde{X}(T) \geq \tilde{F}(\tilde{S}(T)) + \int_0^T \xi(s) d\hat{w}(s)   \text{a.s..}$$

By Theorem 6.14, the initial wealth $X(0) = P_{\max}(0, \mathcal{A}_u) = H(S(0), v(0), 0)$ gives the terminal discounted wealth

$$\tilde{X}(T) = \tilde{F}(\tilde{S}(T)) + \int_0^T \alpha(t) dt - \int_0^T \frac{\partial H}{\partial v}(\tilde{S}(t), v(t), t) b(\tilde{S}(t), Y(t), \eta(t), t) d\hat{w}(t)$$

$$\geq \tilde{F}(\tilde{S}(T)) - \int_0^T \frac{\partial H}{\partial v}(\tilde{S}(t), v(t), t) b(\tilde{S}(t), Y(t), \eta(t), t) d\hat{w}(t).$$

Proofs for Theorems 6.7, 6.10, 6.11, 6.12, and 6.14

Proof of Theorem 6.7. Let us show that the function $H_{BS}(t, \tilde{K}(t), \sigma, r, K)$ is strictly concave in $\sigma > 0$.

We have that $d_-(t, \tilde{K}(t), \sigma, r, K) \equiv -d_+(t, \tilde{K}(t), \sigma, r, K)$. Let

$$D(t, \sigma, r, K) \triangleq \frac{1}{4}(T - t)\sigma^2 = d_+(t, \tilde{K}(t), \sigma, r, K)^2 = d_-(t, \tilde{K}(t), \sigma, r, K)^2.$$

Let $H_{BS} = H_{BS,c}$ or $H_{BS} = H_{BS,p}$. By (6.3), it follows that

$$\frac{\partial H_{BS}}{\partial \sigma}(t, \tilde{K}(t), \sigma, r, K) = \frac{\tilde{K}(t)}{\sqrt{2\pi}} e^{-\frac{D(t,\sigma,r,K)}{2}} \frac{\partial d_+}{\partial \sigma}(t, \tilde{K}(t), \sigma, r, K)$$

$$- \frac{e^{-r(T-t)}K}{\sqrt{2\pi}} e^{-\frac{D(t,\sigma,r,K)}{2}} \frac{\partial d_-}{\partial \sigma}(t, \tilde{K}(t), \sigma, r, K) = \frac{\tilde{K}(t)}{\sqrt{2\pi}} e^{-\frac{D(t,\sigma,r,K)}{2}} \sqrt{T-t}.$$

Then

$$\frac{\partial^2 H_{BS}}{\partial \sigma^2}(t, \tilde{K}(t), \sigma, r, K) = -\frac{\tilde{K}(t)}{\sqrt{2\pi}} e^{-\frac{D(t,\sigma,r,K)}{2}} \frac{1}{2} \frac{\partial D(t,\sigma,r,K)}{\partial \sigma} \sqrt{T-t}$$

$$= -\frac{\tilde{K}(t)}{\sqrt{2\pi}} e^{-\frac{D(t,\sigma,r,K)}{2}} \frac{\sigma}{4}(T-t)^{3/2} < 0.$$

Similarly, we obtain for $H_{BS} = H_{BS,s}$ that

$$\frac{\partial H_{BS,s}}{\partial \sigma}(t, \tilde{K}(t), \sigma, r, K) = \frac{\tilde{K}(t)}{\sqrt{2\pi}}e^{-\frac{D(t,\sigma,r,K)}{2}}\frac{\partial d_+}{\partial \sigma}(t, \tilde{K}(t), \sigma, r, K)$$

$$= \frac{\tilde{K}(t)}{\sqrt{2\pi}}e^{-\frac{D(t,\sigma,r,K)}{2}}\frac{\sqrt{T-t}}{2}.$$

Then

$$\frac{\partial^2 H_{BS,s}}{\partial \sigma^2}(t, \tilde{K}(t), \sigma, r, K) = -\frac{\tilde{K}(t)}{\sqrt{2\pi}}e^{-\frac{D(t,\sigma,r,K)}{2}}\frac{1}{2}\frac{\partial D(\sigma, t)}{\partial \sigma}\frac{\sqrt{T-t}}{2}$$

$$= -\frac{\tilde{K}(t)}{\sqrt{2\pi}}e^{-\frac{D(t,\sigma,r,K)}{2}}\frac{\sigma(T-t)^{3/2}}{8} < 0.$$

Hence, the function $H_{BS}(t, \tilde{K}(t), \sigma, r, K)$ is strictly concave in $\sigma > 0$.

To complete the proof of Theorem 6.7, it suffices to consider the case of $t = 0$. Let us consider $\nu = 1$. By Lemma 6.1, it follows that

$$\mathbf{E}_{Q_{\sigma(\cdot)}}\tilde{F}(\tilde{S}(T), K) = \mathbf{E}_{Q_{\sigma(\cdot)}}\mathbf{E}_{Q_{\sigma(\cdot)}}\{\tilde{F}(\tilde{S}(T), K)|v(0)\}$$

$$= \mathbf{E}_{Q_{\sigma(\cdot)}}H_{BS}\left(0, S(0), \sqrt{v(0)}, r, K\right).$$

We find that $\hat{\sigma}_1(0) = \mathbf{E}\sqrt{v(0)}$ and Var $\sqrt{v(0)} \neq 0$. By Jensen's inequality, it follows that, if $\tilde{S}(0) = \tilde{K}(0)$, then

$$\mathbf{E}_{Q_{\sigma(\cdot)}}\tilde{F}(\tilde{S}(T), K) < H_{BS}(0, S(0), \hat{\sigma}_1(0), r, K). \tag{6.18}$$

This completes the proof for $\nu = 1$.

Let us consider $\nu > 1$. By Hölder's inequality, we obtain

$$\hat{\sigma}_1(0) \leq \hat{\sigma}_\nu(0), \quad \nu > 1. \tag{6.19}$$

Further, $H_{BS}(t, x, \sigma, r, K)$ is strictly increasing in σ for a vanilla put and call, and it is strictly increasing in σ for a share-or-nothing call when $x = e^{-r(T-t)}K$. If $S(0) = \tilde{K}(0)$, then

$$H_{BS}(0, S(0), \hat{\sigma}_1(0), r, K) < H_{BS}(0, S(0), \hat{\sigma}_\nu(0), r, K). \tag{6.20}$$

Therefore, the proof of Theorem 6.7 follows. □

Let $\tilde{P}_{\max}(t, \mathcal{A}) = e^{-rt}P_{\max}(t, \mathcal{A})$ be the corresponding discounted price given a class $\mathcal{A}$.

Proof of Theorem 6.10. It suffices to consider $t = 0$ only. Let $\overline{\mathbf{E}}_Q$ be the risk-neutral measure defined by $\overline{\sigma}(\cdot)$, and let $\overline{v} \triangleq \frac{1}{T}\int_0^T \overline{\sigma}(s)^2 ds$. By the definition,

$$\overline{P}_{\max}(0, \mathcal{A}) \geq \overline{\mathbf{E}}_Q F(\tilde{S}(T), K) = \tilde{H}_{BS}(0, S(0), \sqrt{\overline{v}}, r, K).$$

The second equality here follows from Lemma 6.1.

Further, let $S(0) = \tilde{K}(0)$ and $\mathcal{A} \subset \mathcal{A}^\perp$. By Theorem 6.7, we have that

$$\mathbf{E}_{Q_{\sigma(\cdot)}} \tilde{F}(\tilde{S}(T)) \leq \tilde{H}_{BS}(0, S(0), \sqrt{\bar{v}}, r, K).$$

Similar to the proof of Theorem 6.7, it follows from (6.19) that

$$\tilde{H}_{BS}(0, S(0), \sqrt{\bar{v}}, r, K) \leq \tilde{H}_{BS}(0, S(0), \hat{\sigma}_\nu(0), r, K).$$

It follows that $\tilde{P}_{\max}(0, \mathcal{A}) \leq \tilde{H}_{BS}(0, \tilde{K}, \hat{\sigma}_\nu(0, r, K))$ in that case. Then the proof of Theorem 6.10 follows. $\square$

Proof of Theorem 6.11. Let

$$V(x, y, t) \triangleq H(e^x, y, t), \quad x \in \mathbf{R}. \tag{6.21}$$

Let $x = \ln p$. Formally,

$$\frac{\partial H}{\partial p}(p, y, t) = \frac{1}{p}\frac{\partial V}{\partial x}(x, y, t), \quad \frac{\partial^2 H}{\partial p^2}(p, y, t) = \frac{1}{p^2}\frac{\partial^2 V}{\partial x^2}(x, y, t) - \frac{1}{p^2}\frac{\partial V}{\partial x}(x, y, t).$$

Set $\hat{D} \triangleq \mathbf{R} \times \mathbf{R} \times [0, T)$. Problem (6.14) can be rewritten for $V : \hat{D} \to \mathbf{R}$ as

$$\frac{\partial V}{\partial t}(x, y, t) + \frac{1}{2}f(e^x, y, t)\left[\frac{\partial^2 V}{\partial x^2}(x, y, t) - \frac{\partial V}{\partial x}(x, y, t)\right]$$

$$+ \sup_{u \in \Delta}\left\{g(e^x, y, u, t)\frac{\partial V}{\partial y}(x, y, t) + \frac{1}{2}b(e^x, y, u, t)^2\frac{\partial^2 V}{\partial y^2}(x, y, t)\right\} = 0,$$

$$V(x, y, T) = \Phi(x). \tag{6.22}$$

Let $(\overline{w}(t), \hat{w}(t))$ be a standard Wiener process in $\mathbf{R}^2$. Let $\mathcal{F}_t^w$ be the filtration generated by $(\overline{w}(t), \hat{w}(t))$.

Let $\mathcal{V}$ be the set of all processes $\eta(t)$ that are progressively measurable with respect to $\mathcal{F}_t^w$ and such that $\eta(t) \in \Delta$ for all t a.s. Let $\mathcal{V}_M$ be a subset of $\mathcal{V}$ such that there exists $u(\cdot) \in \mathcal{U}$ such that $\eta(t) = u(\tilde{S}(t), Y(t), t)$ and (6.13) holds.

For $\eta(\cdot) \in \mathcal{V}$, we consider the following controlled diffusion process:

$$d\xi_1(t) = -\frac{1}{2}f\left(e^{\xi_1(t)}, \xi_2(t), t\right)dt + f\left(e^{\xi_1(t)}, \xi_2(t), t\right)^{1/2}d\overline{w}(t),$$

$$d\xi_2(t) = g\left(e^{\xi_1(t)}, \xi_2(t), \eta(t), t\right)dt + b\left(e^{\xi_1(t)}, \xi_2(t), \eta(t), t\right)d\hat{w}(t).$$

Let $\xi^{x,y,s}(t) = [\xi_1^{x,y,s}(t), \xi_2^{x,y,s}(t)]$ be the solution of this equation given $\xi^{x,y,s}(s) = (x, y)$, $x, y \in \mathbf{R}$, $s \in [0, T]$.

We see that (6.22) represents the Bellman equation for the optimal control problem

$$\text{Maximize} \quad \mathbf{E}\Phi(\xi_1(T)) \quad \text{over} \quad \eta(\cdot) \in \mathcal{V}$$

(see e.g., [79]). From the assumptions on f and b, we have that

$$\sup_{u \in \Delta}\left[f(e^x, y, t)z_1^2 + b(e^x, y, u, t)^2 z_2^2\right] > 0 \quad \forall x, y, t, z = (z_1, z_2) \neq 0. \tag{6.23}$$

It follows that conditions of Theorem 5.2.5 from [79], p.225, are satisfied. By this theorem,

$$V(x, y, t) = \sup_{\eta(\cdot) \in \mathcal{V}} \mathbf{E}\Phi(\xi_1^{x,y,t}(T)) = \sup_{\eta(\cdot) \in \mathcal{V}_M} \mathbf{E}\{\Phi(\xi_1^{x,y,t}(T)).$$

By Theorem 3.1.5 from [79], p.132, the function V is continuous.

By Girsanov theorem, it follows that the process $(w_Q(t), \hat{w}(t))$ is a Wiener process under $Q_{\sigma(\cdot)}$, where

$$w_Q(t) \triangleq w(t) + \int_0^t a(s)f(\tilde{S}(s), Y(s), s)^{-1/2} ds.$$

If $\sigma(\cdot) \in \mathcal{A}_\mathcal{U}$ and $\eta(t) = u(\tilde{S}(t), Y(t), t)$ for the corresponding $u(\cdot) \in \mathcal{U}$, then the vector $\xi^{x,y,t}(T)$ has the same distribution under $\mathbf{P}$ as $(\log \tilde{S}(T), Y(T))$ under $Q_{\sigma(\cdot)}$ given that $Y(t) = y$ and $\log \tilde{S}(t) = x$. In this case,

$$\mathbf{E}\Phi(\xi_1^{x,y,t}(T)) = \mathbf{E}_{Q_{\sigma(\cdot)}}\{\tilde{F}(\tilde{S}(T))|Y(t) = y, \tilde{S}(t) = e^x\}.$$

Hence

$$V(x, y, t) = H(e^x, y, t) = \sup_{\sigma(\cdot) \in \mathcal{A}_\mathcal{U}} \mathbf{E}_{Q_{\sigma(\cdot)}}\{\tilde{F}(\tilde{S}(T))|Y(t) = y, \tilde{S}(t) = e^x\}.$$

Further, let (6.17) be satisfied. By Theorem 4.1.1 from [79], p.165, estimate (6.17) implies that the function V has bounded first-order generalized derivatives V_x' and V_y'. It is easy to deduct the required estimates for the first-order derivatives of H. This completes the proof of Theorem 6.11. $\square$

Proof of Theorem 6.12. From the assumptions on f, F, it follows that the function $\Gamma \triangleq (\phi, \Phi)$ is such that

$$|\Gamma(x, y, t)| + |\Gamma_t'(x, y, t)| + |\Gamma_x'(x, y, t)|$$
$$+ |\Gamma_{xy}''(x, y, t)| + |\Gamma_{xx}''(x, y, t)| + |\Gamma_{yy}''(x, y, t)| \leq C(1 + |x| + |y|)^m.$$

By Theorem 4.7.4 from [79], p.206, it follows that there exists a constant $C_1 > 0$ such that

$$|V(x, y, t)| + |V_t'(x, y, t)| + |V_x'(x, y, t)|$$
$$+ |V_{xy}''(x, y, t)| + |V_{xx}''(x, y, t)|$$
$$+ |V_{yy}''(x, y, t)| \leq C_1(1 + |x| + |y|)^{3m} \quad \forall(x, y, t) \in \hat{D}^*,$$

where

$$\hat{D}^* \triangleq \{(x, y, t) \in \hat{D} : \inf_{z \in \mathbf{R}^2 : |z|=1} \sup_{u \in \Delta} \left[f(e^x, y, t)z_1^2 + b(e^x, y, u, t)^2 z_2^2 \right] > 0\}.$$

Here $\hat{D} \triangleq \mathbf{R} \times \mathbf{R} \times [0, T]$. In particular, all generalized derivatives here are locally

bounded in $\hat{D}^*$. Similar to (6.23), we obtain $\hat{D}^* = \hat{D}$. Then the function $H(x, y, t) = V(\ln x, y, t)$ has the required properties. This completes the proof of Theorem 6.12.
$\square$

Proof of Theorem 6.13. Let $V(x, y, t) \triangleq V_1(x, y, t) + M_1 e^x + M_2$. It is easy to see that V is the solution of (6.22). Further,

$$V(x, y, t) = \sup_{\sigma(\cdot) \in \mathcal{A}_u} \mathbf{E}_{Q_{\sigma(\cdot)}} \{\tilde{F}_1(\tilde{S}(T)) | Y(t) = y, \tilde{S}(t) = e^x\} + M_1 e^x + M_2$$

$$= \sup_{\sigma(\cdot) \in \mathcal{A}_u} \mathbf{E}_{Q_{\sigma(\cdot)}} \left\{ \left[\tilde{F}_1(\tilde{S}(T)) + M_1 \tilde{S}(T) + M_2 \right] | Y(t) = y, \tilde{S}(t) = e^x \right\}$$

$$= \sup_{\sigma(\cdot) \in \mathcal{A}_u} \mathbf{E}_{Q_{\sigma(\cdot)}} \{\tilde{F}(\tilde{S}(T)) | Y(t) = y, \tilde{S}(t) = e^x\}. \quad (6.24)$$

Clearly, $H(x, y, t) = V(\ln x, y, t)$ is the solution of (6.14), which has the desired form. This completes the proof of Theorem 6.13. $\square$

Proof of Theorem 6.14. By Ito's formula applied to the process $\tilde{X}(t)$ defined by (6.18), we have that

$$d\tilde{X}(t)$$

$$= \frac{\partial H}{\partial t}(\tilde{S}(t), Y(t), t)dt + \frac{1}{2}f(\tilde{S}(t), Y(t), t)\tilde{S}(t)^2 \frac{\partial^2 H}{\partial x^2}(\tilde{S}(t), Y(t), t)dt$$

$$+ \frac{1}{2}b(\tilde{S}(t), Y(t), \eta(t), t)^2 \frac{\partial^2 H}{\partial y^2}(\tilde{S}(t), Y(t), t)dt + \frac{\partial H}{\partial x}(\tilde{S}(t), Y(t), t)d\tilde{S}(t)$$

$$+ \frac{\partial H}{\partial y}(\tilde{S}(t), Y(t), t)dY(t) + \alpha(t)dt$$

$$- \frac{\partial H}{\partial y}(\tilde{S}(t), Y(t), t)b(\tilde{S}(t), Y(t), \eta(t), t)d\hat{w}(t).$$

It can be rewritten as

$$d\tilde{X}(t) = \frac{\partial H}{\partial t}(\tilde{S}(t), Y(t), t)dt + \frac{1}{2}f(\tilde{S}(t), Y(t), t)\tilde{S}(t)^2 \frac{\partial^2 H}{\partial x^2}(\tilde{S}(t), Y(t), t)dt$$

$$+ \frac{1}{2}b(\tilde{S}(t), Y(t), \eta(t), t)^2 \frac{\partial^2 H}{\partial y^2}(\tilde{S}(t), Y(t), t)dt$$

$$+ g(\tilde{S}(t), Y(t), \eta(t), t)\frac{\partial H}{\partial y}(\tilde{S}(t), Y(t), t)dt$$

$$+ \frac{\partial H}{\partial x}(\tilde{S}(t), Y(t), t)d\tilde{S}(t) + \alpha(t)dt.$$

From the definition of the process $\alpha(t)$, this equation can be rewritten as

$$d\tilde{X}(t) = \frac{\partial H}{\partial t}(\tilde{S}(t), Y(t), t)dt + \frac{1}{2}f(\tilde{S}(t), Y(t), t)\tilde{S}(t)^2 \frac{\partial^2 H}{\partial x^2}(\tilde{S}(t), Y(t), t)dt$$

$$+ \sup_{u \in \Delta} \left\{ g(\tilde{S}(t), Y(t), u, t)\frac{\partial H}{\partial y}(\tilde{S}(t), Y(t), t) \right.$$

$$\left. + \frac{1}{2}b(\tilde{S}(t), Y(t), u, t)^2 \frac{\partial^2 H}{\partial y^2}(\tilde{S}(t), Y(t), t) \right\} dt + \frac{\partial H}{\partial x}(\tilde{S}(t), Y(t), t)d\tilde{S}(t).$$

By (6.14), the selection of the function H ensures that the last equation can be rewritten as

$$d\tilde{X}(t) = \frac{\partial H}{\partial x}(\tilde{S}(t), Y(t), t)d\tilde{S}(t).$$

It follows that $\tilde{X}(t)$ is the discounted wealth for the self-financing strategy such that the quantity of stock shares at time t is $\frac{\partial H}{\partial x}(\tilde{S}(t), Y(t), t)$. By (6.14) again, $\alpha(t) \geq 0$. This implies statement (i).

Let us prove statement (ii). Let $\overline{X}(0)$ be some other initial wealth such that $\overline{X}(0) < X(0)$ and such that, for any $\sigma(\cdot) \in \mathcal{A}_u$, there exists an admissible strategy and an $\mathcal{F}_t$-adapted square integrable process $\overline{\xi}$ such that, for the corresponding discounted wealth $\overline{X}(t)$,

$$\overline{X}(T) \geq \tilde{F}(\tilde{S}(T)) + \int_0^T \overline{\xi}(s)d\hat{w}(s) \quad \text{a.s..} \tag{6.25}$$

From statement (i), it follows that

$$X(0) = H(S(0), Y(0), 0) = \sup_{\sigma(\cdot) \in \mathcal{A}_u} \mathbf{E}_{Q_{\sigma(\cdot)}} \tilde{F}(\tilde{S}(T)).$$

Hence $\overline{X}(0) < \sup_{\sigma(\cdot) \in \mathcal{A}_u} \mathbf{E}_{Q_{\sigma(\cdot)}} \tilde{F}(\tilde{S}(T))$. Therefore, there exists $\overline{\sigma}(\cdot) \in \mathcal{A}_u$ such that $\overline{X}(0) < \mathbf{E}_{Q_{\overline{\sigma}(\cdot)}} \tilde{F}(\tilde{S}(T))$. On the other hand, (6.25) implies that $\mathbf{E}_{Q_{\overline{\sigma}(\cdot)}} \overline{X}(T) = \overline{X}(0) \geq \mathbf{E}_{Q_{\overline{\sigma}(\cdot)}} \tilde{F}(\tilde{S}(T))$. Hence (6.25) does not hold. This completes the proof. $\square$

6.4 Unconditionally implied volatility and risk-free rate

Definition 6.3, the standard definition of implied volatility, ignores the fact that, in reality, r is unknown and needs to be forecasted, because the option price depends on its future (forward) curve. Therefore, the standard implied volatility at time t is a *conditional* one and it depends on the future curve $r(s)|_{s \in [t,T]}$. We shall study the pair $(\sigma_{imp}(t), \rho_{imp}(t))$ of *unconditionally* implied parameters, where $\sigma_{imp}(t)$ is the unconditionally implied volatility, and $\rho_{imp}(t)$ is the unconditionally implied value of $\rho(t)$. This pair of implied parameters can be inferred from a system of two equations for different options.

Definition 6.15 *Assume that we observe two options on the same stock with market prices $P^{(1)}(t)$ and $P^{(2)}(t)$ at time t. These options have the same expiration time $T > 0$. Let $H_{BS}^{(1)}$ and $H_{BS}^{(2)}$ be the Black–Scholes price for the corresponding types of options. Let the pair $(\sigma_{imp}(t), \rho_{imp}(t))$ be such that*

$$H_{BS}^{(1)}(t, S(t), \sigma_{imp}(t), \rho_{imp}(t)) = P^{(1)}(t),$$
$$H_{BS}^{(2)}(t, S(t), \sigma_{imp}(t), \rho_{imp}(t)) = P^{(2)}(t).$$

We say that $\sigma_{imp}(t)$ is the implied volatility and $\rho_{imp}(t)$ is the implied average forward risk-free rate inferred from (6.26).

To avoid technical difficulties, we will assume that the prices and parameters in (6.26) are such that the solution $(\sigma_{imp}(t), \rho_{imp}(t))$ exists and is uniquely defined for all special cases described below. Clearly, Definition 6.15 is model free and does not require any pricing rules and *a priori* assumptions on the evolution law for volatilities and risk-free rates. We need some models and pricing rules only for numerical simulations of $(\sigma_{imp}(t), \rho_{imp}(t))$.

6.4.1   Two calls with different strike prices

Assume that two European call options on the same stock have market prices $P_i(t)$ at time t, $i = 1, 2$. We assume that these options have the same expiration time $T > 0$ and have different strike prices $K_i > 0$, $K_1 \neq K_2$. Let $\sigma_{imp}(t) = \sigma_{imp}(t, K_1, K_2)$ be the *implied volatility* and $\rho_{imp}(t) = \rho_{imp}(t, K_1, K_2)$ be the *implied average forward risk-free rate* given K_1, K_2 at time t, inferred from the system

$$H_{BS,c}(t, S(t), \sigma_{imp}(t), \rho_{imp}(t), K_1) = P_1(t),$$
$$H_{BS,c}(t, S(t), \sigma_{imp}(t), \rho_{imp}(t), K_2) = P_2(t).$$

Remark 6.16 For a solution of system (6.26), the following straightforward algorithm can be applied. Let $\sigma_1(t, K_1 | \rho)$ be such that

$$H_{BS,c}(t, S(t), \sigma_1(t, K_1 | \rho), \rho, K_1) = P_1(t)$$

(i.e., it is the standard (conditional) implied volatility). Consider equation

$$H_{BS,c}(t, S(t), \sigma_1(t, K_1 | \rho), \rho, K_2) = P_2(t).$$

Let $\hat{\rho} = \hat{\rho}(K_1, K_2)$ be the solution of this equation. Then

$$(\sigma_{imp}(t), \rho_{imp}(t)) = (\sigma_1(t, K_1 | \hat{\rho}(K_1, K_2)), \hat{\rho}(K_1, K_2)).$$

6.5   Bond price inferred from option prices

This section considers estimation of the future cumulative short-term interest rate and the related problem of bond pricing. We assume that there is a bond that needs to be priced, and that there is also a pair of put and call options with prices generated by a risk-neutral valuation method for a generalized Black–Scholes model, and they are correlated with the short-term interest rate that generates the bond price. We suggest exploring this connection via $(\sigma_{imp}(t), \rho_{imp}(t))$. We found that the put-call parity ensures that the implied forward risk-free rate inferred from the system of put and call options does not depend on the current stock prices and the strike price and can be explicitly expressed from inferred parameters. Therefore, it can be effectively used for bond pricing (see an example for the Vasicek model).

6.5.1 Definitions

Let $r(t)$ be a random process of the short-term interest rate. Consider a zero coupon bond that pays \$1 at time T. We shall study the problem of pricing of this bond. We assume that the price $B(0)$ of the bond at time $t = 0$ is defined by a modification of (6.2), i.e.,

$$B(0) = \mathbf{E}_{Q_T} \exp\left(-\int_0^T r(s)ds\right), \tag{6.26}$$

where $\mathbf{E}_{Q_T}$ is the expectation with respect to a measure Q_T, and the choice of all measures Q_T for different T is such that the bond market is arbitrage free (see, e.g., [81], Chapter 6).

Further, we assume that there is a stock with price $S(t)$, $t \geq 0$ that evolves as (2.1).

Let $\mathcal{F}_t$ be the filtration generated by the observable data. We assume that $\mathcal{F}_0$ is the P-augmentation of the set $\{\emptyset, \Omega\}$. Further, we assume that $\mathcal{F}_t$ and $(a(t), \sigma(t), r(t))$ do not depend on $\{w(t_2) - w(t_1)\}_{t_2 \geq t_1 \geq t}$. Further, we assume that the process $(S(t), r(t))$ is $\mathcal{F}_t$-adapted (in particular, this means that the process $(S(t), r(t))$ is currently observable). However, we do not assume that the vector $(B(t), w(t), a(t))$ is currently observable, and we do not assume that the prior distribution of (r, a, σ, w) is known. In addition, we assume that there are options on the stock such that their prices are currently observable.

We are going to apply Definition 6.15 to certain models for $(S(t), r(t), B(t))$, and for certain pricing rules for options. To avoid technical difficulties, we will assume that the prices and parameters in (6.26) are such that the solution $(\sigma_{imp}(t), \rho_{imp}(t))$ exists and is uniquely defined for the special case described below.

6.5.2 Inferred ρ from put and call prices

Theorem 6.17 *Consider a market model with pricing rule (6.5) given some Q with random (σ, a, r) such that $\sigma(\cdot)$ does not depend on $w(\cdot)$ under Q. Assume that we observe market prices $P_c(t)$ and $P_p(t)$ at time t for European vanilla call and put options on a stock with current price $S(t)$, with payoff functions $(S(T) - K)^+$ and $(K - S(T))^+$, respectively. We assume that these options have the same expiration time $T > 0$ and have the same strike price $K > 0$. Let the pair $(\sigma_{imp}(t), \rho_{imp}(t)) = (\sigma_{imp}(t, K), \rho_{imp}(t, K))$ be inferred from*

$$H_{BS,c}(t, S(t), \sigma_{imp}(t), \rho_{imp}(t), K) = P_c(t),$$
$$H_{BS,p}(t, S(t), \sigma_{imp}(t), \rho_{imp}(t), K) = P_p(t). \tag{6.27}$$

Then $\rho_{imp}(t)$ does not depend on $(K, S(t))$, and

$$\rho_{imp}(t) = -\frac{1}{T-t} \ln \mathbf{E}_Q \left\{ \exp\left(-\int_t^T r(s)ds\right) \middle| \mathcal{F}_t \right\}. \tag{6.28}$$

In particular, $\rho_{imp}(t)$ is defined solely by the distribution of $r(\cdot)$ given Q and $\mathcal{F}_t$.

Corollary 6.18 *It is easy to see that (6.28) can be rewritten as*

$$\mathbf{E}_Q \left\{ \exp\left(-\int_t^T r(s)ds \right) \Big| \mathcal{F}_t \right\} = e^{-\rho_{imp}(t)(T-t)}.$$

Therefore, the implied forward risk-free rate inferred from system (6.27) can be effectively used as an estimate for the exponent $\exp\left(-\int_t^T r(s)ds \right)$ *presented in the equation for bond price (6.26).*

Clearly, there is an uncertainty in the choice of measures in (6.26) and (6.29), since these measures are not unique for the general case. We discuss this in the next section.

Proof of Theorem 6.17. By the Black–Scholes formula and (6.27), it follows that

$$K \exp\left[-\rho_{imp}(t)(T-t)\right] = H_{BS,p}(t, S(t), \sigma_{imp}(t), \rho_{imp}(t), K)$$
$$-H_{BS,c}(t, S(t), \sigma_{imp}(t), \rho_{imp}(t), K) + S(t). \tag{6.29}$$

By Lemma 6.1 and by the Black–Scholes formula again, it follows that

$$H_{BS,p}(t, S(t), \sigma_{imp}(t), \rho_{imp}(t), K) - H_{BS,c}(t, S(t), \sigma_{imp}(t), \rho_{imp}(t), K) + S(t)$$
$$= P_p(t) - P_c(t) + S(t)$$
$$= \mathbf{E}_Q\{H_{BS,p}(t, S(t), \sqrt{v(t)}, \rho(t), K)|\mathcal{F}_t\}$$
$$-\mathbf{E}_Q\{H_{BS,s}(t, S(t), \sqrt{v(t)}, \rho(t), K)|\mathcal{F}_t\} + S(t)$$
$$= \mathbf{E}_Q\{H_{BS,p}(t, S(t), \sqrt{v(t)}, \rho(t), K) - H_{BS,s}(t, S(t), \sqrt{v(t)}, \rho(t), K)$$
$$+S(t)|\mathcal{F}_t\} = K\mathbf{E}_Q \exp\left(-\int_t^T r(s)ds \right).$$

This completes the proof. □

Note that there is a novelty in this simple proof: the well-known put-call parity for the Black–Scholes prices is applied with the Black–Scholes put and call prices for the case when the Black–Scholes formula is invalid (for random processes $r(t), \sigma(t)$): the true (unknown) values of parameters are replaced by the implied ones.

Corollary 6.19 *It follows from the proof of Theorem 6.17 that*

$$\rho_{imp}(t) = \frac{1}{T-t} \left(\ln K - \ln[P_p(t) - P_c(t) + S(t)] \right).$$

Proof. By (6.29), $K \exp\left[-\rho_{imp}(t)(T-t) \right] = P_p(t) - P_c(t) + S(t).$ □

6.5.3 Application to a special model

Consider a case when the process $r(t)$ is adopted to the filtration $\mathcal{F}_t^w$ generated by the Wiener process $w(t)$. More precisely, assume that $r(t)$ is a solution of the Ito equation

$$dr(t) = f(r(t), t)dt + b(r(t), t)dw(t),$$

where (random) functions $f(x, t, \omega) : \mathbf{R} \times [0, T] \times \Omega \to \mathbf{R}$, $b(x, t, \omega) : \mathbf{R} \times [0, T] \times \Omega \to \mathbf{R}$ are adapted to the filtration $\mathcal{F}_t^w$ for all x and have some regularity as functions of (x, t).

Consider a market with zero coupon bonds with prices $P_k(t)$, where $t \in [0, T_k]$, and where $\{T_k\}$ is a given set of maturing times, $P(T_k, T_k) = 1$. The model for bond prices defined in Chapter 5 implies that

$$P_k(t) = \mathbf{E}\left\{\exp\left(-\int_t^{T_k} r(s)ds + \int_t^{T_k} q(s)dw(s) - \frac{1}{2}\int_t^{T_k} |q(s)|^2 ds\right) \bigg| \mathcal{F}_t\right\}, \quad (6.30)$$

where $q(t)$ is a given $\mathcal{F}_t^w$-adapted function.

To calculate the expectation in (6.30) directly, one needs to know the processes f, b, q. Assume that q is known but the pair (f, b) is unknown. In that case, this expectation cannot be calculated.

An alternative way is to estimate the bond price via option prices using the method described above. Let $t = 0$. It suffices to find a pair of put and call options with the same strike price and with expiration time $T = T_k$, where T_k is the maturity time for a bond under consideration; the options must be on an underlying risky asset $S(t)$ evolving as (2.1) and such that

$$(a(t) - r(t))\sigma(t)^{-1} = -q(t).$$

In that case, the only risk-neutral measure $\mathbf{P}_Q$ is such that $d\mathbf{P}_Q/d\mathbf{P} = Z$, where

$$Z \triangleq \exp\left(\int_0^T q(s)dw(s) - \frac{1}{2}\int_0^T |q(s)|^2 ds\right).$$

By (6.30), it follows that

$$P_k(0) = \mathbf{E}_Q \exp\left(-\int_t^{T_k} r(s)ds\right).$$

This value can be found by (6.29), where $t = 0$, and where $\rho_{imp}(0)$ is the inferred parameter from the pair of the put and call options.

6.6   A dynamically purified option price process

For the dynamic estimation of time-varying implied parameters, one has to separate the impact of the changes of the stock price on the option price from the impact of the change of the values of the stock price parameters. For this, it could be convenient to exclude the current stock price from the system of equations for the implied parameters. To address this, we suggest the following approach.

Let us consider dynamically adjusted parameters $T = t + \tau$ and $K = \kappa S(t)$, where $\kappa \in (0, +\infty)$ is a parameter. In this case, $F(S(T), K) = S(t)F(Y(t+\tau), \kappa)$, where

$$Y(T) = S(t+\tau)/S(t).$$

By rule (6.5), the option price given (a, σ, r), is

$$
\begin{aligned}
P_{RN}(t, \sigma(\cdot), r(\cdot)) &\triangleq \mathbf{E}_Q\{e^{-\int_t^T r(s)ds} F(S(T), K) \,|\, \mathcal{F}_t\} \\
&= S(t)\mathbf{E}_Q\{e^{-\int_t^{t+\tau} r(s)ds} F(Y(t+\tau), \kappa) \,|\, \mathcal{F}_t\},
\end{aligned}
$$

where Q is some risk-neutral measure, and where $\mathbf{E}_Q$ is the corresponding expectation.

Let

$$G(t) \triangleq \frac{P_{RN}(t, \sigma(\cdot), r(\cdot))}{S(t)}.$$

It follows that

$$G(t) = \mathbf{E}_Q\{e^{-\int_t^{t+\tau} r(s)ds} F(Y(t+\tau), \kappa) \,|\, \mathcal{F}_t\}.$$

Therefore, the implied parameters $(\sigma_{imp}(t), \rho_{imp}(t))$ for European options can be calculated using $H_{BS}(t, 1, \sigma, \rho, \delta)$ only with $\kappa = \kappa_i$, $i = 1, 2$, $\kappa_1 \neq \kappa_2$.

Let us consider the following example. Assume that two European call options on the same stock have market prices $P_i(t)$ at time t, $i = 1, 2$. We assume that these options have the same expiration time $T = t + \tau > 0$ and have different strike prices $K_i = \kappa_i S(t) > 0$, $\kappa_1 \neq \kappa_2$. Let $G_i(t) = P_i(t)/S(t)$. In this case, the implied volatility $\sigma_{imp}(t)$ and the implied average forward risk-free rate $\rho_{imp}(t)$ at time t can be inferred from the system

$$
\begin{aligned}
H_{BS,c}(t, 1, \sigma_{imp}(t), \rho_{imp}(t), \kappa_1) &= G_1(t), \\
H_{BS,c}(t, 1, \sigma_{imp}(t), \rho_{imp}(t), \kappa_2) &= G_2(t).
\end{aligned}
$$

In practice, only a finite set of possible strike prices is available. Therefore, it is not possible to collect the prices $P_i(t)$ of the options with the exact strike prices $K_i = \kappa_i S(t)$ with fixed κ_i. To apply the method described above, we have to use the prices $\tilde{P}_i(t)$ of the corresponding options with the closest available strike prices $\tilde{K}_i(t)$. Let

$$\xi_i(t) = \frac{\tilde{K}_i(t)}{S(t)} - \kappa_i.$$

We have that $\tilde{K}_i(t) = (\kappa_i + \xi_i(t))S(t)$; the processes $\tilde{K}_i(t)$ and $\xi_i(t)$ are observable, i.e., adapted to $\mathcal{F}_t$. To take into account the presence of $\xi_i(t) \neq 0$, we suggest using the first-order approximation and approximate $P_i(t)$ by the processes $\tilde{P}_i(t) - \Delta(K_i)\xi_i(t)S(t)$; the error for this approximation has the order of $O(\xi_i^2)$. Here $\Delta(K)$ is the so-called *Delta of the Strike*. For the call options,

$$\Delta(K) = \frac{\partial H_{BS,c}}{\partial K}(t, S(t), \sigma, \rho, K) = e^{-r\tau} \Phi(d_2).$$

For the put options,

$$\Delta(K) = \frac{\partial H_{BS,p}}{\partial K}(t, S(t), \sigma, \rho, K) = e^{-r\tau}\Phi(-d_2).$$

The approach for the calculations of the implied parameters described above has to be adjusted as the following: the processes $G_i(t)$ have to be replaced by

$$\tilde{G}_i(t) = \frac{\tilde{P}_i(t) - \Delta(K_i)\xi_i(t)S(t)}{S(t)} = \frac{\tilde{P}_i(t)}{S(t)} - \Delta(K_i)\xi_i(t),$$

where $\Delta(K_i) = \Delta(K_i, \sigma_{imp}(t), \rho_{imp}(t))$ is considered to be a function of $(\sigma_{imp}(t), \rho_{imp}(t))$ calculated with given $S(t)$ under the assumption that $K_i = (1 + \kappa_i)S(t)$. In other words, the system of equations (6.31) for $(\sigma_{imp}(t), \rho_{imp}(t))$ at time t has to be replaced by the system

$$H_{BS,c}(t, 1, \sigma_{imp}(t), \rho_{imp}(t), \kappa_1) = \frac{\tilde{P}_1(t)}{S(t)} - \Delta(K_1, \sigma_{imp}(t), \rho_{imp}(t))\xi_1(t),$$

$$H_{BS,c}(t, 1, \sigma_{imp}(t), \rho_{imp}(t), \kappa_2) = \frac{\tilde{P}_2(t)}{S(t)} - \Delta(K_2, \sigma_{imp}(t), \rho_{imp}(t))\xi_2(t).$$

Remark 6.20 The observations of option prices with dynamically adjusted strike price $K = \kappa S(t)$ with a fixed κ can be useful for econometrics purposes even without calculation of the implied parameters. In particular, some features of the evolution law for the process $(\sigma(t), r(t))$ can be restored directly from the observations of the process $G(t)$. For instance, the processes $G(t)$ must evolve as a deterministic function of the current historical values of $(\sigma(t), r(t))$ if the process $(\sigma(t), r(t))$ evolves as a Markov process that is independent from $w(\cdot)$. Since the impact of the stock price movements is damped, one may expect that $G(t)$ and $\tilde{G}_i(t)$ are relatively smooth processes; this can make the calculation of implied parameters more sustainable.

6.7 The implied market price of risk with random numéraire

In this section, we reconsider the market model described in Section 3.3 and describe an approach for selection of equivalent martingale measures.

6.7.1 The risk-free bonds for the market with random numéraire

First, it can be noted that the market described in Section 3.3 does not include a risk-free asset since both processes $S(t)$ and $B(t)$ are random. However, one can augment this market with auxiliary tradable risk-free assets representing zero coupon bonds constructed as some options with the payoff \$1 at time T. We represent the price of these bonds as a function of θ. At the next step, we inverse this pricing formula and represent θ via observed market bond prices. This gives a way of selecting implied parameter θ_{imp} consistent with the observed bond prices. It follows the classical approach to the so-called implied volatility where the Black–Scholes formula is reversed.

The model described in Section 3.3 is a continuous time model of a securities market consisting of two tradable assets with the prices $S(t)$ and $B(t)$, $t \geq 0$. The prices evolve as

$$dS(t) = S(t)\big(a(t)dt + \sigma(t)dw(t) + \hat{\sigma}(t)d\hat{w}(t)\big), \quad t > 0,$$

and

$$dB(t) = B(t)\big(\alpha(t)dt + \rho(t)dw(t) + \hat{\rho}(t)d\hat{w}(t)\big). \tag{6.31}$$

We assume that $W(t) = (w(t), \hat{w}(t))$ is a standard Wiener process with independent components on a given standard probability space $(\Omega, \mathcal{F}, \mathbf{P})$, where Ω is a set of elementary events, $\mathbf{P}$ is a probability measure, and $\mathcal{F}$ is a $\mathbf{P}$-complete σ-algebra of events. The initial prices $S(0) > 0$ and $B(0) > 0$ are given constants.

We accept the assumptions of Section 3.3.

6.7.2 The case of a complete market

Let us assume first that

$$\hat{\sigma}(t) \equiv \hat{\rho}(t) \equiv 0. \tag{6.32}$$

In this case, by the definition of $\mathcal{F}_t$, we have that $\mathcal{F}_t = \mathcal{F}_t^w$, and the assumptions of Theorem 3.22 are satisfied. In this case, $\tilde{\rho} \equiv 0$, and (3.38) imply that $\varrho \equiv \rho\theta_1$ and that $\theta_1(t)$ is uniquely defined as $\theta_1(t) = \tilde{\sigma}(t)^{-1}\tilde{a}(t)$. The process θ_1 is called the *marked price of risk process* in this case.

Under these assumptions, the equation for B in (3.39) can rewritten as

$$dB(t) = B(t)([\alpha(t) - \rho(t)\theta_1(t)]dt + \rho(t)dW_{1\theta}(t))). \tag{6.33}$$

Lemma 6.21 *If (6.32) holds, then the claim $\xi \equiv \$1$ is replicable in the following sense: for any $t \in [0, T)$, there exists an $\mathcal{F}_t$-adapted process $\gamma(t)$ such that $\mathbf{E}_\theta \int_0^T \gamma(t)^2 \tilde{S}(t)^2 dt < +\infty$ and*

$$B(T)^{-1} = \mathbf{E}_\theta B(T)^{-1} + \int_0^T \gamma(t)d\tilde{S}(t).$$

Proof of Lemma 6.21 follows from the assumptions of Theorem 3.22 and from the equation

$$B(T)^{-1} = B(t)^{-1} \exp\left([-\alpha + \rho^2/2 + \theta_1](T - t) - \rho[W_{1\theta}(T) - W_{1\theta}(t)]\right).$$

$\square$

Remember that, by the definition of $\mathcal{F}_t$, the processes $a(t), \sigma(t), \alpha(t)$, and $\rho(t)$, are $\mathcal{F}_t^w$-adapted if (6.32) holds.

Under the assumptions of Lemma 6.21, the value $\mathbf{E}_\theta B(T)^{-1}$ represents the price at time $t = 0$ of a zero coupon bond with the payoff \$1 at the maturity time T. The value $\tilde{X}(t) = \mathbf{E}_\theta \{B(T)^{-1}|\mathcal{F}_t\}$ represents the discounted wealth for the hedging (replicating) strategy, and the value

$$P(t, T) = B(t)\tilde{X}(t) = B(t)\mathbf{E}_\theta \{B(T)^{-1}|\mathcal{F}_t\} \tag{6.34}$$

represents the total wealth for the replicating strategy and the price at time t of a zero coupon bond with the payoff \$1 at the maturity time T.

Let us discuss some consequences of these statements.

Lemma 6.21 implies that the value

$$r(t) \overset{\Delta}{=} -(T-t)^{-1}\log P(t,T) = -(T-t)^{-1}\log\left(B(t)\mathbf{E}_\theta \{B(T)^{-1}|\mathcal{F}_t\}\right) \tag{6.35}$$

represents the so-called yield to maturity, or the expected average risk-free rate associated with the zero coupon bond, meaning that the price at time t of a zero coupon bond with the payoff \$1 at the maturity time T is

$$P(t, T) = \exp(-r(t)(T-t)). \tag{6.36}$$

If the processes $a(t), \sigma(t), \alpha(t), \rho(t), \hat{\rho}(t), \theta(t)$ are constant and the assumptions of Lemma 6.21 are satisfied, then

$$B(T)^{-1} \tag{6.37}$$

$$= B(t)^{-1}\exp\left(\left(-\alpha + \frac{\rho^2}{2} + \rho\theta_1\right)[T-t] - \rho\left(W_{1\theta}(T) - W_{1\theta}(t)\right)\right).$$

In this case, a direct calculation of (6.34) gives

$$P(t, T) = B(t)\mathbf{E}_\theta \{B(T)^{-1}|\mathcal{F}_t\} = \exp[-(T-t)(\alpha - \rho^2 - \rho\theta_1)]$$

and

$$-(T-t)^{-1}\log P(t,T) = \alpha - \rho^2 - \rho\theta_1.$$

Since $\theta_1 = \tilde{a}/\tilde{\sigma}$, it gives $-(T-t)^{-1}\log P(t,T) = \alpha - \rho^2 - \rho\tilde{a}/\tilde{\sigma}$. Hence

$$r \overset{\Delta}{=} \alpha - \rho^2 - \rho\tilde{a}/\tilde{\sigma} \tag{6.38}$$

can be interpreted as the "true" risk-free rate for this market. It can be seen that r is close to α if ρ is small. If $\rho = 0$ then $r = \alpha$.

Let us investigate the situation where θ_1 is unknown. Let us consider a scenario where the real market price $P_{market}(0, T)$ of the zero coupon bond with the payoff \$1 at the maturity time T is observable at time $t = 0$. In this case, the corresponding value (6.35) $r_{market} = -T^{-1}\log P_{market}(0, T)$ can be calculated. From the assumptions on the coefficients, (6.32) implies that $\rho \neq 0$. In this case, we can reverse pricing formula (6.34) (or reverse (6.38)) and calculate *implied* $\theta_{1,imp}$ from (6.38) as

$$\theta_{1,imp} = (\alpha - \rho^2 - r_{market})/\rho. \tag{6.39}$$

In this case, equation (6.31) can be rewritten as

$$dB(t) = B(t)\big([r_{market} + \rho^2]dt + \rho dW_{1\theta}(t)\big),$$

where $W_{1\theta}(t) = w(t) + \int_0^t \theta_{1,imp}(s)ds$.

It particular, it follows that a choice of ρ for a market model with given α is consistent with the observed bond prices if $(\alpha - r_{market})/\rho$ is bounded as $\rho \to 0$. This leads to the following heuristic rule: if the observed bond market price is such that $\alpha - r_{market}$ is large, then one should assume a sufficiently large ρ, to avoid overestimation of the market price of risk.

Representation (6.39) follows the classic approach to the so-called implied volatility where the Black–Scholes formula is reversed. Representation of the market price of risk process θ_1 implied from observed bond prices as described above could be a useful addition to the existing methods. Further development of this approach is presented in Section 6.7.3.

Remark 6.22 For the bond pricing model with constant coefficients described above, the choice of (θ_1, r) is independent on T. It follows that a single market price $P_{market}(0, T)$ of a zero-coupon bond for one given maturity time T defines uniquely the prices of similar bonds for all other maturity times $\overline{T} \neq T$ given that these prices are defined by (6.34). This is caused by the fact that this formula has to be applied with the same θ_1 leading to the same r in (6.36).

6.7.3 The case of an incomplete market

We suggested above a method of calculation of the implied market price of risk $\theta_{1,imp}$ from observed bond prices for a case of a complete market with non-zero ρ and $\hat{\rho}$. For this, we established that the value (6.34) represents the price at time t of a zero coupon bond with the payoff $\$1$ at the maturity time T for a case where the claim $\$1$ is replicable. This indicates that it could be reasonable to accept (6.34) as the price of this bond for a more general case for an incomplete market with a non-replicable claim $\$1$ as well (i.e., with non-zero ρ and $\hat{\rho}$).

Up to the end of this section, we assume a market model with non-zero ρ and $\hat{\rho}$, where the price of a zero coupon bond with the payoff $\$1$ at the maturity time T is defined by (6.34) similar to the case of the complete market, regardless of replicability of the corresponding contingent claim.

This leads to a one more important example of selection of ϱ and of the corresponding θ defined by (3.38) that supplements choices described in Section 3.3.3.

Assume that the processes $a(t), \sigma(t), \alpha(t), \rho(t), \hat{\rho}(t), \theta(t)$ are constant, and that $r(t)$ is defined by (6.35). This means that $r(t)$ represents again the expected average risk-free rate associated with the zero coupon bond.

It follows from the Ito formula that

$$B(T)^{-1} = B(t)^{-1} \exp\left(\left(-\alpha + \frac{\rho^2}{2} + \frac{\hat{\rho}^2}{2} + \rho\theta_1 + \hat{\rho}\theta_2\right)(T-t)\right.$$
$$\left.-\rho\left(W_{1\theta}(T) - W_{1\theta}(t)\right) - \hat{\rho}\left(W_{2\theta}(T) - W_{2\theta}(t)\right)\right). \qquad (6.40)$$

In this case, a direct calculation of (6.34) gives

$$P(t,T) = B(t)\mathbf{E}_\theta\left\{B(T)^{-1}|\mathcal{F}_t\right\} = e^{-(T-t)(\alpha-\rho^2-\hat{\rho}^2-\rho\theta_1-\hat{\rho}\theta_2)}$$

and

$$-(T-t)^{-1}\log P(t,T) = \alpha - \rho^2 - \hat{\rho}^2 - \rho\theta_1 - \hat{\rho}\theta_2.$$

Then (6.35) implies that $r(t)$ can be found explicitly; it does not depends on t and T and depends on θ such that $r(t) \equiv r_\theta$, where

$$r_\theta = \alpha - \rho^2 - \hat{\rho}^2 - \rho\theta_1 - \hat{\rho}\theta_2. \qquad (6.41)$$

Using (3.38), equation (6.41) can be rewritten as

$$\varrho = \rho\theta_1 + \hat{\rho}\theta_2 = \alpha - r_\theta - \rho^2 - \hat{\rho}^2. \qquad (6.42)$$

Consider now a scenario where the real market price $P_{market}(0,T)$ of a zero coupon bond with the payoff \$1 at the maturity time T is observed from the market statistics at time $t = 0$, and the corresponding value (6.34) $r_{market} = -T^{-1}\log P_{market}(0,T)$ is calculated. We can reverse pricing formula (6.34) and calculate *implied* ϱ as

$$\varrho_{imp} = \alpha - r_{market} - \rho^2 - \hat{\rho}^2$$

and find implied $\theta = \theta_{imp} = (\theta_{imp,1}, \theta_{imp,2})^\top$ as a solution of the corresponding system (3.38), i.e.,

$$\tilde{\sigma}\theta_{imp,1} + \tilde{\rho}\theta_{imp,2} = \tilde{a},$$
$$\rho\theta_{imp,1} + \hat{\rho}\theta_{imp,2} = \varrho_{imp}. \qquad (6.43)$$

This would follow again the classic approach to the so-called implied volatility where the Black–Scholes formula is reversed.

In particular, equation (3.39) for this $\varrho = \varrho_{imp}$ gives

$$dB(t) = B(t)\left([r_{market} + \rho^2 + \hat{\rho}^2]dt + \rho dW_{1\theta}(t)\right) + \hat{\rho}dW_{2\theta}(t)),$$

where $\theta = \theta_{imp}$.

This model has the same feature as that described in Remark 6.22 for a special model. In particular, the choice of θ is independent of T, and a single market price $P_{market}(0,T)$ of a zero coupon bond for one given maturity time T defines uniquely the prices defined by (6.34) for similar bonds for all other maturity times $\overline{T} \neq T$, since this formula has to be applied with the same θ. For models with time variable coefficients of equations for B and S, the same approach gives a time-dependent solution $\theta(t)$ of (3.38), and the value r defined by (6.38) for a maturity time $\overline{T}$ depends on $\overline{T}$.

6.8 Bibliographic notes

Sections 6.3 and 6.4

Section 6.3 is based on the results from [45]. Section 6.4 is based on the results from [42]. It can also be noted that the case where two parameters are inferred from option historical prices has been addressed by several authors but in different settings. In [30], it was suggested to use two call options with different strike prices for calculation of implied volatility distributions for the case of option prices obtained via the unbiased estimate of option price for random volatility. The mentioned paper addressed the case of implied risk-free rate, but it was focused on the case of the implied stock prices and volatility.

Section 6.5 and 6.6

Section 6.5 is based on the results from [44]. Section 6.6 is based on the results from [90, 89].

Section 6.7

Currently, there are few other implied processes considered in the literature, besides the classic implied volatility. In [115], inference of the implied value $a - \theta_1 \sigma$ from the market option price was considered, in a model corresponding to a special case of our model with $\hat{\sigma} = \rho = \hat{\rho} = 0$, presuming that this value is used as the appreciation rate under the pricing measure; for the Black and Scholes model, this value should be the risk-free rate. In [121], the implied market price of risk for energy prices was estimated as the difference between the observable historical Ornstein-Uhlenbeck long-term mean and the implied long-term mean inferred from the market options prices. Finally, the implied martingale measure for a bond market was introduced in [23]; this construction was based on observation of bond prices for a continuum of maturities. None of these papers considered estimation of the appreciation rate of the stock as an implied parameter inferred from the stock option prices.

7

Inference of implied parameters from option prices

The net present value of a European vanilla call option is the payoff at option contract expiry discounted to the present date. The discount rate reflects the funding cost, liquidity risk, and credit risk, entities that differ among market participants. Since the market participants may use different discount rates to perform valuation on a set of option contracts, some may find these option contracts undervalued and express their views by buying these contracts, while other may find them overvalued and sell these contracts instead.

One may infer from the bid-ask quotes of exchange traded European vanilla call options the aggregate market participants' choice of discount rate. A non-linear inverse problem approach for addressing this problem is to use a European vanilla option pricing formula as a mapping tool to map the bid-ask mid-quote price to the pricing model parameters, and, assuming that the discount rate is also regarded as an unknown parameter, to the model-based discount rate simultaneously. We refer to this model-based discount rate as the implied discount rate.

In this chapter, we use the Black–Scholes option pricing formula (7.1) as the mapping tool to infer the implied volatility and the implied discount rate as the pair of implied parameters of interest from the day-close bid-ask mid-quote prices pertaining to a set of European vanilla call option contracts. In Section 7.1, we first investigate the sensitivity of implied volatility mapping with respect to the uncertainty in discount rate specification using the construct of an under-defined system of nonlinear equations. In Section 7.3.1, we describe a numerical strategy that deploys Algorithm 7.2 to simultaneously infer implied volatility and implied discount rate from option prices using the construct of an over-defined system of nonlinear equations.

Let $(\Omega, \mathcal{F}, \mathbf{P})$ be a standard probability space where Ω is a set of elementary events, $\mathcal{F}$ is a complete sigma algebra of events, and $\mathbf{P}$ is a risk-neutral probability measure. Let $\mathcal{F}_t$ be a complete sigma algebra of events generated by the data observed at time t. Let the stochastic process of the price of a risky asset that evolves under the risk-neutral probability measure be

$$S(t) = S(0) \exp \left\{ \left(\rho(t) - \frac{1}{2}\sigma(t)^2 \right) t + \sigma(t) W(t) \right\}, \quad 0 < t, \qquad (7.1)$$

where $W(t)$ is a standard Wiener process under the risk-neutral probability measure.

Let $T, 0 < t < T$, and let

$$\rho(t) \triangleq \frac{1}{T-t} \int_s^T \widetilde{\rho}(u)du , \quad \text{and} \quad \sigma(t) \triangleq \left(\frac{1}{T-t} \int_t^T \widetilde{\sigma}(u)^2 du\right)^{1/2} , \quad t < T, \quad (7.2)$$

where $\widetilde{\rho}(u)$ and $\widetilde{\sigma}(u)$ are deterministic variables, and where $\widetilde{\rho}(u)$ is the instantaneous, i.e., short, rate at time u under the risk-neutral measure that is related to the price of a risk free zero coupon bond $B(t,T)$ such that

$$B(t,T) = \exp\left(-\int_t^T \widetilde{\rho}(u)du\right)B(T,T),$$

and where $\widetilde{\sigma}(u)$ is the instantaneous volatility, and $S(0) > 0$ is a given deterministic constant. We assume that $S(t)$, $\widetilde{\rho}(t)$, and $\widetilde{\sigma}(t)$ are adapted to the filtration $\mathcal{F}_t$, and do not depend on $\{W(\widetilde{s}) - W(t)\}_{t<\widetilde{s}}$, and we consider the probability measure associated with $\mathcal{F}_t$ to be the risk-neutral probability measure.

Let $K > 0$ be the strike price of a European vanilla call option contract with terminal payoff function $(S(T) - K)^+ = \max(0, S(T) - K)$, let $\sigma(t)$ and $\rho(t)$ be nonrandom at time t, let $0 < t < T$ be fixed, and let $C_{BSLin}(t, T, S(t), \sigma(t), \rho(t), K)$ be the price of a European vanilla call option contract evaluated using the Black–Scholes option pricing formula assuming the dividend rate is zero such that

$$
\begin{aligned}
&C_{BSLin}(t, T, S(t), \sigma(t), \rho(t), K) \\
&= S(t)\Phi(d_+(t, T, S(t), \sigma(t), \rho(t), K)) \\
&\quad - Ke^{-\rho(t)\tau}\Phi(d_-(t, T, S(t), \sigma(t), \rho(t), K)) , \quad (7.3) \\
&d_+(t, T, S(t), \sigma(t), \rho(t), K) \\
&= [\log(S(t)/K) + \rho(t)\tau]/(\sigma(t)\sqrt{\tau}) + (\sigma(t)\sqrt{\tau})/2 , \\
&d_-(t, T, S(t), \sigma(t), \rho(t), K) \\
&= d_+(t, T, S(t), \sigma(t), \rho(t), K) - \sigma(t)\sqrt{\tau} ,
\end{aligned}
$$

where $\tau = T - t$ is the time to maturity, and where $\Phi(d) \triangleq \frac{1}{\sqrt{2\pi}} \int_{-\infty}^d e^{-x^2/2}dx$. In the classical Black–Scholes framework, $\rho(t)$ is assumed to be known, while $\sigma(t)$ is the only unknown parameter to be estimated. In our current model setting, we relax this assumption and regard both $\sigma(t)$ and $\rho(t)$ as unknown model parameters.

7.1 Sensitivity analysis of implied volatility estimation

7.1.1 An under-defined system of nonlinear equations

Let $C(t, T_j, K_{j.\ell}), j = 1, \ldots, m, \ell = 1, \ldots, n_j$, be the prices of $\sum_{j=1}^m n_j$ European vanilla call option contracts written on the same underlying asset observed at time t with times to maturity $\{\tau_j\}_{j=1}^m$, $\tau_j = T_j - t$, and struck at $\{K_{j,\ell}\}_{j=1,\ldots,m,\ell=1,\ldots,n_i}$

respectively. By mapping $C(t, T_j, K_{j.\ell})$ to their corresponding theoretical prices evaluated using the Black–Scholes option pricing formula, we construct a system of nonlinear equations

$$C_{BSLin}(t, T_j, S(t), \sigma_{imp,j,\ell}(t), \rho_j(t), K_{j,\ell}) = C(t, T_j, K_{j,\ell}), \qquad (7.4)$$

where $j = 1, \ldots, m$, $\ell = 1, \ldots, n_j$, $\sigma_{imp,j,\ell}(t)$ is the Black–Scholes model-based implied volatility corresponding to $C(t, T_j, K_{j,\ell})$, $\rho_j(t)$ is the discount rate for tenor τ_j at time t, and where $S(t)$ is the spot price of the underlying asset at time t. By regarding $\{\rho_j(t)\}_{j=1}^m$ as unknown parameters, (7.4) becomes an under-defined system of equations where the solution of $\sum_{j=1}^m n_j$ equations necessitates finding solutions for $m + \sum_{j=1}^m n_j$ unknown parameters, namely $\{\sigma_{imp,j,\ell}(t)\}_{j=1,\ldots,m,\ell=1,\ldots,n_j}$ and $\{\rho_j(t)\}_{j=1}^m$. Under this construction, uncertainty of the discount rate leads to estimation uncertainty of the implied volatility.

We use (7.4) as a convenient working model to quantify the estimation uncertainty of implied volatility based on some pre-defined interval $[\rho_{min}, \rho_{max}]$ that mimics plausible fluctuation of discount rate with respect to two instances in time, one at spot time t, the other at some later time t', $t < t' < T$, where T is the expiry date of the option contract. Let $\rho_{min} = \min(\rho_j(t), \rho_j(t'))$, and $\rho_{max} = \max(\rho_j(t), \rho_j(t'))$, where $\rho_j(t)$ and $\rho_j(t')$ are the discount rates for tenor τ_j at times t and t', respectively. Specifically, we have chosen to construct $\rho_j(t)$ by interpolating at tenor τ_j from the yield curve historical data of a reference rate quoted at time t, and construct $\rho_j(t')$ by interpolating also at tenor τ_j from the yield curve historical data of the same reference rate quoted at time t'. For our current context, we do not construct $\rho_j(t')$ by interpolating at tenor $\tau_j - (t' - t)$ from the yield curve historical data quoted at time t'. Our simple construction of ρ_{min} and ρ_{min} is aimed at capturing the vertical shift of the yield curve in consideration, i.e., the change in constant maturity rate at a horizon τ_j, instead of taking into account the more complicated joint dynamics of the rolldown yield change from tenor τ_j to tenor $\tau_j - (t' - t)$ and the change in constant maturity rate at a horizon $\tau_j - (t' - t)$. Granted, different strategies may be used to construct ρ_{min} and ρ_{max} depending on the context of the application.

For example, in *ex post* study of future discount rate uncertainty, the discount rates at the corresponding instances, $\rho_j(t_1)$ and $\rho_j(t_2)$, $t_1 < t_2$, from the relevant historical interest rates may be used to construct ρ_{min} and ρ_{max}. Additionally, in portfolio scenario simulation, the current discount rate $\rho_j(t_{spot})$, and a simulated future discount rate $\rho_j(t_{future})$ may also be used to construct ρ_{min} and ρ_{max} to facilitate quantification of the corresponding implied volatility uncertainty for risk management purposes. Furthermore, in rare event simulation for stress-testing, one may use the current discount rate $\rho_j(t_{spot})$ and a simulated future discount rate that mimics wide fluctuation in the discount rate comparable to those observed in catastrophic financial events to construct ρ_{min} and ρ_{max}.

7.1.2   Numerical analysis using cross-sectional S&P 500 call options data

We use future London Interbank Offer Rate (Libor) uncertainty, in an *ex post* manner, to mimic plausible future discount rate uncertainty. We construct the intervals of the discount rate uncertainty using the historical Libor time series downloaded from the Economic Research Division, Federal Reserve Bank of St. Louis. For this empirical study, we aim to illustrate the point that *ceteris paribus* future discount rate fluctuation during relatively calm market conditions can still leads to meaningful variability in implied volatility mapping.

Since the commencement of a series of quantitative easing measures by central banks worldwide after the 2008 global financial crisis, the daily variability of the Libor rates have by and large been relatively small. As such we have chosen May 2011 as a subinterval within this time period for our empirical analysis. In this section, we use the day-close prices of the S&P500 index European vanilla call option contracts traded at the Chicago Board Options Exchange (CBOE) on May 5th 2011, obtained from Market Data Express LLC., to analyze the impact of a given, small interval of discount rate uncertainty on the interval of the implied volatility estimation uncertainty for option contracts with different strike prices and times to maturity.

As an example to visualize the potential impact of discount rate uncertainty, four subsets of data corresponding to time-to-maturity of 16, 44, 56, and 72 days, respectively, are depicted in four separate panels to contrast the impact of implied risk-free rates uncertainty on the volatility smile profile across different time-to-maturities. The range of implied risk-free rate used to construct Figure 7.1 for illustrative purposes has lower and upper bounds of $\rho_{min}(t) = 0.001$ and $\rho_{max}(t) = 0.02$, respectively, representing a realistic range of risk-free rates as the Libor term structure quoted on the same trading day ranges between 0.001 and 0.01. For the same range of implied risk-free rate uncertainty, the region of uncertainty of the implied volatility smile profile across strikes becomes wider across increasing magnitude of time-to-maturities. These diagrams highlight the fact that a relatively narrow domain of uncertainty for the implied risk-free rate can lead to a large uncertainty region of the implied volatility smile profile for options with long time-to-maturities.

The impact of different ranges of risk-free rate uncertainty on the uncertainty of the volatility smile caricature is shown in Figure 7.2. For the same set of options data, widening of the risk-free rate uncertainty range leads to a corresponding widening of the implied volatility profile uncertainty. The asymmetric widening of implied volatility profile where in-the-money (ITM) options are subjected to larger extent of uncertainty render the use of the domain of implied volatility surface (IVS) corresponding to ITM call options more prone to mis-pricing.

The two bounding surfaces for the IVS caricature constructed assuming implied risk-free rate as $\rho_{min}(t) = 0.001$ and $\rho_{max}(t) = 0.02$, respectively, are depicted in Figure 7.3. The upper surface corresponds to the IVS constructed assuming the risk-free rate is $\rho_{min}(t) = 0.001$ while the lower surface corresponds to that constructed assuming the risk-free rate is $\rho_{max}(t) = 0.02$. These surfaces represent, given $\rho_{min}(t) = 0.001$ and $\rho_{max}(t) = 0.02$, the upper and lower surface boundaries for the space of uncertainty within which the caricature of the IVS varies in conjunction with the uncertainty of the risk-free rate. The vertical distance between these two surfaces for the long time-to-maturity zone appears in general to be larger than that for the short time-to-maturity zone. This illustrates that the longer time-to-maturity zone in the IVS profile is more sensitive to the uncertainty of the risk-free rate, as expected from differentiating the Black–Scholes formula with respect to risk-free rate. This observation reiterates the concern that long time-to-maturity exotic options that are heavily path dependent are subjected to the danger of a greater extent of being mis-priced even if the range of uncertainty for the risk-free rate remains the same across the entire term structure.

Let $\rho_{min}(t) = 0.001, \rho_{max}(t) = 0.02, \forall t$, where t refers to each of the trading days in the data set spanning May 1st 2011 to May 31st 2011 on which prices for these options are observed. The corresponding bounds of implied volatilities as $\sigma_{min}(t)$ and $\sigma_{max}(t)$ are calculated. The magnitude of the volatility uncertainty range, defined as $\sigma_{max}(t) - \sigma_{min}(t), \forall t$, is plotted against the trading dates t for these European call options with some strike price.

Figure 7.4 depicts the change, across the trading days, in the magnitude of volatility uncertainty range across options of different strike prices for a collection of options that mature on the same day, June 18th 2011. The stochastic behavior in the magnitude of volatility uncertainty range is demonstrated as expected. The call options with lower strike prices are higher in value than call options with higher strike prices. Therefore, the magnitude of volatility uncertainty range for options with lower strike prices is smaller than those for options with higher strike prices on the same trading days. Towards the end of May 2011, these options that mature on June 18th 2011 have a remaining time-to-maturity of less than one month. For the corresponding options across all depicted strikes, the progressive decrease in magnitude of implied volatility uncertainty range, albeit non-monotonic, is marked.

However, the picture is quite different for options with long time-to-maturities as shown in Figure 7.5 which depicts, in four different panels, call options that mature on December 30th 2011, June 16th 2012, December 22nd 2012, and December 21st 2013, respectively. At the end of May 2011, all the options displayed in Figure 7.5 still have remaining time-to-maturity that exceeds six months. The magnitude of uncertainty volatility range for these options with long remaining time-to-maturities at the end of May 2011 do not exhibit sharp, progressive decrease across the 21 trading days considered, in contrast to the comparatively short time-to-maturity options depicted in Figure 7.4. Even among the four long time-to-maturity options, the longer the time-to-maturity, the larger the magnitude of volatility uncertainty in general. This further illustrates that the uncertainty of implied volatility due to the risk-free rate uncertainty has a more pronounced impact on pricing path-dependent contingent claims that depends heavily on the long time-to-maturity zone of the IVS.

One of the primary motivations for profiling the magnitude of the volatility uncertainty range for options with different strike and time-to-maturity characteristics across a consecutive sequence of trading days is to investigate the dynamics of this process. In particular, Figures 7.4 and 7.5 demonstrate that the process is not smooth, exhibiting dynamics that is stochastic in nature resembling the dynamics of implied volatility itself.

We obtain the proxy for the tenor matching discount rates that are used in Sections 7.1.2 and 7.1.3 by first fitting smoothing spline curves to the Libor rates where the smoothing parameters are estimated using generalized cross-validation, and then interpolate the discount rate proxies from the spline curves at the respective tenors that correspond to the times to maturity of the option contracts considered. We use the smoothing spline modeling algorithm implemented as the in-build function smooth.spline() for the R statistical computing environment. This suffices for our current demonstrative purpose of constructing plausible discount rate uncertainty intervals to analyze implied volatility estimation uncertainty in our simple empirical examples. That said, in order to construct arbitrage-free discount rate term structures for coherent pricing and hedging purposes, more sophisticated shape-constrained spline-based techniques may be used instead.

Option contracts	Days to maturity from	Libor rates (in decimal)		
expiry dates	May 5th 2011	May 5th 2011	May 13th 2011	May 27th 2011
June 18th 2011	44	0.002133	0.001967	0.001728
June 30th 2011	56	0.002295	0.002120	0.001881
July 16th 2011	72	0.002513	0.002324	0.002081
Aug. 20th 2011	107	0.003013	0.002790	0.002521
Sep. 17th 2011	135	0.003449	0.003203	0.002903
Sep. 30th 2011	148	0.003663	0.003409	0.003094
Dec. 17th 2011	226	0.005022	0.004752	0.004393
Dec. 30th 2011	239	0.005252	0.004982	0.004623
Mar. 17th 2012	317	0.006625	0.006348	0.006007
Mar. 30th 2012	330	0.006854	0.006576	0.006228

Table 7.1: The Libor rates corresponding to the times to maturity for the S&P 500 European call option contracts with prices quoted on the 5th, 13th, and 27th of May 2011.

Table 7.1 depicts the interpolated Libor rates, obtained based on the historical Libor rates quoted on May 5th 2011, May 13th 2011, and May 27th 2011, at various tenors that correspond to the times to maturity of the option contracts considered. The quoted Libor rates are first converted from the discrete interest rates convention to the continuously compounded convention prior to interpolation. A subset of the interpolated rates depicted in Table 7.1 is used as proxy for the discount rates in the empirical analysis reported in this section.

Using two sets of option prices observed on May 5th 2011 that correspond to the European vanilla call option contracts expiring on either Sept. 30th 2011 or Mar. 30th 2012, we compare the intervals of implied volatility estimation uncertainty in the presence of discount rate uncertainty across the same horizon for these option contracts. The proxy for discount rates pertaining to these two sets of option contracts at the spot time are interpolated from the Libor term structure quoted on May

5th 2011 while the proxy for future discount rates pertaining to the same sets of option contracts are interpolated from the Libor term structure quoted on May 13th 2011. These discount rates are used to construct the future discount rate uncertainty intervals in order to calculate the corresponding intervals of implied volatility estimation uncertainty. For option contracts expiring on Sept. 30th 2011, we obtain $\rho_{min} = 0.003409$, and $\rho_{max} = 0.003663$. For option contracts expiring on Mar. 30th 2012, we obtain $\rho_{min} = 0.006576$, and $\rho_{max} = 0.006854$. We depict the corresponding estimated implied volatilities in Table 7.2. In this sample, the implied volatilities for the into-the-money (ITM) option contracts tend to exhibit a wider interval of estimation uncertainty than those for the out-of-the-money (OTM) option contracts. This trend is more obvious for the set of option contracts expiring on Mar. 30th 2012, but less so for those expiring on Sept. 30th 2011. Additionally, for a given range of discount rate uncertainty and for a given strike price, the option contract with longer time to maturity tends to exhibit a wider interval of implied volatility estimation uncertainty.

We also analyze the impact of future discount rate uncertainty across different horizons ahead of the spot time based on option contracts with the same time to maturity. In the lower panel of Table 7.2, for example, we compare and contrast the intervals of implied volatility estimation uncertainty in the presence of discount rate uncertainty across the two different time intervals considered based on a set of option contracts expiring on Mar. 30th 2012. The shorter time interval mimics the future discount rate uncertainty between the spot date, May 5th 2011, and a subsequent date, May 13th 2011, where we obtain $\rho_{min} = 0.006576$, and $\rho_{max} = 0.006854$ by tenor matching interpolation of the Libor rates quoted on these two dates. The longer time interval mimics the future discount rate uncertainty between the spot date and a subsequent date, May 27th 2011, where we obtain $\rho_{min} = 0.006228$, and $\rho_{max} = 0.006854$ by interpolating the Libor rates quoted on these corresponding dates. The interval of implied volatility estimation uncertainty is wider for the scenario simulating the future discount rate uncertainty across a longer time interval in this particular example as the longer time interval considered is associated with a wider range of discount rate uncertainty. This is a sample-dependent phenomenon as the evolution of interest rates across time is stochastic and non-monotonic, and as such it may turn out that, for some other time frames, the interval between the spot rate and the rate more distant into the future may instead be narrower than that between the spot rate and an interest rate at a time in the nearer future. Henceforth, suffice to note that the range of discount rate uncertainty between the spot time and any two different time points ahead may be different.

In order to provide an idea of the impact of 10% change in the magnitude of the discount rate of choice on the corresponding inferred implied volatilities using the empirical data set analyzed in this section, we repeat the calculations carried out for Table 7.2. Instead of constructing the discount rate uncertainty interval using *ex post* Libor discount rates, we use 90%, 100%, and 110% of the contemporaneously quoted Libor rates for the appropriate tenor on May 5th 2011 to construct the discount rate uncertainty ranges and report the corresponding results in Table 7.3. The findings are qualitatively similar to those of Table 7.2.

Strike price	Moneyness	Implied volatilities estimated from prices observed on May 5th 2011 for option contracts expiring on Sept. 30th 2011 with discount rate proxy interpolated from Libor rates observed on the following dates		
		May 5th 2011 ($\rho = 0.003663$)	May 13th 2011 ($\rho = 0.003409$)	May 27th 2011 ($\rho = 0.003094$)
1000	0.7490	0.040129	0.040074	0.040474
1100	0.8239	0.209290	0.210267	0.211482
1200	0.8988	0.200491	0.200915	0.201419
1300	0.9737	0.177714	0.177956	0.178256
1400	1.0486	0.152567	0.152726	0.152886
1500	1.1235	0.132204	0.132286	0.132387
		Implied volatilities estimated from prices observed on May 5th 2011 for option contracts expiring on Mar. 30th 2012 with discount rate proxy interpolated from Libor rates observed on the following dates		
		May 5th 2011 ($\rho = 0.006854$)	May 13th 2011 ($\rho = 0.006576$)	May 27th 2011 ($\rho = 0.006228$)
1000	0.7490	0.108361	0.126524	0.137647
1100	0.8239	0.186538	0.187506	0.188694
1200	0.8988	0.184604	0.185138	0.185827
1300	0.9737	0.173265	0.173631	0.174092
1400	1.0486	0.158296	0.158559	0.158870
1500	1.1235	0.143280	0.143451	0.143702

Table 7.2: The implied volatility estimated for near-the-money S&P 500 European call options with prices quoted on May 5th 2011, and maturing on either Sep. 30th 2011 or May 5th 2011. The discount rate ρ is assumed to be the Libor rates observed on May 5th, 13th, or 27th 2011, which correspond to the interest rate values of 0.003663, 0.003409, and 0.003094 for options maturing on Sep. 30th 2011, and correspond to the interest rate values of 0.006854, 0.006576, and 0.006228 for options maturing on March 30th 2012. The spot price on May 5th 2011 was 1335.1. Moneyness is the strike price to spot price ratio.

7.1.3 Numerical analysis using longitudinal S&P500 call options data

In this section, we analyze the size of the implied volatility estimation uncertainty interval for a set of option contracts across a number of consecutive trading days with respect to a given, fixed interval of future discount rate uncertainty throughout the trading days considered. Admittedly, the interval of future discount rate uncertainty may change from day to day. However, the combined effect of the fluctuation of future discount rate uncertainty, and the reduction of time to maturity of the option contracts across different trading dates on the implied volatility estimation uncertainty is complicated and difficult to interpret. The current approach, albeit simple, allows us to concentrate on the analysis of the size of implied volatility estimation

Strike price	Moneyness	Implied volatilities estimated from prices observed on May 5th 2011 for option contracts expiring on Sept. 30th 2011 with discount rate proxy interpolated from Libor rates observed on the following dates		
		May 5th 2011	May 13th 2011	May 27th 2011
		($\rho = 0.003663$)	($\rho = 0.003409$)	($\rho = 0.003094$)
1000	0.7490	0.0401293	0.04001317	0.04018039
1100	0.8239	0.2092895	0.21073115	0.20783333
1200	0.8988	0.2004906	0.20110547	0.19985694
1300	0.9737	0.1777135	0.17806266	0.17736667
1400	1.0486	0.1525666	0.15278246	0.15235801
1500	1.1235	0.1322044	0.13232187	0.13204550
		Implied volatilities estimated from prices observed on May 5th 2011 for option contracts expiring on Mar. 30th 2012 with discount rate proxy interpolated from Libor rates observed on the following dates		
		May 5th 2011	May 13th 2011	May 27th 2011
		($\rho = 0.006854$)	($\rho = 0.006576$)	($\rho = 0.006228$)
1000	0.7490	0.1083605	0.1391060	0.03851205
1100	0.8239	0.1865378	0.1889018	0.18414973
1200	0.8988	0.1846038	0.1859554	0.18324797
1300	0.9737	0.1732647	0.1741572	0.17233795
1400	1.0486	0.1582957	0.1589287	0.15762864
1500	1.1235	0.1432796	0.1437429	0.14285416

Table 7.3: The implied volatility estimated for near-the-money S&P500 European call options with prices quoted on May 5th 2011, and maturing on either Sep. 30th 2011 or May 5th 2011. The discount rate ρ is assumed to be the 90%, 100%, and 110% of the Libor rates observed on May 5th 2011, which correspond to the interest rate values of 0.003663, 0.0032967, and 0.0040293 for options maturing on Sep. 30th 2011, and correspond to the interest rate values of 0.006854, 0.0061686, and 0.0075394 for options maturing on March 30th 2012. The spot price on May 5th 2011 was 1335.1.

uncertainty intervals as the times to maturity of the option contracts decrease, conditional on some fixed interval of discount rate uncertainty.

For this purpose, we use the historical data of two sets of S&P500 index European vanilla call option contracts expiring on either Sept. 30th 2011 or Mar. 30th 2012 traded on CBOE between May 3rd 2011 and May 31st 2011 inclusive. The proxy for the discount rates at spot time for the corresponding times to maturity of the option contracts concerned is interpolated from the Libor rates quoted on May 3rd 2011, while the proxy for the discount rates at a future date for the corresponding times to maturity of these option contracts is interpolated from the Libor rates quoted on May 31st 2011. For option contracts expiring on Sept. 30th 2011, we obtain $\rho_{min} = 0.003021$, and $\rho_{max} = 0.003740$. For option contracts expiring on Dec. 21st 2013, we obtain $\rho_{min} = 0.007269$, and $\rho_{max} = 0.007578$.

Let $\sigma_{min}(t)$ and $\sigma_{max}(t)$ be the implied volatilities estimated based on (7.4)

Option contracts expiring on Sept. 30th 2011			Option contracts expiring on Mar. 30th 2012		
Remaining	$\Delta(t) \times 10^{-4}$		Remaining	$\Delta(t) \times 10^{-4}$	
days to expiry	$K = 1300$	$K = 1350$	days to expiry	$K = 1300$	$K = 1350$
150	8.1109	6.1961	963	15.2900	13.6729
149	7.2671	5.5069	962	14.7074	13.1785
148	6.8184	4.8442	961	14.3815	13.0682
147	7.2423	5.2465	960	14.7226	13.2310
144	7.2388	5.1645	957	14.8460	13.5592
143	7.6903	5.4952	956	15.2220	13.4902
142	7.0732	4.8597	955	14.7675	13.0340
141	7.4327	5.1182	954	14.7660	13.2549
140	6.9275	5.1113	953	14.2565	13.2757
137	6.0202	4.5480	950	14.2170	12.7400
136	6.1249	4.9204	949	14.2015	13.0187
135	6.9528	4.9476	948	14.3012	12.8745
134	7.0739	5.4153	947	14.4494	13.1024
133	6.8918	4.5891	946	14.4809	12.5228
130	5.6973	4.1463	943	14.0775	12.4794
129	5.6160	4.1096	942	13.4644	12.4921
128	6.0966	4.2712	941	14.0185	12.7519
127	6.0172	4.6379	940	14.1189	12.7614
126	6.3804	4.3716	939	14.0613	12.5497
122	6.9529	4.9584	935	14.3995	13.1295

Table 7.4: Range of implied volatility uncertainty, $\Delta(t)$, for near-the-money S&P 500 European call options observed for each trading day in May 2011 calculated assuming the discount rate ρ to be the Libor rates observed on May 2nd or 27th 2011, which are 0.003021, and 0.003740 for options maturing on Sept. 30th 2011, and 0.007269, and 0.007578 for options maturing on Dec. 21st 2013, respectively. K is the strike price of the option contract.

by inverting the Black–Scholes option pricing formula assuming that the discount rate is either $\rho_{min} = 0.007269$ or $\rho_{max} = 0.007578$, respectively. Let $\Delta(t) = \sigma_{max}(t) - \sigma_{min}(t)$, where t denotes the trading date on which the option prices $C(t; \tau_j, K_{j,\ell})$ are observed. The time series of $\Delta(t)$ for these two sets of option contracts are tabulated in Table 7.4. The non-monotonic trend of $\Delta(t)$ across different trading dates may, in part, be attributed to the stochastic nature of the underlying asset price evolution. For this sample, the day-to-day variability of $\Delta(t)$ appears to be less marked for longer dated option contracts.

With an aim to provide an idea of the impact of 10% increase in the magnitude of the discount rate on the inferred implied volatilities using the empirical data set analyzed in this section, we repeat the aforementioned calculations based on which Table 7.4 is produced. Instead of constructing the discount rate uncertainty intervals using the *ex post* Libor discount rate fluctuation range, we use 100% and 110% of the contemporaneously quoted Libor rates for the appropriate tenor quoted on on May 2nd 2011 to construct the discount rate uncertainty ranges and report the corresponding results in Table 7.5.

Option contracts expiring on Sept. 30th 2011			Option contracts expiring on Mar. 30th 2012		
Remaining	$\Delta(t) \times 10^{-4}$		Remaining	$\Delta(t) \times 10^{-4}$	
days to expiry	$K = 1300$	$K = 1350$	days to expiry	$K = 1300$	$K = 1350$
150	3.3341	2.6272	963	6.4497	5.8350
149	3.2847	2.3011	962	6.2659	5.3360
148	2.8563	2.1730	961	5.8354	5.4658
147	3.3050	2.3036	960	6.2761	5.4621
144	3.0765	1.7457	957	6.3193	5.5115
143	3.1160	2.2631	956	6.2865	5.5348
142	2.9744	2.1307	955	6.0264	5.5936
141	3.1810	1.9837	954	6.2720	5.5410
140	2.9244	2.3083	953	5.8849	5.7502
137	2.5776	2.1753	950	6.1962	5.2190
136	2.6620	1.9548	949	6.1785	5.4384
135	2.9530	2.2086	948	6.0110	5.6079
134	3.2369	2.2362	947	6.0415	5.5473
133	2.8327	1.8566	946	6.0354	5.1417
130	2.1873	1.7920	943	5.9616	5.3155
129	2.5275	1.6291	942	5.5683	5.3654
128	2.5380	1.9733	941	5.8539	5.4163
127	2.4360	1.7462	940	6.0597	5.4596
126	2.6972	1.9155	939	5.8618	5.3186
122	2.8647	2.3743	935	5.9159	5.7462

Table 7.5: Range of implied volatility uncertainty, $\Delta(t)$, for near-the-money S&P 500 European call options observed for each trading day in May 2011 calculated assuming the discount rate ρ to be 100% or 110% of the Libor rates observed on May 2nd 2011, which are 0.003021 and 0.003740 for options maturing on Sept. 30th 2011, and 0.007269 and 0.007578 for options maturing on Dec. 21st 2013, respectively. K is the strike price of the option contract.

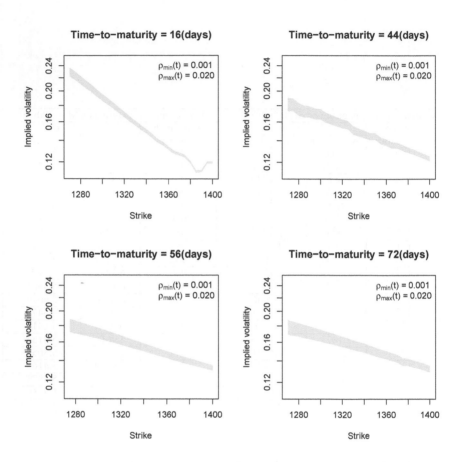

Figure 7.1: The region of implied volatility uncertainty for a given range of implied risk-free rates illustrated using subsets of SPX500 index European call options data observed on May 5th 2011. Underlying spot price was 1361.22. The four panels depict data subsets corresponding to time-to-maturity of 16, 44, 56, and 72 days, respectively. In each panel, the region of uncertainty of implied volatilities corresponding from the same data subset with respect to the same range of uncertainty of implied risk-free rate at spot time t as $\rho_{min}(t) = 0.001$ and $\rho_{max}(t) = 0.02$ is depicted as a shaded region within the corresponding panels.

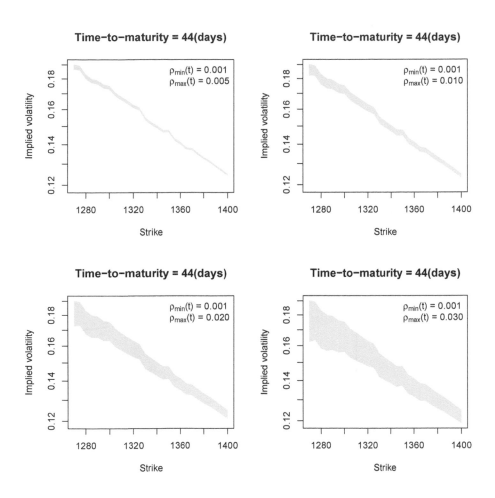

Figure 7.2: The region of implied volatility uncertainty for *different* ranges of implied risk-free rates illustrated using SPX500 index European call options data with time-to-maturity of 44 days observed on May 5th 2011. Underlying spot price was 1361.22. In each panel, the region of uncertainty of implied volatilities corresponding from the same data subset with respect to the different ranges of implied risk-free rate uncertainty at spot time t is a highlighted as shaded region within the corresponding panels.

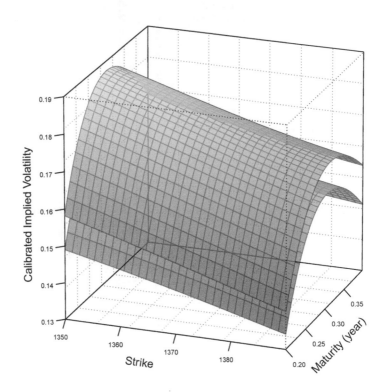

Figure 7.3: A three-dimensional view of the IVS caricature uncertainty region for a given range of implied risk-free rates illustrated using SPX500 index European call options data observed on May 5th 2011. Underlying spot price was 1361.22. The upper surface corresponds to the IVS constructed assuming the risk-free rate is $\rho_{min}(t) = 0.001$ while the lower surface corresponds to the IVS constructed assuming the risk-free rate is $\rho_{max}(t) = 0.02$.

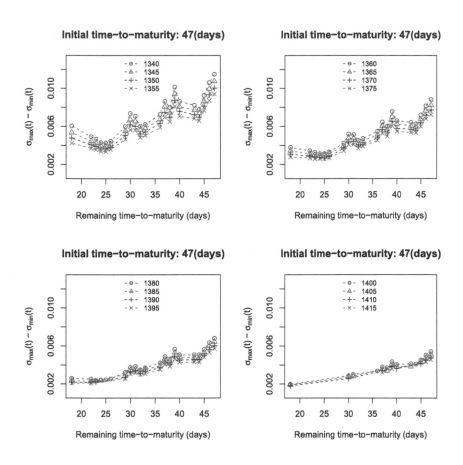

Figure 7.4: The dynamic evolution of the implied volatility uncertainty ranges for S&P 500 index European call options that mature on June 18th 2011 across 21 trading days from May 1st 2011 to May 31st 2011. These options have an initial time-to-maturity of 47 days. In each panel, the vertical axis depicts $\sigma_{max}(t) - \sigma_{min}(t)$ where $\sigma_{max}(t)$ is the value of implied volatility corresponding to the risk-free rate of $\rho_{min} = 0.001$ while $\sigma_{min}(t)$ is the value of implied volatility corresponding to the risk-free rate of $\rho_{max} = 0.02$. The horizontal axis depicts the *remaining* time-to-maturity corresponding to each trading day in May 2011. The legend in each panel enumerates the strike price for each of the options displayed in the corresponding panel.

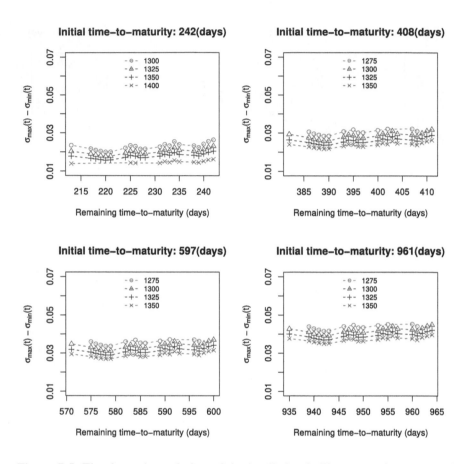

Figure 7.5: The dynamic evolution of the implied volatility uncertainty ranges for S&P 500 index European call options that mature on December 30th 2011, June 16th 2012, December 22nd 2012, and December 21st 2013, respectively, is displayed, in four different panels, across 21 trading days from May 1st 2011 to May 31st 2011. These options have an initial time-to-maturity of 242, 408, 597, and 961 days, respectively. In each panel, the vertical axis depicts $\sigma_{max}(t) - \sigma_{min}(t)$ where $\sigma_{max}(t)$ is the value of implied volatility corresponding to the risk-free rate of $\rho_{min} = 0.001$ while $\sigma_{min}(t)$ is the value of implied volatility corresponding to the risk-free rate of $\rho_{max} = 0.02$. The horizontal axis depicts the *remaining* time-to-maturity corresponding to each trading day in May 2011.

7.2 A brief review of evolutionary optimization

This section contains a brief, self-contained review of evolutionary optimization techniques, in particular a class of multi-point direct search gradient-free optimization procedures known as differential evolution that is central to the numerical implementation of the methodologies described in the current and the next chapter.

Let $\theta = \{\theta_J\}_{J=0}^{D-1}$ be a set of unknown parameters to be calculated by minimization of some user-defined objective function $\Psi(\theta)$, and let $b_{J,L} \leq \theta_J \leq b_{J,U}$ be the constraints of θ_J. The calculation of θ can be casted in the form of a constrained optimzation problem

$$
\begin{aligned}
&\underset{\theta}{\text{minimize}} \quad \Psi(\theta) \\
&\text{subject to} \quad b_{J,L} \leq \theta_J \leq b_{J,U}, \ J = 0, \ldots, D-1.
\end{aligned}
\tag{7.5}
$$

The choice of the optimization procedure for solving (7.5) depends on the differentiability, and uni-modality of the objective function $\Psi(\theta)$.

If the objective function is uni-modal and twice-differentiable, derivative-based procedures such as steepest descent, conjugate gradient, and quasi-Newton methods are computationally efficient. If the uni-modal objective function is non-differentiable, direct search procedures such as random walk and Hooke–Jeeves methods may be used instead.

However, if the objective function is multi-modal, i.e., non-convex, multiple minima may exist. Depending on the starting point, an optimizer may fail to locate the global minimum, instead locating the local minimum, thus posing the starting point problem.

In this situation, derivative-free global optimization procedures such as evolutions strategies, differential evolution, and simulated annealing may be deployed. Both evolution strategies and differential evolution perform an iterative search for the global minimum by mimicking Darwinian evolution of a biological population, and are effective continuous objective function optimizers. Simulated annealing emulates the physical process of slowly cooling a molten substance in search of a global minimum, but its effectiveness to do so is sensitive to a user-defined annealing metaparameter.

Differential evolution is a derivative-free stochastic-based optimizer that addresses the starting point problem by sampling the objective function at multiple starting points that are randomly chosen. Nomenclature assigned to various components and steps in this class of optimization strategies are borrowed from evolutionary biology to highlight their similarities.

We represent θ, the set of D unknown parameters we seek via (7.5), as a point in $\mathbb{R}^d$. The D-dimensional starting points $\theta_{I,g}, I = 1, \ldots, N_p, g = 0$, and the corresponding points evolved from them after g iterations $\theta_{I,g}, g = 1, \ldots, G$, are referred to as individual vectors, while the entire set of N_p starting points or the set of N_p points at any given g is collectively referred to as the population at iteration g. Each iteration consists of three steps, namely mutation, crossover, and selection. In the mutation step, each individual vector $\boldsymbol{\theta}_{I,g}$ is subjected to D-dimensional perturbation to construct a corresponding mutation vector $\mathbf{v}_{I,g}$. In the crossover step, each pair of $\boldsymbol{\theta}_{I,g}$ and $\mathbf{v}_{I,g}$ are used to construct a corresponding crossover vector $\mathbf{u}_{I,g}$ where each of the D elements in $\mathbf{u}_{I,g}$ is randomly selected from between the corresponding elements in $\boldsymbol{\theta}_{I,g}$ and $\mathbf{v}_{I,g}$. In the selection step, if the objective function evaluated using the crossover vector returns a smaller value than that evaluated using the corresponding individual vector it was derived from, the crossover vector is selected to replace the individual vector in the next iteration. Various stopping criteria may be used to terminate the iteration loop, including loop termination upon reaching some fixed G iterations, and a relative change ratio where G is not fixed explicitly.

Most of the differential evolution algorithms in the extant literature can be described under a unified framework as depicted by Algorithm 7.1. They differ in terms of parameter control, i.e., the definition of the mutation parameter $F_{I,g}$ in the mutation step, and the crossover parameter $CR_{I,g}$ in the crossover step. They also differ in perturbation strategy in terms of the construction of the perturbation vector $\Delta_{J,I,g}$, and the reset parameter $\varphi_{J,I,g}$ in the mutation step.

The control parameters $F_{I,g}$ and $CR_{I,g}$ may be fixed throughout G iterations, or may be specified to evolve based on either a deterministic or a stochastic rule. Specifically, deterministic evolution of these control parameters is carried out by a deterministic rule without taking into account any feedback from the evolutionary search. On the other hand, stochastic evolution of the control parameters is carried out by incorporating the feedback from the evolutionary search and pseudo-random numbers generated based on some user-defined probability distributions in order to dynamically change the control parameters.

The perturbation vector $\Delta_{J,I,g}$ may be constructed in many different ways. One may construct $\Delta_{J,I,g}$ as the difference of one or multiple pairs of individual vectors chosen randomly from the current population excluding an individual vector indexed by I. Alternatively, one may construct $\Delta_{J,I,g}$ based on random individual vectors from the current population, and random individual vectors meeting some user-defined criteria from the past populations that are stored in an archive. Differential evolution strategies using the latter approach are known as the archive-assisted differential evolution algorithms.

The boundary constraint handled operation $\varphi_{J,I,g}$ may be defined using the resetting scheme, or the penalty scheme. Variants of the resetting scheme include resetting the corresponding $v_{J,I,g}$ in question to the mid-point of the interval $[b_{J,L}, b_{J,U}]$, or resetting it to some random point in $[b_{J,L}, b_{J,U}]$. Penalty schemes include the brick wall penalty that sets the objective function value to such a high value that guarantees it not to be selected, the caveat of which is that when the global minimum lies near the parameter bounds, convergence may be slowed down by this scheme.

For comparative purposes, we describe the original differential evolution algorithm, and the Zhang–Sanderson algorithm, an adaptive version of a differential evolution algorithm that will be used in this book to obtain a numerical solution of sets of non-linear equations.

7.2.1 The original differential evolution algorithm

The original differential evolution algorithm sets $\Delta_{J,I,g} = \theta_{I_1,g} - \theta_{I_2,g}$, where $I_1, I_2 \in \{0, \ldots, N_p - 1\}, I_1 \neq I_2 \neq I$. While this perturbation strategy encourages search in different directions in the parameter space, convergence may be slow.

Additionally, the original differential evolution algorithm sets

$$F_{I,g} = F, CR_{I,g} = CR, \forall I = 0, \ldots, N_p - 1, g = 0, \ldots, G - 1,$$

and recommends setting $F \in [0.5, 1.0], CR \in [0.8, 1.0], N_p = 10D$ as a rule of thumb. By fixing the control parameters across all iterations, they cannot be adjusted based on the feedback from the evolutionary search.

The optimization performance of differential evolution algorithms depends on the specification of the control parameters. Different function profiles of $\Psi(\theta)$ may require different values of F and CR to be specified in order to achieve rapid convergence. Keeping these parameters fixed may not achieve a faster convergence compared to putting a feedback mechanism in place to automatically evolve and update the control parameters based on the feedback from the evolutionary search.

7.2.2 The Zhang–Sanderson adaptive differential evolution algorithms

The Zhang–Sanderson algorithms are a set of adaptive differential evolution algorithms that take the feedback from the evolutionary search into account in the construction of a stochastic evolution process for the control parameters, and the construction of the perturbation vectors. While some variants in this set of algorithms make use of an archive, the other variants do not. In this book, we use the variant of a Zhang–Sanderson algorithm without archive that constructs the perturbation vector as

$$\Delta_{I,g} = F_{I,g} \left(\theta^p_{best,g} - \theta_{I,g} \right) + F_{I,g} \left(\theta_{I_1,g} - \theta_{I_2,g} \right), \tag{7.6}$$

where $\theta^p_{best,g}$ is randomly chosen from the subset of $\{\theta_{I,g}, I = 1, \ldots, N_p\}$ such that $\Psi\left(\theta^p_{best,g}\right)$ is in the lowest p-percentiles of $\{\Psi\left(\theta_{I,g}\right), I = 1, \ldots, N_p\}$. The boundary constraint handled operation $\varphi_{J,I,g}$ is implemented using a random reinitialization scheme where

$$\varphi_{J,I,g} = b_{J,L} + \xi_{J,I}(b_{J,U} - b_{J,L})$$

and where $\xi_{J,I}$ is a pseudo-random number in the uniform distribution $U[0,1]$.

The control parameter $F_{I,g}$ for the mutation step at iteration g is constructed by first generating a pseudo-random variable from Cauchy$(\mu_{F,g}, 0.1)$, the Cauchy distribution with location parameter $\mu_{F,g}$ and scale parameter 0.1, and then truncating it to 1 if $F_{I,g} \geq 1$ or regenerated if $F_{I,g} \leq 0$. The location parameter of the Cauchy distribution is initialized as $\mu_{F,0} = 0.5$, and is evolved from one iteration to the next by the rule

$$\mu_{F,g+1} = (1 - c)\mu_{F,g} + c\mu_{L,S_{F,g},g}, \quad \mu_{L,S_{F,g},g} = \frac{\sum_{F_{I,g} \in S_{F,g}} F^2_{I,g}}{\sum_{F_{I,g} \in S_{F,g}} F_{I,g}},$$

where c is a user-specified parameter, and $S_{F,g} = \{F_{I,g} : \theta_{I,g} \neq \theta_{I,g+1}\}$ is the set of control parameters $F_{I,g}$ corresponding to the individual vectors that are replaced by the crossover vector in the selection step.

The control parameter $CR_{I,g}$ for the crossover step at iteration g is constructed by first generating a pseudo-random number from $\mathcal{N}(\mu_{CR,g}, 0.01)$, the Gaussian distribution mean $\mu_{CR,g}$, and variance 0.01, and then truncating it to 1 if $CR_{I,g} > 1$ or truncating it to 0 if $CR_{I,g} < 1$. The location parameter of the Gaussian distribution is initialized as $\mu_{CR,0} = 0.5$, and is evolved from one iteration to the next by the rule

$$\mu_{CR,g+1} = (1 - c)\mu_{CR,g} + c\mu_{A,CR,g} ,$$

where c is the same user-specified parameter involved in the evolution of the control parameter $F_{I,g}$, and where $\mu_{A,CR,g}$ is the arithmetic mean of the set of all successful crossover probabilities $S_{F,g}$. For comparison with the generic differential evolution algorithm, the Zhang–Sanderson algorithm described herein is depicted by Algorithm 7.2.

Algorithm 7.1: A generic differential evolution algorithm.

Input: (i) $\Psi(\boldsymbol{\theta})$, (ii) $\{b_{J,L}, b_{J,U}\}_{J=0}^{D-1}$, (iii) N_p, (iv) $F_{I,g=0}$, (v) $CR_{I,g=0}$.

Output: $\{\boldsymbol{\theta}_{I,G-1}\}_{I=0}^{N_p}$.

begin Initialization stage

> **for** $I = 0, \ldots, N_p - 1$, **do**
>
> > **for** $J = 0, \ldots, D - 1$, **do**
> >
> > > Generate $\eta_{J,I,g}, \eta_{J,I,g} \sim U[0,1]$
> > >
> > > $\theta_{J,I,0} \leftarrow b_{J,L} + \eta_{J,I,g}(b_{J,U} - b_{J,L})$
>
> $\boldsymbol{\theta}_{I,0} \leftarrow \{\theta_{J,I,0}\}_{J=0}^{D-1}$

begin Iteration stage

> **for** $g = 0, \ldots, G - 1$, **do**
>
> > **begin** Mutation step
> >
> > > $\mathbf{v}_{I,g} = \{v_{J,I,g}\}_{J=1}^{D} \leftarrow \boldsymbol{\theta}_{I,g} + F_{I,g}\, \Delta_{J,I,g}$
> > >
> > > **for** $I = 0, \ldots, N_p - 1$, **do**
> > >
> > > > **for** $J = 0, \ldots, D - 1$, **do**
> > > >
> > > > > **if** $v_{J,I,g} < b_{J,L}$ *or* $v_{J,I,g} > b_{J,U}$ **then**
> > > > >
> > > > > > $v_{J,I,g} \leftarrow \varphi_{J,I,g}, \quad \varphi_{J,I,g} \in [b_{J,L}, b_{J,U}]$
> >
> > **begin** Crossover step
> >
> > > **for** $I = 0, \ldots, N_p - 1$, **do**
> > >
> > > > **for** $J = 0, \ldots, D - 1$, **do**
> > > >
> > > > > Generate $\zeta_{J,I,g}, \zeta_{J,I,g} \sim U[0,1]$
> > > > >
> > > > > **if** $\xi_{J,I,g} \le CR_{I,g}$ **then**
> > > > >
> > > > > > $u_{J,I,g} \leftarrow v_{J,I,g}$
> > > > >
> > > > > **else**
> > > > >
> > > > > > $u_{J,I,g} \leftarrow \theta_{J,I,g}$
> > >
> > > $\mathbf{u}_{I,g} \leftarrow \{u_{J,I,g}\}_{J=1}^{D}$
> >
> > **begin** Selection step
> >
> > > **for** $I = 0, \ldots, N_p - 1$, **do**
> > >
> > > > **if** $\Psi(\mathbf{u}_{I,g}) < \Psi(\boldsymbol{\theta}_{I,g})$ **then**
> > > >
> > > > > $\boldsymbol{\theta}_{I,g+1} \leftarrow \mathbf{u}_{I,g}$

Algorithm 7.2: Zhang–Sanderson algorithm without archive.

Input: (i) $\Psi(\boldsymbol{\theta})$, (ii) $\{b_{J,L}, b_{J,U}\}_{J=0}^{D-1}$, (iii) N_p, (iv) c, (v) p.

Output: $\{\boldsymbol{\theta}_{I,G-1}\}_{I=0}^{N_p}$.

begin Initialization stage

    for $I = 0, \ldots, N_p - 1, J = 0, \ldots, D - 1,$ **do**

        Generate $\eta_{J,I}, \eta_{J,I} \sim U[0,1]$

        $\theta_{J,I,0} \leftarrow b_{J,L} + \eta_{J,I}(b_{J,U} - b_{J,L})$

    $\boldsymbol{\theta}_{I,0} \leftarrow \{\theta_{J,I,0}\}_{J=0}^{D-1}$, $\mu_{F,0} = \mu_{CR,0} \leftarrow 0.5$

begin Iteration stage

    for $g = 0, \ldots, G - 1,$ **do**

        begin Mutation step

            for $I = 0, \ldots, N_p - 1,$ **do**

                Generate $F_{I,g}, F_{I,g} \sim \text{Cauchy}(\mu_{F,g}, 0.1)$

                $\Delta_{J,I,g} \leftarrow \left(\boldsymbol{\theta}_{best,g}^p - \boldsymbol{\theta}_{I,g}\right) + \left(\boldsymbol{\theta}_{I_1,g} - \boldsymbol{\theta}_{I_2,g}\right)$

                $\mathbf{v}_{I,g} = \{v_{J,I,g}\}_{J=1}^{D} \leftarrow \boldsymbol{\theta}_{I,g} + F_{I,g}\,\Delta_{J,I,g}$

                for $J = 0, \ldots, D - 1,$ **do**

                    if $v_{J,I,g} < b_{J,L}$ *or* $v_{J,I,g} > b_{J,U}$ **then**

                        Generate $\xi_{J,I,g}, \xi_{J,I,g} \sim U[0,1]$

                        $v_{J,I,g} \leftarrow b_{J,L} + \xi_{J,I,g}(b_{J,U} - b_{J,L})$

        begin Crossover step

            for $I = 0, \ldots, N_p - 1,$ **do**

                Generate $CR_{I,g}, CR_{I,g} \sim \mathcal{N}\left(\mu_{CR,g}, 0.01\right)$

                for $J = 0, \ldots, D - 1,$ **do**

                    Generate $\zeta_{J,I,g}, \zeta_{J,I,g} \sim U[0,1]$

                    if $\xi_{J,I,g} \leq CR_{I,g}$ **then**

                      $u_{J,I,g} \leftarrow v_{J,I,g}$

                  else

                      $u_{J,I,g} \leftarrow \theta_{J,I,g}$

            $\mathbf{u}_{I,g} \leftarrow \{u_{J,I,g}\}_{J=1}^{D}$

        begin Selection step

            for $I = 0, \ldots, N_p - 1,$ **do**

                if $\Psi(\mathbf{u}_{I,g}) < \Psi(\boldsymbol{\theta}_{I,g})$ **then**

                    $\boldsymbol{\theta}_{I,g+1} \leftarrow \mathbf{u}_{I,g}$

        $\mu_{F,g+1} \leftarrow (1 - c)\,\mu_{F,g} + c\mu_{L,S_{F,g},g}$

        $\mu_{CR,g+1} \leftarrow (1 - c)\mu_{CR,g} + c\mu_{A,CR,g}$

7.3 Inference of implied parameters from over-defined systems

7.3.1 An over-defined system of nonlinear equations

Let $C(t_i, T_j, K_{j,\ell}), i = 1, \ldots, p, j = 1, \ldots, m, \ell = 1, \ldots, n_j$, be the prices of a set of European vanilla call option contracts that expire at times T_j and are struck at $K_{j,\ell}$, respectively, where $p, m,$ and n_j, are some given positive integers. These option prices are observed at times t_i where $t_{i-1} < t_i$. We regard both the implied discount rates and the implied volatilities as unknown parameters that have to be estimated from the set of option prices $\{C(t_i, T_j, K_{j,\ell})\}_{i=1,\ldots,p,j=1,\ldots,m,\ell=1,\ldots,n_j}$. In order to proceed, we make several assumptions.

We assume that the prices for this set of European call option contracts with different $(T_j - t_i, K_{j,\ell})$ pairs are associated with different implied volatilities $\sigma_{imp,j,\ell}(t_i)$, and that different times to maturity $\tau_j = T_j - t_i$ are associated with different implied discount rates $\rho_j(t_i)$. We also assume that, within some reasonable short time intervals $[t_1, t_p]$, there is some stability for the implied volatility and the implied discount rate within these time intervals. Specifically, we assume that

$$\sigma_{imp,j,\ell,[t_1,t_p]} = \sigma_{imp,j,\ell}(t_1) = \cdots = \sigma_{imp,j,\ell}(t_p) ,$$

and

$$\rho_{j,[t_1,t_p]} = \rho_j(t_1) = \cdots = \rho_j(t_p) .$$

In practice, an implicit assumption on the stability in the implied volatility of an option contract over a short time interval is commonly made in the execution of risk management activities in the financial industry. Dynamic hedging of positions in option contracts cannot be carried out continuously due to practical constraints such as transaction costs, and has to be carried out at discrete time intervals. The discrete rebalancing of dynamic hedge positions for option contracts implicitly assumes that, for practical purposes, the implied volatility remains the same within the time period between two successive rebalancing activities. Additionally, the reference interest rates used as proxy for the risk-free rates are usually quoted on a daily basis, except on weekends and business holidays. This implies that the risk-free rate, or discount rate, term structure may be assumed to remain the same between two successive announcements of this set of reference rates.

Based on these assumptions, we construct an over-defined system of nonlinear equations by mapping a set of observed option prices $C(t_i, T_j, K_{j,\ell})$ to $\sigma_{imp,j,\ell,[t_1,t_p]}$ and $\rho_{j,[t_1,t_p]}$ via the Black–Scholes option pricing formula (7.3)

$$C_{BSLin}(t_i, T_j, S(t_i), \sigma_{imp,j,\ell,[t_1,t_p]}, \rho_{j,[t_1,t_p]}, K_{j,\ell}) = C(t_i, T_j, K_{j,\ell}) , \qquad (7.7)$$

where $i = 1, \ldots, p, \quad j = 1, \ldots, m, \quad \ell = 1, \ldots, n_j$. This over-defined system of equations has $m + \sum_{j=1}^m n_j$ unknown variables to be solved based on $p \times \sum_{j=1}^m n_j$ equations. While it is possible to construct this system of nonlinear equations using a time series of prices pertaining to a set of option contracts observed at uniform or non-uniform intervals for some $p > 1$ across some short time span, we construct the over-defined systems of nonlinear equations, for simplicity, using day-close prices of option contracts on two consecutive trading days where we set $p = 2$.

We calculate $\sigma_{imp,j,\ell,[t_1,t_p]}$ and $\rho_{j,[t_1,t_p]}$ by seeking approximate numerical solution of (7.7). We cast the numerical solution problem in the form of an optimization problem as a special case of (7.5)

$$\begin{array}{c} \underset{\boldsymbol{\theta}}{\text{minimize}} \quad \Psi(\boldsymbol{\theta}) \\[2mm] \text{subject to} \quad 0 \leq \theta_J < \infty, \ J = 0, \ldots, D-1. \end{array} \tag{7.8}$$

where the objective function $\Psi(\boldsymbol{\theta})$ may be defined as an L_1 loss metric

$$\Psi_{L_1}(\boldsymbol{\theta}) = \sum_{i=1}^{p}\sum_{j=1}^{m}\sum_{\ell=1}^{n_j} \Delta_{i,j,\ell} \,, \tag{7.9}$$

where

$$\Delta_{i,j,\ell} = \left| C_{BSLin}(t_i, T_j, S(t_i), \sigma_{imp,j,\ell,[t_1,t_p]}, \rho_{j,[t_1,t_p]}, K_{j,\ell}) - C(t_i, T_j, K_{j,\ell}) \right| \,,$$

an L_2 loss metric

$$\Psi_{L_2}(\boldsymbol{\theta}) = \sum_{i=1}^{p}\sum_{j=1}^{m}\sum_{\ell=1}^{n_j} \Delta_{i,j,\ell}^2 \,, \tag{7.10}$$

or some other loss metric, where the D-dimensional unknown parameters are

$$\boldsymbol{\theta} = \{\theta_J\}_{J=0}^{D-1} = (\boldsymbol{\sigma}, \boldsymbol{\rho}) \,,$$

where

$$\boldsymbol{\sigma} = \left(\sigma_{imp,1,1,[t_1,t_p]}, \ldots, \sigma_{imp,m,n_m,[t_1,t_p]} \right) \,,$$
$$\boldsymbol{\rho} = \left(\rho_{1,[t_1,t_p]}, \ldots, \rho_{m,[t_1,t_p]} \right) \,,$$

with $D = m + \sum_{j=1}^{m} n_j$ and where the parameter boundaries are

$$\sigma_{imp,j,\ell,[t_1,t_p]} > 0 \,, \quad \rho_{j,[t_1,t_p]} > 0, \quad j = 1, \ldots, m, \ell = 1, \ldots, n_j$$

via the Zhang–Sanderson non-archive assisted differential evolution algorithm as detailed in Algorithm 7.2. Let $\boldsymbol{\theta}_{best,G-1} = \left(\boldsymbol{\sigma}_{best,G-1}, \boldsymbol{\rho}_{best,G-1} \right)$ be the individual vector in the population $\{\boldsymbol{\theta}_{I,G-1}, I = 1, \ldots, N_p\}$ at algorithm termination after G iterations.

7.3.2 Computational implementation

We perform the numerical experiment reported in the rest of this chapter using the implementation of the Zhang–Sanderson algorithm without archive described in Algorithm 7.2 implemented in the `DEoptim` package for the R computing environment

[12]. We used 48 parallel cores provided by the National eResearch Collaboration Tools and Resources (NeCTAR) cloud computing infrastructure to carry out all the calculations reported in this paper to achieve considerable reduction in computation time. For example, completion of the two numerical experiments described in Section 7.3.4 took 5.68 and 5.74 hours, respectively, on a single core, but only took 1.23 and 1.24 hours, respectively, instead on 48 parallel cores, leading to an approximately 78% reduction in computation time in this particular example. Due to the substantial inter-core communication at the end of each iteration, the extent of computational acceleration does not approach that of the classic Amdahl's law to be expected from that of an embarrassingly parallel algorithm such as a parallel Monte Carlo simulation.

7.3.3 Construction of the estimation uncertainty bounds for the estimated implied discount rates and implied volatilities

Since the estimated parameters $\sigma_{best,G-1}$ and $\rho_{best,G-1}$ are approximate solutions to (7.7), some measures of in-sample goodness-of-fit assessment can give us an idea of the solution accuracy of the system of nonlinear equations. The N_p D-dimensional member vectors $S_{\theta,G-1} = \{\theta_{I,G-1}, I = 0, \ldots, N_p - 1\}$ are the result of iterative stochastic perturbation of the random initialization vectors $\{\theta_{I,0}, I = 0, \ldots, N_p-1\}$ via Zhang–Sanderson's algorithm. In this sense, $S_{\theta,G-1}$ may be regarded as a set of D-dimensional random variables.

If the number of iterative loops G is sufficiently large for the set of member vectors to converge to, hopefully, the global minimum, then one would expect the corresponding collection of objective function values $S_{\Psi(\theta),G-1} = \{\Psi(\theta_{I,G-1}), I = 0, \ldots, N_p - 1\}$ to be small in absolute terms. Various standard measures of dispersion may be used as an empirical gauge to quantify the dispersion of $S_{\Psi(\theta),G-1}$. For example, the narrower the interval $\left[\min S_{\Psi(\theta),G-1}, \max S_{\Psi(\theta),G-1}\right]$, the more likely that the member vectors are converging towards a minimum.

By regarding $S_{\rho,G-1} = \{\rho_{I,G:j,[t_1,t_p]}, I = 1, \ldots, N_p\}$ as the set of random possible solutions of $\rho_{j,[t_1,t_p]}, j = 1, \ldots, m$, and $S_{\sigma,G-1} = \{\sigma_{I,G:imp,j,\ell,[t_1,t_p]}, I = 1, \ldots, N_p\}$ as the set of random possible solutions of $\sigma_{imp,j,\ell,[t_1,t_p]}$, for each (j, ℓ) pair where $j = 1, \ldots, m, \ell = 1, \ldots, n_j$, we can use $[\min S_{\rho,G-1}, \max S_{\rho,G-1}]$ as a gauge to quantify the parameter estimation uncertainty interval of $\rho_{best,G:j,[t_1,t_p]}$ as an estimator of $\rho_{j,[t_1,t_p]}$ for each j where $j = 1, \ldots, m$, and we can also use $[\min S_{\sigma,G-1}, \max S_{\sigma,G-1}]$ as a gauge to quantify the parameter estimation uncertainty interval of $\sigma_{best,G:imp,1,1,[t_1,t_p]}$ as an estimator of $\sigma_{imp,j,\ell,[t_1,t_p]}$ for each (j, ℓ) pair where $j = 1, \ldots, m, \ell = 1, \ldots, n_j$.

Our construction of the aforementioned parameter calculation uncertainty intervals is inspired by [19] which suggested using the probabilistic approach to the model calibration problem within the stochastic-based differential evolution optimization framework similar to the Zhang–Sanderson algorithm we use. The rationale is that the population converges to the set of global minima of pricing errors after a sufficiently large number of iterations. That being the case, the individual vectors may still be associated with different magnitudes of pricing error, leading to the existence of a set of pricing model parameters that may be compatible with the contingent claim prices in question up to some in-sample goodness-of-fit error. Our empirical parameter estimation uncertainty interval builds on this notion of model parameter calculation uncertainty. This strategy relies on neither the large sample results nor the assumption of independently and identically distributed errors across options.

7.3.4   Numerical experiment with synthetic test data

In this section, we construct two sets of synthetic test data sets to assess the performance of our proposed parameter estimation strategy in terms of convergence profile, the parameter estimation performance of the proposed strategy, and the effect of the choice of objective function on the parameter estimation performance.

Construction of synthetic test data sets

Two sets of synthetic test data sets are constructed. Each of these data sets simulate the prices of a set of European vanilla call option contracts quoted on two successive days. The hypothetical prices in both test data sets are synthesized using the same set of hypothetical times to maturity, spot, and strike prices, and implied volatilities as input values to (7.3), setting the dividend rate to be zero. They differ only in the input of the hypothetical discount rates to (7.3).

For the set of synthetic test data that simulate an up-sloping discount rate term structure, the hypothetical discount rates corresponding to the unique times to maturity at increasing tenors are 0.002, 0.004, 0.006, 0.012, 0.015, and 0.019, respectively. For the set of test data that simulate the inverted discount rate term structure, the hypothetical discount rates corresponding to the unique times to maturity at increasing tenors are 0.002, 0.004, 0.019, 0.015, 0.012, and 0.006, respectively.

We set the hypothetical times to maturity corresponding to the hypothetical prices observed at the first instance to be 0.17, 0.42, 0.69, 0.94, 1, and 1.5 years and set those observed at the second instance to be $0.17 - \delta_\tau$, $0.42 - \delta_\tau$, $0.69 - \delta_\tau$, $0.94 - \delta_\tau$, $1 - \delta_\tau$, and $1.5 - \delta_\tau$, where $\delta_\tau = 1/365$. For simplicity, we set the spot underlying asset price to be \$590 on both instances, and set the strike-to-spot ratio to be 0.85, 0.9, 0.95, 1, 1.05, 1.1, 1.15, 1.2, 1.3, and 1.4 on both instances as well. The hypothetical implied volatilities are adapted from the implied volatilities of the European call options on the S&P500 equity index quoted in October 1995 and reported in Table 1 of [5]. They are chosen to provide a volatility surface with smile that constitutes a system of nonlinear equations that is numerically challenging to solve. Figure 7.6 depicts a graphical profile of this set of implied volatilities.

Each data set consists of 120 synthetic data points; 60 of them are regarded as the hypothetical prices of 60 option contracts observed at one instance in time while the remaining 60 are regarded as the hypothetical prices of the same set of hypothetical option contracts quoted one day later. For each synthetic test data set, we construct an over-defined system of 120 nonlinear equations based on (7.7). In each of these systems, there are 66 unknown parameters, i.e., 60 implied volatilities and 6 implied

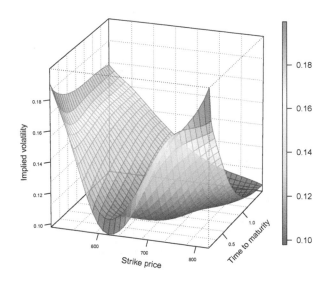

Figure 7.6: A graphical profile of the set of test-implied volatilities, adapted from Table 1 of [5], that is used to construct the test data set.

discount rates, to be solved for based on the corresponding hypothetical option prices where $p = 2, m = 6$, and $n_j = 10$.

The convergence profile and the parameter calculation accuracy profile pertaining to the synthetic test data sets

We compare the convergence profile and parameter calculation accuracy for the numerical approximate solution to the two aforementioned over-defined systems of nonlinear equations formulated using the synthetic test data sets by minimizing either the L_1 loss function $\Phi_{L_1}(\boldsymbol{\theta})$ or the L_2 loss function $\Phi_{L_2}(\boldsymbol{\theta})$ using Algorithm 7.2. We set $c = 0.2, p = 0.01, N_p = 50 \times D$, and $G = 5,000$.

We use $\Psi(\boldsymbol{\theta}_{best,g})/(p \times \sum_{j=1}^{m} n_j), g = 0, \ldots, G - 1$, as a gauge of the best in-sample goodness-of-fit at each iteration step of Algorithm 7.2. For $\Phi_{L_2}(\boldsymbol{\theta})$, this gauge may be regarded as the average residual sum of squares, while for $\Phi_{L_1}(\boldsymbol{\theta})$, it is the mean absolute deviation. Figure 7.7 compares the convergence profiles of using $\Psi_{L_1}(\boldsymbol{\theta})$ versus $\Psi_{L_2}(\boldsymbol{\theta})$ as the objective function in Algorithm 7.2 in terms of $\Psi(\boldsymbol{\theta}_{best,g})/(p \times \sum_{j=1}^{m} n_j)$ against $g = 0, \ldots, G - 1$. Since the convergence profiles for both test data sets are similar, only that for the up-sloping term structure is

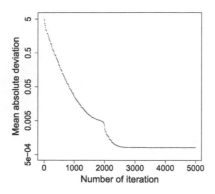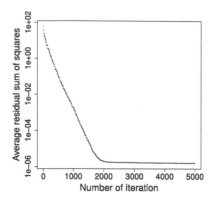

Figure 7.7: **Left**: Convergence profile for the average residual sum of squares $\Psi_{L_1}(\boldsymbol{\theta}_{best,g})/(p \times \sum_{j=1}^{m} n_j)$. **Right**: Convergence profile for the mean absolute deviation $\Psi_{L_2}(\boldsymbol{\theta}_{best,g})/(p \times \sum_{j=1}^{m} n_j)$. The number of iterations indexes g, $g = 0, \ldots, G-1$, $G = 5000$. For this synthetic test data set, $p \times \sum_{j=1}^{m} n_j = 120$.

depicted in Figure 7.7. It appears that the convergence rate tends to be faster for L_2 minimization and the eventual residual error is smaller for L_2 minimization too.

Table 7.6 depicts the absolute error between the interest rates used to simulate the European vanilla call option prices of the synthetic test data sets and the corresponding implied discount inferred from these prices. It also depicts the ratio of the absolute error for the implied discount rates calculated by minimization of $\Psi_{L_2}(\boldsymbol{\theta})$ to the absolute error for the implied discount rates calculated by minimization of $\Psi_{L_1}(\boldsymbol{\theta})$. It is interesting to observe that while using L_1 and L_2 metrics produce somewhat different implied discount rates in the test data set with up-sloping interest rate term structure, they arrive at implied discount rates that are similar up to three decimal places for the inverted interest rate term structure.

Table 7.7 depicts the absolute error between the implied volatilities used to simulate the European vanilla call option prices of the synthetic test data sets and the corresponding implied volatilities inferred from these prices. It also depicts the ratio of the absolute error for the implied volatilities calculated by minimization of $\Psi_{L_2}(\boldsymbol{\theta})$ to the absolute error for the implied volatilities calculated by minimization of $\Psi_{L_1}(\boldsymbol{\theta})$. For these sets of synthetic data sets, the absolute error in calculating the implied volatilities by minimization of $\Psi_{L_1}(\boldsymbol{\theta})$ and $\Psi_{L_2}(\boldsymbol{\theta})$ metric objective functions are comparable, except for far OTM options near expiry where $\Psi_{L_1}(\boldsymbol{\theta})$ appears to perform better. Additionally, the uncertainty bounds for the calculated implied discount rates and the implied volatilities for both test data sets are narrow, and less than one basis points on either side of $\rho_{best,G-1}$ less than 10 basis points on either side of $\sigma_{best,G-1}$.

Time to maturity at first spot time (years)	Up-sloping term structure	Inverted term structure
Absolute error for $\Psi_{L_1}(\boldsymbol{\theta})$ ($\times 10^{-3}$)		
0.17	3.347	2.990
0.42	8.809	8.902
0.69	10.615	4.650
0.94	7.603	7.182
1	4.765	7.580
1.5	4.139	8.761
Ratio of absolute error for $\Psi_{L_2}(\boldsymbol{\theta})$ to absolute error for $\Psi_{L_1}(\boldsymbol{\theta})$		
0.17	0.879	1.000
0.42	1.027	1.000
0.69	0.881	1.000
0.94	0.776	1.000
1	1.298	1.000
1.5	0.590	1.000

Table 7.6: **Upper panel**: Absolute error profiles for the implied discount rates calculated using $\Psi_{L_1}(\boldsymbol{\theta})$ from the two synthetic test data sets; one simulating up-sloping discount rate term structure, the other simulating an inverted discount rate term structure. **Lower panel**: Ratio of absolute error for $\Psi_{L_2}(\boldsymbol{\theta})$ to absolute error for $\Psi_{L_2}(\boldsymbol{\theta})$.

7.3.5 Numerical analysis using historical S&P500 call options data

For the empirical analysis reported in this section, we use the historical day-close prices of S&P500 index European vanilla call option contracts traded at the CBOE.[1] We consider OTM options with moneyness between 1 and 1.2 because this set of OTM option contracts are not "too far" out of the money as they are suitable for implied volatility and risk-free rate analysis because far OTM options are low in liquidity and are associated with higher numerical uncertainty on implied volatility, while ITM option contracts are more expensive and often less liquid than the OTM option contracts.

We use the option contracts with remaining times to maturity between 150 and 320 days in our empirical analysis. Near expiry option contracts are excluded on grounds that their prices as the information content of these options in terms of volatility is questionable, while long dated options are potentially less liquid and thus may contain less contemporaneous information on the market participants' aggregate choice of discount rate, and aggregate expectation of the underlying asset price volatility for the remaining lifespan of the option contracts.

The estimated implied discount rates inferred from the historical option prices are compared with the contemporaneously quoted USD Libor[2] and OIS[3] rates. The USD Libor rates are quoted uncollateralized interbank reference rates designed to

---

[1] Source: Market Data Express LLC

[2] Source: Federal Reserve Bank of St. Louis

[3] Source: Thomson Reuters Tick History (TRTH) supplied by the Securities Industry Research Centre of Asia-Pacific (SIRCA)

Time to maturity at first instance (years)	Moneyness									
	0.85	0.9	0.95	1	1.05	1.1	1.15	1.2	1.3	1.4
	Synthetic test data mimicking up-sloping discount rate term structure									
	Absolute error for $\Psi_{L_1}(\theta)$ $(\times 10^{-3})$									
0.17	44.202	9.581	3.944	1.527	0.733	0.444	0.439	0.371	0.133	54.220
0.42	55.243	23.881	12.567	7.253	4.491	3.067	2.272	2.031	1.731	1.599
0.69	18.802	11.763	7.651	5.168	3.606	2.546	1.937	1.588	1.218	1.121
0.94	27.962	18.391	12.739	9.248	6.870	5.143	3.997	3.259	2.362	2.079
1	28.599	19.196	13.452	9.905	7.444	5.650	4.410	3.593	2.615	2.308
1.5	31.282	22.783	17.348	13.576	10.819	8.734	7.223	6.043	4.538	3.589
	Ratio of absolute error for $\Psi_{L_2}(\theta)$ to absolute error for $\Psi_{L_1}(\theta)$									
0.17	0.958	0.982	0.983	0.989	0.997	0.995	0.834	0.631	18.789	1.894
0.42	1.038	1.020	1.017	1.017	1.016	1.015	1.016	1.032	1.068	1.166
0.69	2.139	1.964	1.917	1.907	1.906	1.912	1.923	1.946	2.000	1.930
0.94	0.770	0.789	0.798	0.803	0.806	0.808	0.810	0.812	0.823	0.774
1	0.808	0.820	0.825	0.827	0.827	0.827	0.826	0.825	0.817	0.823
1.5	0.278	0.289	0.295	0.297	0.298	0.298	0.298	0.297	0.295	0.290
	Synthetic test data mimicking inverted discount rate term structure									
	Absolute error for $\Psi_{L_1}(\theta)$ $(\times 10^{-3})$									
0.17	63.516	10.764	4.428	1.714	0.825	0.505	0.436	0.401	0.339	0.325
0.42	53.647	23.564	12.433	7.186	4.449	3.038	2.264	2.029	1.666	1.512
0.69	48.938	26.790	16.797	11.235	7.822	5.536	4.236	3.502	2.743	2.485
0.94	29.114	19.126	13.256	9.630	7.166	5.372	4.183	3.414	2.501	2.181
1	17.171	11.908	8.464	6.271	4.728	3.593	2.805	2.284	1.662	1.437
1.5	15.250	11.410	8.783	6.902	5.503	4.436	3.651	3.042	2.266	1.781
	Ratio of absolute error for $\Psi_{L_2}(\theta)$ to absolute error for $\Psi_{L_1}(\theta)$									
0.17	0.696	0.890	0.891	0.891	0.888	0.879	1.007	0.925	0.392	166.831
0.42	1.030	1.013	1.011	1.009	1.009	1.010	1.004	1.001	1.039	1.058
0.69	0.384	0.439	0.455	0.460	0.461	0.460	0.457	0.453	0.444	0.451
0.94	0.960	0.962	0.961	0.960	0.959	0.957	0.956	0.955	0.944	0.953
1	1.666	1.612	1.589	1.579	1.574	1.573	1.572	1.573	1.573	1.606
1.5	2.051	1.997	1.975	1.967	1.966	1.969	1.978	1.987	2.003	2.015

Table 7.7: **Upper panel**: Absolute error profiles for the implied volatilities calculated using $\Psi_{L_1}(\theta)$ from the two synthetic test data sets; one simulating up-sloping discount rate term structure, the other simulating an inverted discount rate term structure. **Lower panel**: Ratio of absolute error for $\Psi_{L_2}(\theta)$ to absolute error for $\Psi_{L_2}(\theta)$.

reflect the average credit risk among different banks, while the USD OIS rates are OTC-quoted interest rate swap rates, the floating legs of which are based on discrete daily compounded effective Federal Funds rate and are virtually free of credit risk.

We select two sets of option contracts, the prices of which are quoted on May 18th and 19th 2010, and May 18th and 19th 2011, respectively. We construct, for each of the two samples, a system of over-defined nonlinear equations using (7.7), and apply Algorithm 7.2 to calculate the implied discount rates and implied volatilities by minimization of $\Psi_{L_2}(\theta)$. Here we set $c = 0.15$, $p = 0.01$, and $N_p = 25 \times D$ for the empirical data analysis reported in Section 7.3.5.

We depict in Table 7.8 the estimated implied discount rates and the contemporaneously quoted Libor and OIS rates. Due to the fact that the tenors corresponding to the implied discount rates differ from sample to sample, and are in general dif-

	I-(a). **May 18th 2010**				I-(b). **May 19th 2010**			
Time to maturity (year)	0.5863	0.6219	0.8356	0.8685	0.5836	0.6192	0.8329	0.8658
Implied discount rate	0.02525	0.04829	0.02793	0.04628	0.02525	0.04829	0.02793	0.04628
Libor	0.00730	0.00764	0.00965	0.00995	0.00746	0.00779	0.00979	0.01009
OIS rate	0.00276	0.00282	0.00323	0.00329	0.00258	0.00263	0.00301	0.00307
	II-(a). **May 18th 2011**				II-(b). **May 19th 2011**			
Times to maturity (year)	0.5836	0.6192	0.8329	0.8685	0.5808	0.6164	0.8301	0.8658
Implied discount rate	0.01244	0.02441	0.02286	0.03348	0.01244	0.02441	0.02286	0.03348
Libor	0.00463	0.00486	0.00623	0.00646	0.00459	0.00482	0.00622	0.00646
OIS rate	0.00133	0.00137	0.00165	0.00170	0.00143	0.00147	0.00178	0.00183

Table 7.8: The implied discount rates depicted are estimated based on four samples of historical S&P 500 call option prices indexed by I and II. The two consecutive dates in each sample are indexed by (a) and (b) in the same row. Tenor matching interpolated values of contemporaneously quoted Libor and OIS rates are depicted for comparison.

ferent from the fixed tenors of the contemporaneously quoted Libor and OIS rates, we apply the interest rate interpolation technique previously used in Section 7.1 in order to obtain the interpolated Libor and OIS rates corresponding to the tenors of the implied discount rates for comparison depicted in Table 7.8.

Worthy of note is that the differences between the implied discount rates and the tenor matching interpolated values of the contemporaneously quoted OIS rates depicted in Table 7.8 appear to be comparable in magnitude to the credit default swap (CDS) spreads of US firms with an average Moody's and S&P ratings between BBB and BB, as depicted in Table II of [87] and Table 1 of [120] reflecting, to some extent, the premium paid by agents to fund their investment.

I. Option chain strike price: 1150	I-(a). **May 18th 2010**				I-(b). **May 19th 2010**			
Times to maturity (year)	0.5863	0.6219	0.8356	0.8685	0.5836	0.6192	0.8329	0.8658
(i) Proxy: Implied discount rate	0.22473	0.20314	0.21405	0.19427	0.22473	0.20314	0.21405	0.19427
(ii) Proxy: Libor	0.23705	0.23675	0.22968	0.23005	0.24170	0.24105	0.23511	0.23526
(iii) Proxy: OIS rate	0.24076	0.24082	0.23603	0.23670	0.24562	0.24528	0.24165	0.24219
II. Option chain strike price: 1400	II-(a). **May 18th 2011**				II-(b). **May 19th 2011**			
Times to maturity (year)	0.5836	0.6192	0.8329	0.8685	0.5808	0.6164	0.8301	0.8658
(i) Proxy: Implied discount rate	0.14229	0.13283	0.13493	0.12592	0.14229	0.13283	0.13493	0.12592
(ii) Proxy: Libor	0.14904	0.14830	0.15050	0.15236	0.14680	0.14708	0.14987	0.15152
(iii) Proxy: OIS rate	0.15135	0.15085	0.15452	0.15669	0.14906	0.14955	0.15384	0.15575

Table 7.9: The implied volatilities calculated jointly with the implied discount rates, and the implied volatilities calculated by assuming the discount to be either the contemporaneously quoted Libor or OIS rates based on four separate samples of historical S&P 500 call options data indexed by I and II. The two consecutive dates in each sample are indexed by (a) and (b) in the corresponding rows.

Table 7.9 illustrates the sensitivity of implied volatility estimation with respect to the choice of discount rate from among the implied discount rate, and the contemporaneously quoted Libor and OIS rates pertaining to the samples of option prices considered. We calculate the implied volatilities from the same sets of option prices by inverting (7.3), using the estimated implied discount rates, or the tenor matching interpolated values with respect to the contemporaneously quoted Libor or OIS rates as the discount rate proxy.

The implied volatilities calculated jointly with the implied discount rates are smaller than the implied volatilities calculated by inverting (7.3) and using either the tenor matching interpolated values of the contemporaneously quoted Libor or OIS rates as risk-free rate proxy. Since higher discount rates map a given European vanilla call option price to lower implied volatilities, the comparatively higher implied discount rates relative to the tenor matching interpolated contemporaneous Libor and OIS rates in these samples map the option prices to lower implied volatilities. This is a sample dependent phenomenon. If the implied discount rates are lower than the tenor matching interpolated contemporaneous Libor and OIS rates instead, then the implied volatilities estimated jointly with the implied discount rates will be higher than those estimated using the contemporaneous Libor and OIS rates as discount rates.

In an ideal world, one would like to verify whether the implied discount rate is an accurate estimator of the true market participants' aggregate choice of the discount rate, or, equivalently, whether the implied volatility estimated jointly with the implied discount rate is an accurate estimator of the market participants' aggregate expectation of the underlying asset price volatility for the remaining lifespan of the option contract. However, both the market participants' aggregate choice of discount rate and aggregate expectation of future realized volatility are latent, unobservable state variables, and it is unclear how one should construct an econometric test to verify the accuracy of these estimates. That said, from the numerical experiment results reported in Section 7.3.4, it appears that our proposed estimation strategy is capable of estimating the implied discount rates and implied volatilities used to generate the synthetic option prices with reasonable accuracy. While these experimental results raise the prospect that the application of our estimation strategy to historical data may be capable of estimating the market participants' aggregate choice of discount rates and aggregate expectation of future realized volatilities, these parameter estimates need to be examined carefully in terms of the proposed in-sample goodness-of-fit measures, and in terms of the econometric plausibility of the magnitude of these estimates.

7.4   Bibliographic notes and literature review

Implied volatility – a representation of market sentiment inferred from option prices

In 1973, the Chicago Board Options Exchange (CBOE) was founded and became the first platform for trading listed options. In the same year, [24] proposed a framework for the valuation of European vanilla option contracts. Assuming that the underlying asset price process follows a log-normal stochastic process, they proposed the celebrated Black–Scholes option pricing formula that maps six parameters, namely the spot underlying asset price, the contract strike price, the time to maturity of the option contract, the dividend rate, the short term interest rate, and the implied volatility, i.e., the volatility of the underlying asset for the remaining lifespan of the option contract, to a theoretical net present value of the option contract. The first five parameters are regarded as known quantities; the last is not.

In the spirit that the option price contains information of the market participants' aggregate expectation of the underlying asset volatility for the remaining lifespan of the option contract, financial economist and financial industry practitioners alike have tried to infer the implied volatility from the option prices. The notion of implied volatility as a value that matches the Black–Scholes model price to the market price as closely as possible was first introduced by [83]; an errata to this paper was subsequently published by [82].

From the econometric perspective, one may utilize the implied volatility inferred from the option price to provide forecast of the future realized volatility of the underlying asset for the remaining lifespan of the option contract. From the arbitrage-free pricing perspective, one may utilize the implied volatilities inferred from a cross-sectional set of option prices via the Black–Scholes option pricing formula, along with the values of other input parameters, to infer the prices of more complex contingent claims that have the same underlying asset but are not exchange traded, i.e., traded "over-the-counter" (OTC), and European vanilla options with strike-expiry pairs that are OTC traded. This is to ensure that the same set of modeling assumptions of asset price dynamics is applied to all exchange traded and OTC traded contingent claims with the same underlying asset. In so doing, the asset price dynamics assumptions and parameter values used as input to price the OTC contingent claims can recover, at least to a close approximation, the exchanged traded European vanilla options, thereby abrogating the possibility of constructing an arbitrage position that exploits the pricing difference between these exchange traded and OTC traded contingent claims. This practice of consistent pricing of exchange traded and OTC contingent claims is not confined to the Black–Scholes pricing framework. A wide range of models that are more sophisticated than the Black–Scholes model can be calibrated with this approach.

A crucial assumption of the Black–Scholes option pricing formula is that the implied volatility is assumed to be constant. This assumption is not met in reality. Before the Oct. 19th 1987 stock market crash, the relationship between the implied volatility and the strike price for otherwise identical options from samples of S&P 500 index option prices is close to a horizontal line. However, the relationship between the implied volatility and the strike price or the time to maturity for option prices sampled at dates subsequent to the 1987 crash are found to be markedly non-horizontal, and non-linear [104]. They exhibit a set of features not previously observed from option prices sampled prior to 1987. The non-linear relationship between the implied volatility and the strike price, referred to the "volatility smile," is more pronounced for options with shorter time to maturity. The contingent claims for different asset classes exhibit different features of volatility smile; for example, the smile is negatively skewed for equity and index options, while the smile exhibits as less skewed for foreign exchange options. The option contracts with shorter time to maturity exhibit a larger extent of fluctuation of the smile across time. The term structure of implied volatility is typically upward sloping in calm times, while in times of market turmoil it is downward sloping with the short dated options having higher levels of implied volatilities then the longer dated ones. Additionally, the volatility smile and the volatility term structure fluctuates with time.

These empirical findings that demonstrate non-negligible violation of the constant volatility assumption of the Black–Scholes option pricing formula fueled active research in several directions. Various generalizations of the constant volatility log-normal process assumed in the Black–Scholes option pricing formula have been made to better capture salient features of the volatility smile and the implied volatility term structure. In this monograph, we are concerned with using the implied volatility as a representation of market consensus of option underlying asset volatility for the remaining lifespan of the contract, and the extraction of this information from a set of option prices.

The estimation of implied volatilities from historical option prices is an area of active research as the implied volatility is a financial parameters that has many practical applications. For example, the implied volatility estimated from the European vanilla options can be used to forecast future realized volatility of the underlying asset to construct market benchmarks that reflect the market sentiment of the future realized volatility of certain asset classes, to provide input for the coherent pricing of exotic contingent claims with the same underlying asset for trading and risk management purposes, and to carry out pricing of credit default swap (CDS) contracts that are related to the same underlying asset of the option contracts from which the implied volatilities are estimated from.

Thus far, the extant literature appears to suggest that the implied volatilities, i.e., both the Black–Scholes [24] model-based implied volatility, and the model-free implied volatility [29], appear to be better predictors of the future realized volatility of the underlying asset for the remaining lifespan of the option contract than the historical volatility of the underlying asset. While the Heterogeneous Autoregressive model of Realized Volatility (HAR-RV) proposed by [35] appears to be a serious contender of the implied volatility in the forecast of future realized volatility, there is currently no definitive empirical evidence pointing one way or the other.

It has long been reported in the literature that there exists a noticeable difference between the implied and realized volatilities. This difference is larger when the index return distribution is more negatively skewed and leptokurtic in the presence of investor risk aversion [17]. A plethora of theories have been put forth in the attempt to explain this difference, including, among others, the existence of negative variance risk premium that rationalizes the willingness of risk-adverse investors to pay a premium for protection against an increase in volatility of asset prices. Theoretical underpinning aside, the numerical implementation of the estimators for the implied and realized volatilities may have contributed to the estimated difference between implied and realized volatilities.

The squared-root of the sum of squared log return of an asset is widely regarded as a robust estimator of its historical volatility [2, 25]. This estimator is independent of the risk-free rate proxy used in the pricing of the contingent claims that regard this asset as the underlying asset of these claims. On the contrary, the estimation of both the Black–Scholes model-based implied volatility and the model-free implied volatility require making an assumption on the risk-free rate. Given an option price, other things being equal, different choices of risk-free rate proxy map the same option price to different values of implied volatility. Therefore, for a given estimate of the realized volatility, different choices of risk-free rate proxy will result in different estimates of the difference between implied and realized volatilities. In order to infer the market participants' aggregate expectation of the underlying asset price volatility for the remaining lifespan of the option contract, we have to choose a risk-free rate proxy that matches the market participants' aggregate choice of risk-free rate proxy. This motivates the subject of study in this chapter, i.e., to estimate the discount rate and implied volatility jointly from a set of option prices.

Implied discount rate – an aggregate of discount rates chosen by market participants with different funding costs

References [42] and [30] proposed a theoretical framework that relaxes the assumption of *a priori* known risk-free rate, and regards the discount rate and implied volatility in the Black–Scholes option pricing formula as unknown parameters. This idea provides impetus to our development of a strategy to jointly estimate the implied discount rates and the implied volatilities from a set of option prices.

The choice of the risk-free rate proxy made by market participants in the context of contingent claim valuation tend to reflect their corresponding credit and liquidity risk. The demand from market participants for the flexibility of being able to choose different reference rates as proxy to risk-free rate corresponding to different tenors for different applications is recognized in a recent report by the Bank of International Settlement. The said report goes further to indicate the need to make available more diverse reference interest rates that better match the individual needs of the market participants. In practice, a combination of reference rates that reflect different levels of risk premia at different tenors may be used to construct the discount rate term structure for the appropriate context.

One such example is the calculation of volatility indexes. In the calculation of the volatility index based on S&P 500 equity index (VIX) quoted by CBOE, the bond-equivalent yields of the US Treasury bill, which contain near zero credit risk, are used as proxy for the risk-free rates. In the calculation of the volatility index based on DAX equity index (VDAX) quoted by Deutsche Börse, the proxy to risk-free rates are obtained by linear interpolation between Euro OverNight Index Average (EONIA) rate and 1 month Euribor, the London Interbank Offer Rate for the Euro denomination. In the calculation of the volatility index based on the ASX 200 equity index (S&P/ASX 200 VIX) quoted by the Australian Stock Exchange, the proxy for the risk-free rates are obtained by linear interpolation between overnight Reserve Bank of Australia (RBA) rate and 1 month Australian Financial Markets Association (AFMA) Bank Bill Swap (BBSW) benchmark rate. While EONIA and RBA rates are virtually free of bank credit risk, one month Euribor and the AFMA BBSW reflect the average bank credit risk premia.

While [97] reported that the risk-free rate has relevant predictive information with respect to the conditional variance of individual stock return, and that interest rate changes have different effects for different assets, [27], among others, emphasized the importance of the choice of risk-free rate proxy, or discount rate, in contingent claim valuation. In fact, not only would the choice of discount rate affect the calculation of net present value of contingent claims from a pricing perspective, it would affect the magnitude of the implied volatility inferred from the historical price of an option contract from an econometric perspective.

Joint inference of implied volatility and implied discount rate from option prices

Chapter 7.3 offers a discussion on the relationship between the discount rate and the implied volatility estimated jointly from the price of a European vanilla call option contract. It concerns the utilization of an over-defined system of equations to develop a strategy in order to infer the discount rates and the implied volatilities simultaneously from a set of option prices, thereby circumventing the need to make *a priori*, and potentially erroneous, assumptions on the discount rate that may affect the accuracy of implied volatility calibration from option prices.

We propose, therein, a strategy to infer, from the prices of a set of European vanilla call option contracts, the market participants' aggregate choice of the discount rate, and the market participants' aggregate expectation of the underlying asset price volatility for the remaining lifespan of these option contracts. Working within the Black–Scholes option pricing framework, we relax the assumption of *a priori* known risk-free rate, and assume that both the discount rate and the implied volatility are unknown parameters. We use the Black–Scholes option pricing formula as a mapping tool to construct an over-defined system of nonlinear equations. Approximate numerical solution of this system allows us to map a set of option prices to their corresponding pairs of discount rates and implied volatilities. The set of discount rates may be regarded as the implied discount rates with respect to this set of option prices. We suggest regarding the set of implied discount rates as model-based estimates of the market participants' aggregate choice of discount rate, and the set of implied volatilities estimated jointly with the implied discount rates as model-based estimates of the market participants' aggregate expectation of the underlying asset price volatility for the remaining lifespan of these option contracts.

A recent paper by [22] proposed a strategy to jointly estimate the implied risk-free rates and the implied volatilities from option prices. Although both [65] and [22] propose strategies to estimate the implied discount rates or implied risk-free rates, and the implied volatilities from option prices, they differ in terms of model assumptions, numerical implementation, and empirical application.

Firstly, [22] arrange a cross-sectional set of prices of European vanilla call option contracts written on the same underlying asset with the same expiry date in ascending magnitudes of strike, and group the adjacent option contracts into multiple non-overlapping pairs. They assume that each pair of these option contracts with different strike prices but the same expiry date are associated with the same implied risk-free rate and implied volatility. For each pair of option contracts, a set of two nonlinear equations is formulated by mapping both option prices to two unknown parameters, i.e., one implied risk-free rate and one implied volatility, via the Black–Scholes option pricing formula. Specifically, they assume that option contracts with different strike prices that are grouped in the same pair share the same implied volatility, while option contracts with the same expiry date but grouped in different pairs may have different implied risk-free rates.

In contrast, our proposed strategy utilizes prices of a set of European vanilla call option contracts written on the same underlying asset with different strike prices and different expiry dates observed on two successive trading days. We assume that the option contracts with the same time to maturity and expiry date are associated with the same implied discount rate. In this sense, the definition of implied risk-free rate in [22] is different from the definition of implied discount rate in our proposed strategy. We believe our assumption that option contracts sharing the same expiry date are associated with the same discount rate is more in line with the current practice of coherent contingent claim pricing where cash flows at the same time horizon are discounted with the same discount rate. On the other hand, the assumption used by [22] is somewhat at odds with this practice as it seems to allow non-uniqueness of discount rate at the same tenor for the same contingent claim. Additionally, we assume that option contracts with different strike and expiry characteristics may be associated with different implied volatilities. We believe this assumption is less restrictive than requiring option contracts with different strike prices to share the same value of implied volatility on the basis of user-specified pairing of option contracts.

Secondly, [22] utilize a pair of option prices observed at the same instance to formulate a system of two nonlinear equations with two unknown parameters to be solved. In contrast, we utilize prices of a set of option contracts observed at two instances in time to formulate an over-defined system of nonlinear equations and seek numerical approximate solution to this system.

Additionally, [22] seek numerical solution to their system of nonlinear equations by a three-stage optimization procedure that minimizes a scaled L_2 objective function. Interior point search is first performed based on a pair of user-defined starting values to obtain a first set of solutions for implied risk-free rate and implied volatility. One then use this first set of solutions as input to the *patternsearch* algorithm, a derivative-free method in the optimization tool box of MATLAB (MathWorks, Inc., Natick, MA) to obtain a second set of solution values. One finally uses this second set of solutions as input to the interior point search algorithm again to obtain a third set of solution values. The set of solutions among the three that corresponds to the smallest value of the objective function is regarded as the solution to the system of equations.

In contrast, we seek numerical solution to our over-defined system of nonlinear equations by the Zhang–Sanderson algorithm [125], a stochastic-based multi-point direct-search global optimization algorithm that belongs to the differential evolution family of optimization techniques implemented in the DEoptim [12] package for the R computing environment [102]. Our numerical implementation strategy can accommodate minimization of both L_1 and L_2 metric objective functions. Provision of user-defined starting values is not required as the algorithm is initialized using pseudo-random numbers.

Thirdly, [22] use the implied risk-free rates and implied volatilities estimated as solutions to their system of equations based on the 2007 to 2008 historical prices of the S&P500 index call option contracts to interpolate at-the-money (ATM) implied volatilities based on a seemingly unrelated regression framework. Then the interpolated ATM implied volatilities are used to forecast the VIX index.

In contrast, we apply our proposed strategy to samples from the 2004 to 2013 historical prices of S&P 500 index call option contracts in order to analyze the trend of the implied discount rate across time. Specifically, we highlight the difference between the implied discount rates and contemporaneously quoted Libor rates and Overnight Indexed Swap (OIS) rates in these samples.

Various numerical strategies have been proposed in the extant literature to address the inverse problem of mapping a cross-sectional set of European vanilla option prices to the local, i.e., instantaneous, volatilities. Due to the finite number of option prices available for the calibration of the entire local volatility surface, many numerical strategies utilize interpolation techniques, such as B-splines [38], in the numerical solution of the partial differential equation pertaining to this calibration problem.

From a numerical point of view, the problem of seeking approximate solutions to systems of nonlinear equations can be recast into an optimization problem where the objective function is a metric measuring the discrepancy between the left-hand side and the right side of the system of equations. Due to the fact that the objective function may not be convex with respect to the unknown variables in question, the use of non-gradient-dependent optimization strategies prevents trapping in the local minima. We adopt Zhang–Sanderson's differential evolution algorithm [125] to implement the numerical solutions carried out in [64] due to the adaptive nature of this algorithm that ensures convergence towards global minimum [66]. As such, the numerical strategies implemented to address the problems considered in this monograph sit in the same strand of literature as [19].

Although smile consistent models such as the local volatility model can provide an excellent fit to a given set of cross-sectional option prices, there are some shortcomings. Among them is that the local volatility model predicts an unrealistic flat future smile; the longer the forecast horizon, the flatter the predicted implied volatility and local volatility surfaces. Therefore, even if it can be calibrated perfectly to current option prices, it is not suitable for pricing contingent claims in the likes of forward starting options and cliquet options that require a model that can offer a realistic future smile prediction.

8

Forecast of short rate based on the CIR model

The classic interest rate term structure modeling approach constructs a single yield curve for interpolation and forecast purposes. In contrast, the multi-curve approach constructs separate yield curves for different tenors. In this chapter, we propose a strategy to model different segments of the yield curve that correspond to different tenor intervals using different short rate processes. While it is well known that the single-factor CIR process does not perform well in forecasting future interest rates in the classic single-curve framework, we demonstrate that even this model can be improved when cast into a multi-curve framework.

We implement the single-factor Cox–Ingersoll–Ross [36] process in a multi-curve framework to extract the market participants' aggregate expectation of the future short rate at various available tenors from a cross-sectional set of zero coupon bond prices. More precisely, we utilize the single-factor CIR model zero coupon bond pricing formula as a mapping tool to map the market mid-quote prices of a triplet of zero coupon bonds to the three model parameters of a single-factor CIR model in order to construct systems of well-defined non-linear equations.

We then seek approximate solution to the systems of non-linear equations via Algorithm 7.2 to capture the information contained in the prices of the triplets of zero coupon bonds and represent them as the corresponding inferred CIR model parameters. We use the various sets of inferred CIR model parameters inferred from zero coupons of different tenors to construct predictors to forecast short rates at different tenors. We then compare its forecast performance against a predictor constructed based on the classic single-curve framework. In this sense, the numerical strategies used in this context also follow the strand of model calibration literature for equity options and interest rate model.

For numerical demonstration purposes, we apply our proposed algorithm to infer the market participants' aggregate view of the future short rate from the historical prices of the United States Separately Traded Registered Interest and Principal Securities (US STRIPS) sampled between 2001 and 2014 inclusive, and use this information to forecast the US effective Federal Funds rate.

8.1   The model framework

Let $(\Omega, \mathcal{F}, \mathbf{P})$ be a standard probability space where Ω is a set of elementary events, $\mathcal{F}$ is a complete sigma algebra of events, and $\mathbf{P}$ is a probability measure. Let $\mathcal{F}_t$ be a complete sigma algebra of events generated by the data observed at time t. Let the short rate process $r(u), u \in [t, T], t < T$, be a stochastic process adapted to the filtration $\{\mathcal{F}_t\}$, where $\{\mathcal{F}_t\}$ is some filtration generated by the flow of the currently observed market data, $r(u)$ is non-negative, or at least that the process $\min(r(u), 0)$ is bounded, $\mathbf{E}\left[\int_t^T r(u)^2 du \,\Big|\, \mathcal{F}_t\right] < +\infty, t < T$, where $\mathbf{E}$ is the expectation taken over $r(u)$ under some probability measure $\mathbf{P}$, and the case when the filtration $\{\mathcal{F}_t\}$ is generated by the process $r(u)$ is not excluded. The price of a zero coupon bond at the current time t and with maturity time T is defined by the conditional expectation

$$P(t, T) = \mathbf{E}\left[e^{-\int_t^T r(u)du} \,\Big|\, \mathcal{F}_t\right]. \tag{8.1}$$

8.1.1   General setting

Let

$$\rho(t; s, T) \triangleq \mathbf{E}\left[\frac{1}{T-s} \int_s^T r(u)du \,\Big|\, \mathcal{F}_t\right], \tag{8.2}$$

where $u \in [s, T], s < T$, be the 'expected average integrated short rate' observed at time t for the time interval $[s, T]$, and let $\gamma(t; s, T) = (T - s)\rho(t; T_1, T_2)$. Let

$$\eta(t, T) = -\frac{1}{T-t} \log P(t, T) \tag{8.3}$$

be the yield-to-maturity of the zero coupon bond maturing at the tenor $T - t$, and let

$$J(t, T) = \rho(t; t, T) - \eta(t, T) \tag{8.4}$$

be the corresponding 'convexity adjustment' term. Since $e^{-x}, x \in \mathbb{R}$, is a convex function, we obtain, by Jensen's inequality,

$$\mathbf{E}\left[e^{-\int_t^T r(u)du} \,\Big|\, \mathcal{F}_t\right] \geq e^{-\mathbf{E}\left[\int_t^T r(u)du \,\big|\, \mathcal{F}_t\right]},$$

and it follows from (8.2) and (8.3) that $J(t, T) \geq 0$. If the short rate process $r(u), u \in [t, T]$, is non-random, then $J(t, T) = 0$.

Let $\eta(t, T_1)$ and $\eta(t, T_2)$ be the yield-to-maturities with respect to two different zero coupon bonds $P(t, T_1)$ and $P(t, T_2)$, respectively, where $t < T_1 < T_2$. Based on $P(t, T_1)$ and $P(t, T_2)$, we can infer the forward rate in the interval $[T_1, T_2]$ by

$$F(t; T_1, T_2) = \frac{-\log P(t, T_2) + \log P(t, T_1)}{T_2 - T_1}$$
$$= \frac{(T_2 - t)\, \eta(t, T_2) - (T_1 - t)\, \eta(t, T_1)}{T_2 - T_1}. \qquad (8.5)$$

It follows from (8.4) and (8.5) that

$$\rho(t; T_1, T_2) - F(t; T_1, T_2) = \frac{(T_2 - t)\, J(t, T_2) - (T_1 - t)\, J(t, T_1)}{T_2 - T_1}. \qquad (8.6)$$

Further, let $\gamma(t; s, T) = (T - s)\rho(t; s, T)$, and let

$$\Gamma(t; s, T) = \mathbf{E}\left[\left(\int_s^T r(u)\,du \right)^2 \Bigg| \mathscr{F}_t \right],$$

where $u \in [s, T]$, and $t \le s < T$. We expand $e^{-\int_t^{T_q} r(u)\,du}, q = 1, 2$, up to the second order to obtain an approximation expression of the zero coupon bond pricing formula (8.1)

$$P(t, T_q) \approx 1 - \gamma(t; t, T_q) + \frac{1}{2}\Gamma(t; t, T_q), \qquad q = 1, 2,$$

and then obtain the approximation expression of $\log P(t, T_q)$ using this expression by expanding the log term up to the second order and discarding the terms higher than the second order to obtain

$$\log P(t, T_q) \approx \log\left(1 - \gamma(t; t, T_q) + \frac{1}{2}\Gamma(t; t, T_q) \right)$$

$$\approx -\gamma(t; t, T_q) + \frac{1}{2}\Gamma(t; t, T_q) - \frac{1}{2}\left(-\gamma(t; t, T_q) + \frac{1}{2}\Gamma(t; t, T_q) \right)^2$$

$$\approx -\gamma(t; t, T_q) + \frac{1}{2}\left(\Gamma(t; t, T_q) - \gamma(t; t, T_q)^2 \right)$$

$$= -\gamma(t; t, T_q) + \frac{1}{2}\left(\mathbf{E}\left[\left(\int_t^{T_q} r(u)\,du \right)^2 \Bigg| \mathscr{F}_t \right] \right.$$

$$\left. - \mathbf{E}\left[\int_t^{T_q} r(u)\,du \Bigg| \mathscr{F}_t \right]^2 \right)$$

$$= -\gamma(t; t, T_q) + \frac{1}{2}\mathrm{Var}\left[\int_t^{T_q} r(u)\,du \Bigg| \mathscr{F}_t \right], \qquad q = 1, 2, \quad (8.7)$$

where, in the last two lines, we have used the identity $\mathrm{Var}\,[x] = \mathbf{E}[x^2] - \mathbf{E}[x]^2, x \in$

$\mathbb{R}$. Substituting (8.7) into (8.5), we obtain

$$F(t; T_1, T_2) = \rho(t; T_1, T_2) - \frac{\delta(t; T_1, T_2)}{2(T_2 - T_1)} + \mathscr{R}\left(P(t, T_1) - P(t, T_2)\right)$$

$$\approx \rho(t; T_1, T_2) - \frac{\delta(t; T_1, T_2)}{2(T_2 - T_1)} , \qquad (8.8)$$

where

$$\delta(t; T_1, T_2) = \mathrm{Var}\left[\int_t^{T_2} r(u)du \,\middle|\, \mathscr{F}_t\right] - \mathrm{Var}\left[\int_t^{T_1} r(u)du \,\middle|\, \mathscr{F}_t\right] ,$$

and $\mathscr{R}\left(P(t, T_1) - P(t, T_2)\right)$ represents the collection of all the higher-order terms in the exponential expansions of $P(t, T_q), q = 1, 2$, and the higher-order terms in the logarithmic expansions of $\log P(t, T_q), q = 1, 2$, which we disregard in obtaining the approximation (8.8). Let

$$\Delta(t; T_1, T_2) = 2\left(\rho(t; T_1, T_2) - F(t; T_1, T_2)\right) \approx \frac{\delta(t; T_1, T_2)}{T_2 - T_1} , \qquad (8.9)$$

be the 'approximate average forward variance,' an expression that we may regard as a gauge of the average volatility of the integrated short rate in the interval $[T_1, T_2]$ for some $t < T_1 < T_2$ that may be approximated using $\delta(t; T_1, T_2)$.

8.1.2   The CIR model

The expressions (8.1) to (8.9) do not rely on a particular model for the short rate process. Although we do not assume that a particular short rate model can capture all the salient features of the interest rate term structure, for the reasons outlined in Section 8.7, we choose to use the one-factor CIR process to model the dynamics of the tenor-specific short rate processes in a multi-curve framework.

Let $r_{CIR}(u), u \in [t, T], t < T$, be a one-factor CIR process where

$$r_{CIR}(T) = r_{CIR}(t) + \kappa\beta(T - t) - \kappa \int_t^T r_{CIR}(u)du$$

$$+ \sigma \int_t^T r_{CIR}(u)^{1/2}dW(u) , \qquad (8.10)$$

where κ is the speed of mean-reversion of the short rate, β is the long-range short rate, σ is the volatility of the process, $r_{CIR}(t) = r(t)$ where $r(t)$ is the short rate at time t, and $W(u)$ is the standard Wiener process. If $\kappa = 0$ or $\beta = 0$, $r_{CIR}(T)$ reaches zero almost surely and the point zero is absorbing. If $2\kappa\beta \geq \sigma^2$, $r_{CIR}(T)$ is a transient process that stays positive and never reaches zero. If $0 < 2\kappa\beta < \sigma^2$, $r_{CIR}(T)$ is instantaneously reflective at point zero.

Taking expectation across (8.10) conditional on $\mathscr{F}_t$ with respect to $r_{CIR}(T)$, we obtain

$$\mathbf{E}\left[r_{CIR}(T) \,|\mathscr{F}_t\right] = r_{CIR}(t) + \kappa\beta(T - t) - \kappa \int_t^T \mathbf{E}\left[r_{CIR}(u) \,|\mathscr{F}_t\right] du .$$

Differentiating $\mathbf{E}\left[r_{CIR}(T)\,|\mathcal{F}_t\right]$ with respect to T, we obtain

$$\frac{d\mathbf{E}\left[r_{CIR}(T)\,|\mathcal{F}_t\right]}{dT} = \kappa\left(\beta - \mathbf{E}\left[r_{CIR}(T)\,|\mathcal{F}_t\right]\right)\,.$$

Let $\mathbf{E}\left[r_{CIR}(T)\,|\mathcal{F}_t\right] = \Phi(T)$, and we solve the variable separable first-order first-degree ordinary differential equation

$$\frac{d\Phi(T)}{dT} = \kappa\left(\beta - \Phi(T)\right)\,,$$

and since $\Phi(t) = r(t)$, we obtain

$$\mathbf{E}[r_{CIR}(T)|\mathcal{F}_t] = \beta + (r(t) - \beta)\,e^{-\kappa(T-t)}\,. \tag{8.11}$$

It follows from (8.11) that

$$\begin{aligned}
\gamma_{CIR}(t;t,T) &= \mathbf{E}\left[\left.\int_t^T r_{CIR}(u)du\,\right|\,\mathcal{F}_t\right] \\
&= \int_t^T \mathbf{E}[r_{CIR}(u)|\mathcal{F}_t]\,du \\
&= \beta(T - t) + \left(\frac{r(t) - \beta}{\kappa}\right)\left(1 - e^{-\kappa(T-t)}\right) \tag{8.12}
\end{aligned}$$

and, using (8.2), we obtain

$$\rho_{CIR}(t;t,T) = \frac{\gamma_{CIR}(t;t,T)}{T - t} = \theta + \frac{r(t) - \beta}{\kappa(T - t)}\left(1 - e^{-\kappa(T-t)}\right)\,, \tag{8.13}$$

and it follows that, for some $t < T_1 < T_2$,

$$\rho_{CIR}(t;T_1,T_2) = \beta + \left(\frac{r(t) - \beta}{\kappa(T_2 - T_1)}\right)\left(e^{-\kappa(T_1-t)} - e^{-\kappa(T_2-t)}\right)\,, \tag{8.14}$$

where, as $T_1 \to \infty$ and $T_2 \to \infty$, $\rho_{CIR}(t;T_1,T_2) \to \beta$. Using (8.4), we may define $\rho_{CIR}(t;t,T) = \eta(t,T) + J_{CIR}(t,T)$, and express the relation between $\rho_{CIR}(t;T_1,T_2)$ and $F(t;T_1,T_2)$ as

$$\rho_{CIR}(t;T_1,T_2) - F(t;T_1,T_2) = \frac{(T_2-t)\,J_{CIR}(t,T_2)-(T_1-t)\,J_{CIR}(t,T_1)}{T_2-T_1}\,. \tag{8.15}$$

Using (8.9), and (8.15), we define

$$\Delta_{CIR}(t;T_1,T_2) = 2\left(\rho_{CIR}(t;T_1,T_2) - F(t;T_1,T_2)\right) \approx \frac{\delta_{CIR}(t;T_1,T_2)}{T_2 - T_1}\,, \tag{8.16}$$

and may use this expression as an estimator of the approximate average forward variance when the short rate dynamics is modelled using the CIR process where

$$\delta_{CIR}(t;T_1,T_2) = \mathrm{Var}_{\,CIR}\left[\left.\int_t^{T_2} r_{CIR}(u)du\,\right|\,\mathcal{F}_t\right] - \mathrm{Var}_{\,CIR}\left[\left.\int_t^{T_1} r_{CIR}(u)du\,\right|\,\mathcal{F}_t\right]\,,$$

and $\mathrm{Var}_{\,CIR}(\cdot)$ denote the variance expressions for the integrated CIR process.

8.2   Inference of the implied CIR model parameters based on cross-sectional zero coupon bond prices

We arrange a set of cross-sectional zero coupon bonds in ascending order of tenor, and organize the adjacent bonds into multiple triplets that correspond to different non-overlapping tenor intervals. We assume each triplet that corresponds to a different tenor interval is associated with a different short rate process, and model each of these short rate processes using a separate one-factor CIR process specific for each of the tenor intervals considered.

Since the zero coupon bond pricing formula for a one-factor CIR process under the risk-neutral measure is given by

$$P_{CIR}(t, T; \kappa, \beta, \sigma, r(t)) = e^{B_1 - B_2 r(t)}, \tag{8.17}$$

where

$$B_1 = \frac{\kappa \beta}{\sigma^2} \left[(\kappa + B_3)(T - t) - 2 \log \left(1 + \frac{(\kappa + B_3)(e^{B_3(T-t)} - 1)}{2B_3} \right) \right],$$

$$B_2 = \frac{2 \left(e^{B_3(T-t)} - 1 \right)}{(\kappa + B_3) \left(e^{B_3(T-t)} + 2B_3 \right)},$$

$$B_3 = \left(\kappa^2 + 2\sigma^2 \right)^{1/2},$$

and $r(t)$ is the initial short rate, we construct, for each triplet, a set of three nonlinear equations with three unknown parameters by mapping the corresponding triplet of cross-sectional zero coupon bond prices to their respective set of implied CIR model parameters via (8.17). Based on this mapping, we extract the information on the market participants' aggregate view of the future short rate at various forecast horizons from the triplets with comparable tenors, and represent this information as sets of implied CIR parameters for the one-factor CIR processes associated with different tenors.

8.3   Numerical framework for the inference

Let $\{P(t, T_{j,k})\}_{j=1,\ldots,n,k=1,2,3}$ denote the prices of a set of zero coupon bonds observed at time t maturing at $T_{j,k}$ where $T_{j,k+1} > T_{j,k} > t$, and $T_{j+1,1} > T_{j,3} > t$, $j = 1, \ldots, n$. We assume that each triplet of prices $\{P(t, T_{j,k})\}_{k=1}^{3}$, indexed by $j = 1, \ldots, n$, that corresponds to a different non-overlapping tenor interval $(T_{j-1,3} - t, T_{j,3} - t]$, $j = 1, \ldots, n$, is associated with a different short rate process, while $T_{j-1,3} - t = 0$ for $j - 1 = 0$, and this triplet can be mapped to a set of CIR model parameters $(\kappa_j(t), \beta_j(t), \sigma_j(t))$. Using (8.17) as a mapping tool, we construct a well-defined system of nonlinear equations

$$P_{CIR}(t, T_{j,k}; \kappa_j(t), \beta_j(t), \sigma_j(t), r(t)) = P(t, T_{j,k}), \quad j = 1, \ldots, n, \quad k = 1, 2, 3, \tag{8.18}$$

where $(\kappa_j(t), \beta_j(t), \sigma_j(t))$ is the set of CIR model parameters inferred from $\{P(t, T_{j,k})\}_{k=1}^3$. Specifically, there are n sets of equations in this system, each set comprising 3 nonlinear equations and three unknown parameters.

We use an observable overnight interest rate in the same economy as $P(t, T_{j,k})$ that reflects the credit and liquidity risk levels that are comparable to those of $P(t, T_{j,k})$ as a proxy for $r(t)$. Specifically, we assume that, in a multi-curve framework, the proxy short rate process can be used to model the true short rate process in the tenor interval $[T_{j,1} - t, T_{j,3} - t]$ corresponding to the triplet $\{P(t, T_{j,k})\}_{k=1}^3$.

We calculate $\boldsymbol{\kappa}(t) = (\kappa_1(t), \ldots, \kappa_n(t))$, $\boldsymbol{\beta}(t) = (\beta_1(t), \ldots, \beta_n(t))$, $\boldsymbol{\sigma}(t) = (\sigma_1(t), \ldots, \sigma_n(t))$, by seeking approximate numerical solution of (8.18). We cast the numerical solution problem in the form of an optimization problem as a special case of (7.5)

$$\underset{\boldsymbol{\theta}}{\text{minimize}} \quad \Psi(\boldsymbol{\theta}) \tag{8.19}$$

$$\text{subject to} \quad 0 < \theta_J < \infty, \quad J = 0, \ldots, D - 1$$

where, in the present context, we define the objective function $\Psi(\boldsymbol{\theta})$ as an L_2 loss metric

$$\Psi(\boldsymbol{\theta}(t)) = \sum_{j=1}^n \sum_{k=1}^3 (P(t, T_{j,k}) - P_{CIR}(t, T_{j,k}; \kappa_j(t), \beta_j(t), \sigma_j(t), r(t)))^2 \tag{8.20}$$

where the D-dimensional unknown parameters are $\boldsymbol{\theta}(t) = \{\theta_J(t)\}_{J=0}^{D-1} = (\boldsymbol{\kappa}(t), \boldsymbol{\beta}(t), \boldsymbol{\sigma}(t))$, $\boldsymbol{\kappa}(t) = (\kappa_1(t), \ldots, \kappa_n(t))$, $\boldsymbol{\beta}(t) = (\beta_1(t), \ldots, \beta_n(t))$, $\boldsymbol{\sigma}(t) = (\sigma_1(t), \ldots, \sigma_n(t))$, with $D = 3n$, and where the parameter boundaries are

$$\kappa_j(t) > 0, \quad \beta_j(t) > 0, \quad \sigma_j(t) > 0, \quad j = 1, \ldots, 3n,$$

via the Zhang–Sanderson non-archive assisted differential evolution algorithm as detailed in Algorithm 7.2. Let $\boldsymbol{\theta}_{best,G-1} = (\boldsymbol{\kappa}_{best,G-1}, \boldsymbol{\beta}_{best,G-1}, \boldsymbol{\sigma}_{best,G-1})$ be the individual vector in the population $\{\boldsymbol{\theta}_{I,G-1}, I = 0, \ldots, N_p - 1\}$ at algorithm termination after G iterations.

More precisely,

$$\boldsymbol{\theta}_{best,G-1}(t) = \{\theta_{best,G-1,j}(t)\}_{j=1}^{3n} = (\boldsymbol{\kappa}_{best,G-1}(t), \boldsymbol{\beta}_{best,G-1}(t), \boldsymbol{\sigma}_{best,G-1}(t)),$$

where $\boldsymbol{\kappa}_{best,G-1}(t) = \{\kappa_{best,G-1,j}(t)\}_{j=1}^n$, $\boldsymbol{\beta}_{best,G-1}(t) = \{\beta_{best,G-1,j}(t)\}_{j=1}^n$, $\boldsymbol{\sigma}_{best,G-1}(t) = \{\sigma_{best,G-1,j}(t)\}_{j=1}^n$, are the implied CIR model parameters estimated based on the corresponding zero coupon prices along the respective tenors $\{T_{j,3} - t\}_{j=1}^n$. The triplets $(\kappa_{best,G-1,j}(t), \beta_{best,G-1,j}(t), \sigma_{best,G-1,j}(t))$ are regarded as the sets of estimated CIR model parameters for the short rate process that span the tenors $T_{j,3} - t$. The estimated parameters $\boldsymbol{\theta}_{best,G-1}(t)$ are approximate solutions to (8.18). The uncertainty bound for estimated parameters may be constructed based on $\{\kappa_{I,G-1,j}(t)\}_{I=0}^{N_p-1}$, $\{\beta_{I,G-1,j}(t)\}_{I=0}^{N_p-1}$, and $\{\sigma_{I,G-1,j}(t)\}_{I=0}^{N_p-1}$ for each $j, j = 1, \ldots, n$, based on the strategy detailed in Section 7.3.3.

8.4 Computational implementation

A typical set of cross-sectional US STRIPS zero coupon bond prices used in the numerical analysis reported in Section 8.5.1 contains more than 120 data points with which we can construct approximately 40 sets of nonlinear equations, each based on a non-overlapping triplet of zero coupon bonds. In order to accelerate the numerical computation involved in the simultaneous approximate solution of such a large system of equations, we use the Zhang–Sanderson algorithm implemented in the DEoptim [12] package for the R computing environment [102]. We used 48 parallel cores provided by the NeCTAR cloud computing infrastructure to carry out all the calculations in order to accelerate the speed of computation. The computational acceleration achieved is considerable. As an example to give an idea of the extent of computational acceleration, we estimated the CIR model parameters based on day-close US STRIPS prices quoted on Jan. 2nd 2001. There are 168 parameters to be estimated from the system of equations constructed from that set of zero coupon bond prices. Using one processor, it took 3 hours and 10 minutes to carry out 2000 iterations. Using 48 parallel processors, it took 46 minutes instead, leading to a 76% reduction in computation time. This is comparable to the computational acceleration achieved in the numerical experiment reported in Section 7.3.

8.5 Forecast of short rate using the implied CIR model parameters

We propose a set of predictors in Section 8.5.1 to forecast the short rate. They are designed for the multi-curve framework where different sets of implied CIR model parameters that correspond to different triplets of zero coupon bond prices are used to predict short rate at different forecast horizons.

In order to compare the performance of our proposed predictors with that of a set of short rate predictors implemented in the single-curve framework, we construct, in Section 8.5.2, a separate set of predictors that utilize only the set of implied CIR model parameters corresponding to the shortest tenor to forecast short rate at horizons beyond its tenor, which may be regarded as a 'naive' short rate predictor within the single-curve framework.

8.5.1 Forecast within the multi-curve framework

Assuming that the estimated implied CIR model parameters $\{\kappa_{best,G-1,j}(t), \beta_{best,G-1,j}(t), \sigma_{best,G-1,j}(t)\}$ contain some information on the market participants' expectation of the future short rate at forecast horizon $T_{j,3} - t$, we propose to formulate the forecast strategy by using the estimated implied CIR model parameters $\{\kappa_{best,G-1,j}(t), \beta_{best,G-1,j}(t), \sigma_{best,G-1,j}(t)\}$ to model the short rate process associated with the triplet $\{P(t, T_{j,k})\}_{k=1}^{3}$. We then use the non-linear projection of the future short rate based on the corresponding tenor-specific short rate curve in order to forecast $r(T_{j,3})$, the realized short rate at time $T_{j,3}$.

Based on (8.12), we may predict $\gamma_{CIR}(t; t, T_{j,3})$ at the forecast horizons $\{T_{j,3} - t\}_{j=1}^{n-1}$ using

$$\hat{\gamma}_{CIR}(t; t, T_{j,3}) = \beta_{best,G-1,j}(t) (T_{j,3} - t)$$
$$+ \frac{r(t) - \beta_{best,G-1,j}(t)}{\kappa_{best,G-1,j}(t)} \left(1 - e^{-\kappa_{best,G-1,j}(t)}\right), \quad (8.21)$$

where $j = 1, \ldots, n$. Using (8.14) and (8.21), we may predict $\rho_{CIR}(t; T_{j,3}, T_{j+1,3})$ within the time intervals $[T_{j+1,3}, T_{j,3}], j = 1, \ldots, n-1$, using

$$\hat{\rho}_{CIR}(t; T_{j,3}, T_{j+1,3}) = \frac{\hat{\gamma}_{CIR}(t; t, T_{j+1,3}) - \hat{\gamma}_{CIR}(t; t, T_{j,3})}{T_{j+1,3} - T_{j,3}}, \quad (8.22)$$

where $j = 1, \ldots, n$. We suggest using (8.22) to forecast short rates $\{r(T_{j,3})\}_{j=1}^{n-1}$ at forecast horizons $\{T_{j,3} - t\}_{j=1}^{n-1}$ for each $j, j = 1, \ldots, n-1$, respectively. Additionally, based on (8.5), we may calculate

$$F(t; T_{j,3}, T_{j+1,3}) = \frac{\log P(t, T_{j,3}) - \log P(t, T_{j+1,3})}{T_{j+1,3} - T_{j,3}}, \quad (8.23)$$

where $j = 1, \ldots, n$. as the forward rates in the intervals $[T_{j,3}, T_{j+1,3}]$, respectively.

8.5.2 Forecast within the single-curve framework

We construct a separate set of predictors based on a one-factor CIR process in a single-curve framework to compare with the aforementioned set of predictors. They are different from our proposed multi-curve strategy implemented in (8.21) to (8.32). In formulating this set of predictors, we use the estimated CIR model parameters $\{\kappa_{best,G-1,1}(t), \beta_{best,G-1,1}(t), \sigma_{best,G-1,1}(t)\}$ implied from the triplet $\{P(t, T_{1,k})\}_{k=1}^{3}$ corresponding to the shortest available tenors in the set of cross-sectional zero coupon bond prices to model the short rate process and to forecast the short rate at different forecast horizons that may span beyond $T_{1,3} - t$.

Based on (8.12), we construct the single-curve predictor for $\gamma_{CIR}(t; t, T_{j,3})$ at the forecast horizons $\{T_{j,3} - t\}_{j=1}^{n-1}$ using

$$\hat{\gamma}_{CIR,1}(t; t, T_{j,3}) = \beta_{best,C,1}(t) (T_{j,3} - t)$$
$$+ \frac{r(t) - \beta_{best,G-1,1}(t)}{\kappa_{best,G-1,1}(t)} \left(1 - e^{-\kappa_{best,G-1,1}(t)}\right), \quad (8.24)$$

where $j = 1, \ldots, n$. Using (8.14) and (8.24), we construct the single-curve predictor for $\rho_{CIR}(t; T_{j,3}, T_{j+1,3})$ for the time intervals $[T_{j+1,3}, T_{j,3}], j = 1, \ldots, n-1$, as

$$\hat{\rho}_{CIR,1}(t; T_{j,3}, T_{j+1,3}) = \frac{\hat{\gamma}_{CIR,1}(t; t, T_{j+1,3}) - \hat{\gamma}_{CIR,1}(t; t, T_{j,3})}{T_{j+1,3} - T_{j,3}}, \quad (8.25)$$

where $j = 1, \ldots, n$. We use (8.25) as a naive predictor of future short rate $\{r(T_{j,3})\}_{j=1}^{n-1}$ at the forecast horizons $\{T_{j,3} - t\}_{j=1}^{n-1}$ in the single-curve framework for comparison purposes only.

8.6　Numerical analysis using historical data

We use the day-close prices of the US STRIPS[1] zero coupon bonds from Jan. 2nd 2001 to Apr. 28th 2014 for the empirical assessment of our proposed forecast algorithm. The US STRIPS are zero coupon bonds that are created by stripping the Treasury notes and bonds. If so desired, they can be reconstituted to form the original Treasury notes and bonds. The market for the US STRIPS zero coupons is highly liquid. Additionally, these zero coupons contain negligible credit risk. Specifically, we use the average of the day-close bid and ask prices to construct the system of nonlinear equations in order to estimate the implied CIR model parameters.

The historical data of the US effective Federal Funds rate[2] is converted to continuously compound convention to be used as proxy for the short rate process to provide input for $r(t)$ in the CIR model. We use this as the proxy for short rate on grounds that the credit and the liquidity risk profiles of the effective Federal Funds rate, a weighted average of the uncollateralized overnight borrowing rate for a group of federal funds brokers, are comparable to those of the US STRIPS, a highly liquid financial instrument derived from the uncollateralized US Treasuries.

Empirical findings reported in the extant literature suggest that the effective Federal Funds rate and the daily three-month T-bills rate are cointegrated, and exhibit bidirectional Granger causality at various lags. Additionally, the effective Federal Funds rate and US Treasury bill (T-bill) rates tend to move together, and the T-bill rates have been used in some studies to predict the effective Federal Funds rates. We envisage that, since the US STRIPS are derived from the US T-bills and T-bonds, this instrument may contain information for the prediction of the effective Federal Funds rate as well.

Let $t = t_i, i = 1, \ldots, 3234$, index the date of the cross-section day-close prices of the US STRIPS for each of the 3,234 trading days between Jan. 2nd 2001 and Apr. 28th 2014 denoted by $\{P(t_i, T_{i,j,k})\}_{k=1,2,3}, i = 1, \ldots, 3234, j = 1, \ldots, n_i$, where $T_{i,j,k}$ is the maturity date of the corresponding zero coupon bond that is observed at t_i, and $r(T_{i,j,k})$ is the effective Federal Funds rate on the same day. For each t_i, we construct a system of nonlinear equations using (8.18), and estimate the term structure of CIR model parameters using the numerical strategy described in Section 8.3.

In the empirical example data set analyzed and reported in this section, the day-close prices of the US STRIPS zero coupon bonds from Jan. 2nd 2001 to Apr. 28th

---

[1] Source: Thomson Reuters Tick History (TRTH) supplied by the Securities Industry Research Centre of Asia-Pacific (SIRCA)

[2] Source: Federal Reserve Bank of St. Louis

2014 are used. These zero coupon bonds mature at various different tenors, ranging from less than one year all the way out to 30 years. However, since the most recent historical data of the US effective Federal Funds rate available at the date of study is Apr. 28th 2014, the empirical analysis reported herein can only include US STRIPS zero coupon bonds with tenors spanning approximately 10 years out. Therefore, the results reported herein reflect the performance of the fixed-income instruments, in this case US STRIPS zero coupon bonds, having maturities of more than one year. In fact, they span maturities up to 10 years.

The effective Federal Funds rates are a weighted average of the overnight borrowing rates for federal funds brokers guided by the target Federal Funds rates set by the Federal Open Market Committee (FOMC) that meets eight times a year, and whenever necessary. There is a tendency for the effective Federal Funds rates to oscillate around the target Federal Funds rates. It is possible that at forecast horizons shorter than 0.4 years, the mean reverting nature of the effective Federal Funds at such short time intervals may, for this data sample, in general, behave like a random walk due to the oscillatory nature of the effective Federal Funds rate.

8.6.1   Short rate prediction in the multi-curve framework

Let the term structure of implied CIR model parameters estimated from the US STRIPS day-close prices from Jan. 2nd 2001 to Apr. 28th 2014 be indexed by their respective observation dates t_i as $\kappa_{best,G-1}(t_i) = \{\kappa_{best,G-1,j}(t_i)\}_{j=1}^{n_i}$, $\beta_{best,G-1}(t_i) = \{\beta_{best,G-1,j}(t_i)\}_{j=1}^{n_i}$, and $\sigma_{best,G-1}(t_i) = \{\sigma_{best,G-1,j}(t_i)\}_{j=1}^{n_i}$. The sets of parameters $(\kappa_{best,G-1,j}(t_i), \beta_{best,G-1,j}(t_i), \sigma_{best,G-1,j}(t_i))$ are regarded as the sets of implied CIR model parameters for the one-factor CIR processes estimated from $\{P(t_i, T_{i,j,k})\}_{k=1,2,3}$ that span the tenors $\{T_{i,j,3} - t_i\}_{j=1}^{n_i}$.

We consider the forecast performance of using $\hat{\rho}_{CIR}(t_i; T_{i,j,3}, T_{i,j+1,3})$, defined in (8.22), to forecast $r(T_{i,j,3})$, $j = 1, \ldots, n_i - 1$, for each t_i, $i = 1, \ldots, 3234$. Let

$$\text{Error}\,(t_i; r(T_{i,j,3}), \hat{\rho}_{CIR}(t_i; T_{i,j,3}, T_{i,j+1,3})) = r(T_{i,j,3}) - \hat{\rho}_{CIR}(t_i; T_{i,j,3}, T_{i,j+1,3}) , \qquad (8.26)$$

where $j = 1, \ldots, n_i - 1$ is the differences between the realized short rates, $r(T_{i,j,3})$, and the predicted short rates, $\hat{\rho}_{CIR}(t_i; T_{i,j,3}, T_{i,j+1,3})$. Additionally, let

$$\text{Error}\,(t_i; r(T_{i,j,3}), F(t_i; T_{i,j,3}, T_{i,j+1,3})) = r(T_{i,j,3}) - F(t_i; T_{i,j,3}, T_{i,j+1,3}) , \qquad (8.27)$$

where $j = 1, \ldots, n_i - 1$ is the differences between the $r(T_{i,j,3})$ and the forward rates, $F(t_i; T_{i,j,3}, T_{i,j+1,3})$, at forecast horizons $\{T_{i,j,3} - t_i\}_{j=1}^{n-1}$. We use the random walk as a benchmark to evaluate the short rate forecast accuracy where the short rate $r(t_i)$ is used as the forecast of the future short rate at the forecast horizon. Let

$$\text{Error}\,(t_i; r(T_{i,j,3}), RW) = r(T_{i,j,3}) - r(t_i) , \qquad (8.28)$$

where $j = 1, \ldots, n_i - 1$ is the differences between the realized short rates, $r(T_{i,j,3})$, and the random walk prediction of future short rate, which is simply $r(t_i)$, at forecast horizons $\{T_{i,j,3} - t_i\}_{j=1}^{n-1}$. The random walk benchmark is known to be a tough benchmark to match, and it is known to perform better than the one-factor affine term structure models in the single-curve framework in some of the extant literature.

We group the results evaluated using (8.26) and (8.27) into 100 non-overlapping intervals, or 'buckets,' with respect to the forecast horizon. Specifically, we choose 100 intervals of forecast horizons $(\tau_\ell, \tau_{\ell+1}]$, where τ_ℓ is the ℓ-th percentile of the set of forecast horizons denoted by $\{T_{i,j,3} - t_i, i = 1, \ldots, 3234, j = 1, \ldots, n_i - 1\}$ and $\ell = 0, \ldots, 99$. For each $(\tau_\ell, \tau_{\ell+1}]$, we compute the forecast horizons interval-specific, i.e., bucket-specific, root mean squared error (RMSE) for $\hat{\rho}_{CIR}(t_i; T_{i,j,3}, T_{i,j+1,3})$ as

$$\text{RMSE}_\ell \left(\hat{\rho}_{CIR}\right)^2 = \frac{1}{n_\ell} \sum_{i=1}^{3234} \sum_{j}^{n_i} \mathbf{1}_{\tau_\ell < T_{i,j,3}-t_i \leq \tau_{\ell+1}}$$

$$\times \text{Error}\left(t_i; r(T_{i,j,3}), \hat{\rho}_{CIR}(t_i; T_{i,j,3}, T_{i,j+1,3})\right)^2, \quad (8.29)$$

for $F(t_i; T_{i,j,3}, T_{i,j+1,3})$ as

$$\text{RMSE}_\ell \left(F\right)^2 = \frac{1}{n_\ell} \sum_{i=1}^{3234} \sum_{j}^{n_i} \mathbf{1}_{\tau_\ell < T_{i,j,3}-t_i \leq \tau_{\ell+1}}$$

$$\times \text{Error}\left(t_i; r(T_{i,j,3}), F(t_i; T_{i,j,3}, T_{i,j+1,3})\right)^2, \quad (8.30)$$

and for the random walk benchmark as

$$\text{RMSE}_\ell \left(RW\right)^2 = \frac{1}{n_\ell} \sum_{i=1}^{3234} \sum_{j}^{n_i} \mathbf{1}_{\tau_\ell < T_{i,j,3}-t_i \leq \tau_{\ell+1}} \text{Error}\left(t_i; r(T_{i,j,3}), RW\right)^2.$$

$$(8.31)$$

We depict in Table 8.1 and Figure 8.1 the forecast horizon specific performance of $\hat{\rho}_{CIR}(t_i; T_{i,j,3}, T_{i,j+1,3})$, the predictor for the multi-curve framework, and $F(t_i; T_{i,j,3}, T_{i,j+1,3})$, the forward rate for each interval $(\tau_\ell, \tau_{\ell+1}], \ell = 0, \ldots, 99$, and, for comparison, $\hat{\rho}_{CIR,1}(t_i; T_{i,j,3}, T_{i,j+1,3})$, the predictor for the single-curve framework. Reference [51] has remarked that the random walk is a tough benchmark for the standard class of affine models to beat. The forward rate in general performs less favorably against the random walk. The predictor for the multi-curve framework $\hat{\rho}_{CIR}(t_i; T_{i,j,3}, T_{i,j+1,3})$ performs marginally better than the random walk at forecast horizons of 0.8 to 1.6 years, but not for forecast horizons shorter than 0.4 years and longer than 1.6 years.

Both the predictors $\hat{\rho}_{CIR}(t_i; T_{i,j,3}, T_{i,j+1,3})$ and $F(t_i; T_{i,j,3}, T_{i,j+1,3})$ appear to be reasonably good predictors of future short rate up to a forecast horizon of one year. However, progressive deterioration of the forecasting performance becomes noticeable beyond a forecast horizon of one year, possibly due to factors such as the effect of term premium on forecasting ability of both $\hat{\rho}_{CIR}(t_i; T_{i,j,3}, T_{i,j+1,3})$ and $F(t_i; T_{i,j,3}, T_{i,j+1,3})$.

The values of the forecast horizon interval-specific RMSE and MAE for the multi-curve predictor $\hat{\rho}_{CIR}(t_i; T_{i,j,3}, T_{i,j+1,3})$ and the forward rate

Forecast horizon (years)	$\mathrm{RMSE}_\ell\,(\hat{\rho}_{CIR})$ $\times 10^{-3}$	$\mathrm{RMSE}_\ell\,(\hat{\rho}_{CIR,1})$ $\times 10^{-3}$	$\mathrm{RMSE}_\ell\,(F)$ $\times 10^{-3}$	$\mathrm{RMSE}_\ell\,(RW)$ $\times 10^{-3}$
0.4	6.49	13.12	4.01	4.31
0.8	13.96	23.69	17.46	14.08
1.2	7.75	22.97	10.87	15.11
1.6	24.41	34.69	28.43	25.40
1.8	28.54	39.88	31.92	20.89
2.2	28.80	44.85	25.38	19.40
2.4	36.13	40.89	31.51	30.72
2.8	30.07	38.17	26.11	30.09
3.2	38.10	48.07	33.63	35.17
3.4	33.61	42.16	28.14	28.09
3.8	40.33	51.77	31.38	33.39
4.2	42.13	52.69	32.41	31.79
4.6	51.49	57.52	42.42	34.98
5.0	45.67	54.96	35.97	24.95
5.4	56.65	61.53	46.46	36.02
5.8	51.15	57.94	40.95	26.43
6.2	60.89	64.39	49.26	37.96
6.6	59.20	63.09	47.46	25.97
7.1	62.97	68.88	51.73	32.70
7.6	66.66	65.96	54.01	28.33
8.1	68.41	69.79	57.32	29.37
8.5	69.85	69.21	57.89	22.98
9.0	75.04	64.22	61.29	27.87
9.6	75.17	69.21	63.20	22.20

Table 8.1: Forecasting performance of effective Federal Funds rate at different forecast horizons. $\mathrm{RMSE}\,(\hat{\rho}_{CIR})$, $\mathrm{RMSE}\,(\hat{\rho}_{CIR,1})$, $\mathrm{RMSE}\,(F)$, and $\mathrm{RMSE}\,(RW)$ are the forecast horizon-specific MSE for the multi-curve and single-curve CIR predictors, forward rate, and the random walk benchmark, respectively.

$F(t_i; T_{i,j,3}, T_{i,j+1,3})$ for the corresponding tenors are comparable up to forecast horizons of 2 years, beyond which those of $\hat{\rho}_{CIR}(t_i; T_{i,j,3}, T_{i,j+1,3})$ become noticeably larger than those of $F(t_i; T_{i,j,3}, T_{i,j+1,3})$. Based on (8.22) and $F(t_i; T_{i,j,3}, T_{i,j+1,3})$, the estimated average variance of the integrated CIR process within the time interval $[T_{i,j,3}, T_{i,j+1,3}]$ can be expressed as

$$\hat{\Delta}_{CIR}(t_i; T_{i,j,3}, T_{i,j+1,3}) = 2\left(\hat{\rho}_{CIR}(t_i; T_{i,j,3}, T_{i,j+1,3}) - F(t_i; T_{i,j,3}, T_{i,j+1,3})\right)$$

$$\approx \frac{\hat{\delta}_{CIR}(t_i; T_{i,j,3}, T_{i,j+1,3})}{T_{i,j+1,3} - T_{i,j,3}}, \tag{8.32}$$

where

$$\hat{\delta}_{CIR}(t_i; T_{i,j,3}, T_{i,j+1,3}) = \widehat{\text{Var}}_{CIR}\left[\int_{t_i}^{T_{i,j+1,3}} r_{CIR}(u)du \,\Bigg|\, \mathscr{F}_{t_i}\right]$$
$$- \widehat{\text{Var}}_{CIR}\left[\int_{t_i}^{T_{i,j,3}} r_{CIR}(u)du \,\Bigg|\, \mathscr{F}_{t_i}\right],$$

and where $\widehat{\text{Var}}_{CIR}(\cdot)$ represents the variance of the integrated short rate within the multi-curve construct where the short rate process $r_{CIR}(u)$ in the time interval $[T_{i,j,3}, T_{i,j+1,3}]$ is associated with the one-factor CIR process model using the estimated implied parameters $\theta_{best,G,j}(t_i)$, $\kappa_{best,G,j}(t_i)$, and $\sigma_{best,G,j}(t_i)$. From (8.32), we obtain

$$\frac{1}{2}\hat{\Delta}_{CIR}(t_i; T_{i,j,3}, T_{i,j+1,3}) = \hat{\rho}_{CIR}(t_i; T_{i,j,3}, T_{i,j+1,3}) - F(t_i; T_{i,j,3}, T_{i,j+1,3})$$
$$= (\hat{\rho}_{CIR}(t_i; T_{i,j,3}, T_{i,j+1,3}) - r(T_{i,j,3})) - (F(t_i; T_{i,j,3}, T_{i,j+1,3}) - r(T_{i,j,3}))$$
$$= \text{Error}\,(t_i; r(T_{i,j,3}), \hat{\rho}_{CIR}(t_i; T_{i,j,3}, T_{i,j+1,3}))$$
$$- \text{Error}\,(t_i; r(T_{i,j,3}), F(t_i; T_{i,j,3}, T_{i,j+1,3}))$$
$$\approx \frac{\hat{\delta}_{CIR}(t_i; T_{i,j,3}, T_{i,j+1,3})}{2(T_{i,j+1,3} - T_{i,j,3})}. \tag{8.33}$$

Take squares on both sides of

$$\text{Error}\,(t_i; r(T_{i,j,3}), \hat{\rho}_{CIR}(t_i; T_{i,j,3}, T_{i,j+1,3}))$$
$$\approx \text{Error}\,(t_i; r(T_{i,j,3}), F(t_i; T_{i,j,3}, T_{i,j+1,3})) + \frac{\hat{\delta}_{CIR}(t_i; T_{i,j,3}, T_{i,j+1,3})}{2(T_{i,j+1,3} - T_{i,j,3})},$$

then multiply both sides by $\mathbf{1}_{\tau_\ell < T_{i,j,3} - t_i \leq \tau_{\ell+1}}$ and sum across $i, i = 1, \ldots, 3234$, and $j, j = 1, \ldots, n_i$, and finally divide both sides by n_ℓ before rearranging, we obtain

$$\text{RMSE}_\ell\,(\hat{\rho}_{CIR})^2 - \text{RMSE}_\ell\,(F)^2$$
$$\approx \frac{1}{n_\ell}\sum_{i=1}^{3234}\sum_{j}^{n_i}\mathbf{1}_{\tau_\ell < T_{i,j,3} - t_i \leq \tau_{\ell+1}}$$
$$\times \text{Error}\,(t_i; r(T_{i,j,3}), F(t_i; T_{i,j,3}, T_{i,j+1,3}))\frac{\hat{\delta}(t_i; T_{i,j,3}, T_{i,j+1,3})}{T_{i,j+1,3} - T_{i,j,3}}$$
$$+ \frac{1}{n_\ell}\sum_{i=1}^{3234}\sum_{j}^{n_i}\mathbf{1}_{\tau_\ell < T_{i,j,3} - t_i \leq \tau_{\ell+1}}\left(\frac{\hat{\delta}(t_i; T_{i,j,3}, T_{i,j+1,3})}{2\,(T_{i,j+1,3} - T_{i,j,3})}\right)^2. \tag{8.34}$$

Since $\hat{\delta}(t_i; T_{i,j,3}, T_{i,j+1,3}) > 0$, the sign of $\text{RMSE}_\ell\,(\hat{\rho}_{CIR})^2 - \text{RMSE}_\ell\,(F)^2$ depends on the sign and magnitude of $\text{Error}\,(t_i; r(T_{i,j,3}), F(t_i; T_{i,j,3}, T_{i,j+1,3}))$.

The RMSE for both $\hat{\rho}_{CIR}(t_i; T_{i,j,3}, T_{i,j+1,3})$ and $F(t_i; T_{i,j,3}, T_{i,j+1,3})$ increases along the forecast horizon, as depicted in Table 8.1. We envisage that the forecast performance deterioration at horizons beyond two years is the result of, among other factors, the noticeable effect of the term premium.

Other factors may affect the forecast performance of $\hat{\rho}_{CIR}(t_i; T_{i,j,3}, T_{i,j+1,3})$. The forecast performance assessment of $\hat{\rho}_{CIR}(t_i; T_{i,j,3}, T_{i,j+1,3})$ is a joint test of whether $\rho_{CIR}(t_i; T_{i,j,3}, T_{i,j+1,3})$ is a good predictor of the future short rate, and whether $\hat{\rho}_{CIR}(t_i; T_{i,j,3}, T_{i,j+1,3})$ is a good estimator of $\rho_{CIR}(t_i; T_{i,j,3}, T_{i,j+1,3})$. The former is affected by the model specification risk, the latter by the model estimation risk.

Model specification risk arises as $\rho_{CIR}(t_i; T_{i,j,3}, T_{i,j+1,3})$ is a model-based predictor. While the extant literature suggests that the drift and diffusion coefficients of the short rate process may not be linear, the one-factor CIR process is nonetheless a relatively good approximation if the short rate process being modeled lies within the range between 0 and 0.09 [1]. The time series of effective Federal Funds rate, our short rate proxy, for the period we consider lies within this boundary. While this lends support to our choice of using the CIR process to model this short rate dynamics, the CIR model is nonetheless only an approximate working model.

Model estimation risk arises as $\hat{\rho}_{CIR}(t_i; T_{i,j,3}, T_{i,j+1,3})$ is evaluated using the implied CIR model parameters estimated based on zero coupon bond prices as approximate solutions to a system of nonlinear equations (8.18), thus incurring some uncertainty in the point estimation of $\rho_{CIR,1}(t_i; T_{i,j,3}, T_{i,j+1,3})$. That said, the model estimation risk is an issue faced by all parametric models constructed to model the short rate dynamics due to the necessity for numerical estimation of the model parameters.

8.6.2 Short rate prediction in the single-curve framework

The objective of using $\hat{\rho}_{CIR,1}(t_i; T_{i,j,3}, T_{i,j+1,3})$, as a naive one-factor CIR model-based predictor constructed in the single-curve framework to predict $r(T_{i,j,3})$, the future short rate at various forecasting horizons, is to perform a simple comparison between the forecast performance of the one-factor CIR model-based short rate predictor implemented in the multi-curve framework and that implemented in the single-curve framework.

The evaluation of $\hat{\rho}_{CIR,1}(t_i; T_{i,j,3}, T_{i,j+1,3})$ at each $t_i, i = 1, \ldots, 3234$, is based on the set of implied CIR model parameters $(\kappa_{best,G,1}(t_i), \theta_{best,G,1}(t_i), \sigma_{best,G,1}(t_i))$ inferred from $\{P(t_i, T_{i,1,k})\}_{k=1,2,3}$. Let

$$\text{Error}\,(t_i; r(T_{i,j,3}), \hat{\rho}_{CIR,1}(t_i; T_{i,j,3}, T_{i,j+1,3})) = r(T_{i,j,3}) - \hat{\rho}_{CIR,1}(t_i; T_{i,j,3}, T_{i,j+1,3}),$$
$$(8.35)$$

where $j = 1, \ldots, n_i - 1$ is the differences between the realized short rates, $r(T_{i,j,3})$, and the predicted short rates $\hat{\rho}_{CIR,1}(t_i; T_{i,j,3}, T_{i,j+1,3})$ at forecast horizons $\{T_{i,j,3} - t_i\}_{j=1}^{n-1}$.

We group the results evaluated using (8.35) into 100 intervals of forecast horizons $(\tau_\ell, \tau_{\ell+1}]$, where τ_ℓ is the ℓ-th percentiles of the set of forecast horizons $\{T_{i,j,3} - t_i\}_{i=1,\ldots,3234, j=1,\ldots,n_i-1}$, and where $\ell = 0, \ldots, 99$. For each $(\tau_\ell, \tau_{\ell+1}]$, we use (8.26) to compute the forecast horizons interval-specific RMSE

$$
\text{RMSE}_\ell\left(\hat{\rho}_{CIR,1}\right) = \left\{ \frac{1}{n_\ell} \sum_{i=1}^{3234} \sum_j \mathbf{1}_{\tau_\ell < T_{i,j,3} - t_i \leq \tau_{\ell+1}} \right.
$$

$$
\left. \times \text{Error}\left(t_i; r(T_{i,j,3}), \hat{\rho}_{CIR,1}(t_i; T_{i,j,3}, T_{i,j+1,3})\right)^2 \right\}^{1/2} .
$$

$$(8.36)$$

Table 8.1 indicates that the forecast performance of $\hat{\rho}_{CIR,1}(t_i; T_{i,j,3}, T_{i,j+1,3})$ is inferior to $\hat{\rho}_{CIR}(t_i; T_{i,j,3}, T_{i,j+1,3})$, $F(t_i; T_{i,j,3}, T_{i,j+1,3})$, and the random walk.

This comparison verifies the weak forecast performance of the one-factor CIR model implemented in a single-curve framework. However, the forecast performance of the one-factor CIR model implemented can be improved if it is implemented in a multi-curve framework instead.

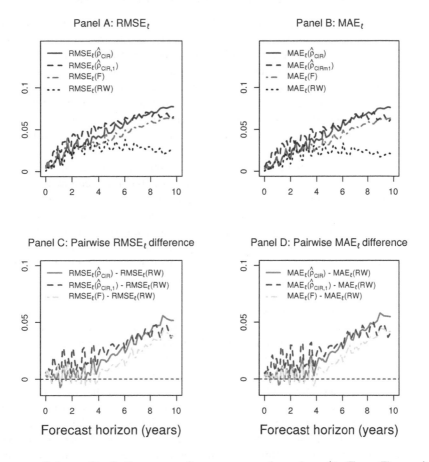

Figure 8.1: Prediction performance of $\hat{\rho}_{CIR}(t_i; T_{i,j,3}, T_{i,j+1,3})$, $\hat{\rho}_{CIR,1}(t_i; T_{i,j,3}, T_{i,j+1,3})$, $F(t_i; T_{i,j,3}, T_{i,j+1,3})$, and the random walk benchmark at various forecast horizons expressed in terms of forecast horizon interval-specific RMSE and MAE.

8.7 Bibliographic notes and literature review

Implied CIR model parameters – a representation of market sentiment of future interest rate inferred from zero coupon prices

While the option contracts are arguably some of the most actively traded financial instruments in the equity sphere, the zero coupon bonds are some of the most liquid instruments traded in the fixed-income market. Notwithstanding their apparent differences, there is an intricate link between these two classes of contingent claims [31, 28]. The net present value of an option contract is the expected discounted payoff. If it pays one dollar at terminal time only, the net present value of the option is equivalent to that of a zero coupon bond paying one dollar at maturity. If it delivers a fixed cash flow at fixed, pre-defined dates, and pays a terminal cash flow, then it replicates a coupon-bearing bond. In fact, the portfolio replication argument [24] invoked to derive the celebrated Black–Scholes option pricing formula is to use cash and zero coupon bonds to replicate a European vanilla option contract.

While the majority of the option pricing literature focuses primarily on modeling the asset price dynamics and assume deterministic discount rate for ease of exposition, the primary focus of the interest rate modeling is to capture the stochastic nature of the interest rate dynamics observed in the market. While one of the main objectives of asset price dynamics is to develop a working model that can adequately simulate realistic features of implied volatility surface at some forecast horizon for applications such as interpolation of OTC traded equity or index linked option prices and exotic contingent claims valuation, one of the main objectives of interest rate dynamics modeling is to develop a working model that can adequately simulate realistic features of the yield curve, i.e., the term structure of interest rates, at some forecast horizon so that cash flows of interest rate linked contingent claims can be determined and discounted in a coherent pricing framework.

After the subprime mortgage crisis of 2007, financial market practitioners observed the empirical phenomenon from the interest rate market where the quoted interest rates that were once very close to each other, such as the overnight indexed swap (OIS) rate and deposit rate with the same maturity or the swap rates with the same tenor but based on different floating-leg frequencies, started to diverge substantially, displaying non-negligible interest rate spreads [95]. This immediately posed the problem of the consistent definition of a zero coupon, i.e., yield, curve. In order to construct a working model for consistent pricing and discounting in the face of this new empirical phenomenon, financial market practitioners adopted an empirical approach of constructing as many curves as possible rate tenors so that future values of interest rates are projected through the associated forward (or projection) curves, whereas future cash flows are discounted by a possibly different discount curve. References [95], [91], and [96] are among the first authors working in this direction to capture the salient features corresponding to different tenors using different yield curves. This approach is commonly referred to as the multi-curve approach, as opposed to the classic single-curve approach that models the entire interest rate term structure using a single yield curve.

Typically, the vast literature of interest rate forecast builds on the practice of using a single yield curve to model the entire interest rate term structure. It begs the question of whether constructing predictors to forecast short rate in the multi-curve framework would improve the forecast performance.

In order to analyze the potential role of the multi-curve term structure modeling framework on interest rate forecasting, we may choose an interest rate model that has well understood forecast performance in the single-curve framework, and implement this interest rate model in the multi-curve framework for interest rate forecasting so that the interest rate forecast performance of the same model implemented in the two different frameworks, i.e., single-curve and multi-curve frameworks, may be compared.

Admittedly, the long held notion that interest rates should be non-negative has been violated in recent times. In June 2014, the European Central Bank (ECB) set the first negative deposit rate; in March 2016, the ECB further reduced the deposit rate to -0.4%. In January 2016, Bank of Japan followed the ECB in adopting negative interest rates. This renders the investors in shorter-terms Japanese and German government bonds as receiving negative yield [103]. That said, the interest rate dynamics that allow possibility of negative interest rates cannot be modelled by straightforward application of Gaussian short-rate models in the likes of the Ho Lee model. Instead, more sophisticated models in the likes of the arbitrage-free SABR model proposed by [61] and the modified SABR model [11] need to be used in order to capture the salient features of the short rate transition density. The interest rate modeling that allow for negative rates is yet an unsettled research question, and more research results are expected to emerge to try to merge this empirical phenomenon into a coherent pricing framework. In the present dissertation, we restrict our consideration to non-negative interest rates for ease of exposition.

In this chapter, we use the single-factor CIR process to model the short rate process associated with each triplet of zero coupon bonds because this is the simplest possible model that guarantees positivity of the short rate, and its zero coupon bond pricing formula is available in closed form. Reference [91] also used the single-factor CIR model in the multi-curve framework for similar reasons, but in the context of a two-price economy instead. We also draw support for our choice from the empirical results in the single-curve framework reported in [1] that for short rate of magnitude less than 0.09, the CIR model is a good approximation model for the seven-day Eurodollar deposit short rate dynamics. We conjecture that this benefit may carry through to the multi-curve setting.

The role of the CIR model in interest rate modeling bears resemblance to the role of the Black–Scholes option pricing formula in the pricing of contingent claim pertaining to the equity asset class in that it has been used by many researchers as a starting point in the single-curve framework from which more sophisticated models are constructed with an aim to better capture the salient features of the interest rate term structure. As far as we are aware, the segmented CIR model [106] that models the long range interest rate parameter as a piecewise-continuous constant is probably one of the earliest reports in the literature aimed at capturing the salient features of different segments of the yield curve by the use of tenor-specific short rate model parameters. That said, our proposed approach differs from the segmented CIR model in several aspects.

Firstly, our approach is designed for the multi-curve framework, while the segmented CIR model is designed for the single-curve framework. Secondly, we model different segments of the yield curve corresponding to different tenors using different single-factor CIR processes that give rise to different sets of implied CIR model parameters, whereas the segmented CIR models model the entire yield curve using one single-factor CIR process, keeping the same speed of mean reversion fixed across all tenors, specifying the volatilities as exogenous input based on the volatility parameters of some contingent claims of interest, and estimating the long-range mean short rate as a piecewise constant that spans the time intervals between the maturity dates of two adjacent zero coupon bonds arranged in increasing tenor. Thirdly, while our parameter estimation procedure ensures that all the estimated CIR model parameters are positive, the segmented CIR model encounters some numerical instability issue where strongly downward sloping initial forward rate curves can lead to negative values of long-range mean short rate.

As far as we are aware, there are no previously published reports in the literature on interest rate forecast based on the multi-curve framework except [65], a recently accepted paper derived from the work reported in this chapter. We attempt to contribute to the literature by proposing a strategy to forecast short rate within the multi-curve framework using the information of the market participants' aggregate view on future short rate inferred from a cross-sectional set of zero coupon bond prices. The performance of this strategy on some historical data sets may give us some idea of the feasibility of constructing short rate forecast strategy within the multi-curve framework, and the effect of term risk premium on the forecast performance of such a strategy.

The extant literature on interest rate forecast in the single-curve framework appears to suggest that the effect of term premium on interest rate forecast performance becomes noticeable beyond some forecast horizon. On the one hand, [53] noted that the forward rates calculated from T-Bill rates do not perform well in predicting short rate at forecast horizons beyond two years unless adjustment is made to account for the term premium. On the other hand, [86] reported that the effect of term premium for short-term repurchase rates is negligible for tenors up to three months. Since these empirical studies considered two different sets of interest rates, the forecast horizon beyond which the term premium starts to show a noticeable effect on the performance of future short rate predictors constructed from the same set of interest rates is not entirely clear. This motivates us to investigate the effect of term premium on short rate forecast in the multi-curve framework using our proposed algorithm as a test model.

Although some of the existing literature recommends disentangling the expected future short rate from the expected term premium to improve the forecast performance of short rate predictors, the difficulty in the practical implementation of this recommendation lies in the lack of consensus on the functional form of the term premium. In fact, many different definitions and parametric expressions of the term premium have been suggested, and numerous different numerical strategies have proposed to estimate them. To make the matter even more challenging, the estimation of the term premium is sensitive to the choice of estimator and is data dependent. Reference [76] pointed out that it is this lack of robustness in existing methodologies that renders it less appealing to practitioners to adjust for the term premium in their analysis of interest rates.

The current chapter focuses on the construction of a strategy to forecast the short rate, and does not consider the proposition of a strategy to forecast volatility of the short rate. This is because the one-factor CIR model that we have chosen as our working model is not capable of capturing potential jumps and stochastic components in the volatility of the effective Federal Funds rate [99]. Our focus of study in this chapter is to propose, using a simple working model, a strategy to extract information on the market participants' expectation of the future short rate in a multi-curve framework.

Additionally, recent empirical studies [52] suggest that the cross-sectional interest rate term structure does not contain information on the short rate volatility. In this light, our model framework may not be an appropriate construct to analyze short rate volatility. Instead, one should extract this information from the time series of the interest rates in order to forecast the realized volatility of the short rate.

In order to estimate the realized variance of a short rate process, one may need to construct a more sophisticated model that takes into account stochastic volatility and jump of the short rate process or adopt time series modeling strategies such as the extended EGARCH-type model.

As an alternative, one may, instead, use the Heterogeneous Autoregressive model of Realized Volatility (HAR-RV, [35]) to forecast the realized volatility of yield by linear projection of the historical volatilities that are estimated at different lags. The out-of-sample forecast performance of the HAR-RV model is comparable to that of the extended EGARCH-type model [8]. Additionally, it is superior to that of the lag-1 and the lag-3 Autoregressive models, and the Autoregressive Fractional Integrated Moving Average (ARFIMA) model [35]. Refinements of the HAR-RV model in the extant literature include the modification by [6] to forecast the realized volatility of affine jump-diffusion stochastic processes, and the extension proposed by [94] to capture nonlinearities and long-range dependence in the time series dynamics via a flexible multiple regime smooth transition model.

Bibliography

[1] Ait-Sahalia, Y. (1996). Nonparametric pricing of interest rate derivative securities. *Econometrica* 64, 527–560.

[2] Ait-Sahalia, Y., Mykland, P.A. and Zhang, L. (2005). A tale of two time scales: Determining integrated volatility with noisy high-frequency data. *Journal of the American Statistical Association* 100, 1394–1441.

[3] Ait-Sahalia, Y., Mykland, P.A. and Zhang, L. (2005). How often to sample a continuous-time process in the presence of market microstructure noise. *Review of Financial Studies* 18, 351–416.

[4] Andersen T.G. and Bollerslev, T. (1998). Answering the skeptics: Yes, standard volatility models do provide accurate forecasts. *International Economic Review* 39, 885-905.

[5] Andersen, L.B.G. and Brotherton-Ratcliffe, R. (1997). The Equity Option Volatility Smile: An Implicit Finite-Difference Approach. *Journal of Computational Finance* 1, 5–37.

[6] Andersen, T.G., Bollerslev, T. and Diebold, F.X. (2007). Roughing it up: Including Jump Components in the Measurement, Modeling, and Forecasting of Return Volatility. *The Review of Economics and Statistics* 89, 701–720.

[7] Andersen, T.G., Bollerslev, T., Diebold, F. and Labys, P. (2003). Modeling and forecasting realized volatility. *Econometrica* 71, 579-625.

[8] Andersen, T.G. and Benzoni, L. (2010). Do bonds span volatility risk in the U.S. treasury market? A specification test for affine term structure models. *The Journal of Finance* 65, 603–653.

[9] Andersen, T.G., Dobrev, D. and Schaumburg, E. (2012). Jump-robust volatility estimation using nearest neighbor truncation. *Journal of Econometrics* 169, 75–93.

[10] Andersen, T.G. and Lund, J. (1997). Estimating continuous-time stochastic volatility models of the short-term interest rate. *Journal of Econometrics* 77, 343–377.

[11] Antonov, A., Konikov, M. and Spector, M. (2015). The free boundary SABR: Natural extension to negative rates. *Risk*, September.

[12] Ardia, D., Mullen, K.M., Peterson, B.G. and Ulrich, J. (2012). *DEoptim: Differential Evolution in 'R'* version 2.2-2.

[13] Arriojas, M., Hu, Y., Mohammed, S-E A. and Pap, G. (2007). A delayed Black and Scholes formula. *Stochastic Analysis and Applications* 25, 471–492.

[14] Bandi, F.M. and Russell, J.R. (2008). Microstructure noise, realized variance, and optimal sampling. *The Review of Economic Studies* 75(2), 339–369.

[15] Arnold, L. (1973). *Stochastic Differential Equations. Theory and Applications*. Wiley-Inter-Science, New York.

[16] Back, K. (2010). Martingale pricing. *Annual Review of Financial Economics* 2, 235–250.

[17] Bakshi, G. and Madan, D. (2006). A Theory of Volatility Spreads. *Management Science* 52, 1945–1956.

[18] Becherer, D. (2010). The numeraire portfolio for unbounded semimartingales. *Finance and Stochastics* 5, 327–341.

[19] Ben Hamida, S. and Cont, R. (2005). Recovering Volatility from Option Prices by Evolutionary Optimization. *Journal of Computational Finance* 8, 1–34,

[20] Benninga, S. Bjork, T. and Wiener, Z. (2002). On the use of numeraires in option pricing. *Journal of Derivatives* 10, 43–58.

[21] Biagini, F. and Pratelli, M. (1999). Local risk minimization and numeraire. *Journal of Applied Probability* 36, 1126–1139.

[22] Bianconi, M., MacLachlan, S. and Sammon, M. (2015). Implied volatility and the risk-free rate of return in options markets. *The North American Journal of Economics and Finance* 31:1–26.

[23] Bielecki, T.R., Jeanblanc, M. and Rutkowski, M. (2009). *Credit Risk Modeling* Osaka University Press, Japan.

[24] Black, F. and Scholes, M. (1973). The pricing of options and corporate liabilities. *Journal of Political Economics* 81, 637–659.

[25] Bollerslev, T.,Gibson, M. and Zhou, H. (2011). Dynamic estimation of volatility risk premia and investor risk aversion from option-implied and realized volatilities. *Journal of Econometrics* 160, 235–245.

[26] Brennan, M.J. (1998). The role of learning in dynamic portfolio decisions. *European Finance Review* 1, 295–306.

[27] Brenner. M. and Galai, D. (1986). Implied Interest Rates. *The Journal of Business* 59,493–507.

[28] Brigo, D. and Mercurio, F. (2006). *Interest Rate Models – Theory and Practice*. Springer-Verlag, Berlin Heidelberg.

[29] Britten-Jones, M. and Neuberger, A.J. (2000). Option Prices, Implied Price Processes, and Stochastic Volatility. *Journal of Finance* 55, 839–866,

[30] Butler, J.S. and Schachter, B. (1996). Statistical properties of parameters Inferred from the Black–Scholes formula. *International Review of Financial Analysis* 5, 223–235.

[31] Cairns, A. (2004). *Interest Rate Models: An Introduction*. Princeton University Press, Princeton.

[32] Carr, P. and Sun, J. (2007). A new approach for option pricing under stochastic volatility. *Review of Derivatives Research* 10, 87–250.

[33] Chan, K.C., Karolyi, G.A., Longstaff, F.A. and Sanders, A.B. (1992). An empirical investigation of alternative models of the short-term interest rate. *Journal of Finance* 47, 1209–1227.

[34] Cheng, S.T. (1991). On the feasibility of arbitrage-based option pricing when stochastic bond price processes are involved. *Journal of Economic Theory* 53, 185–198.

[35] Corsi, F. (2009). A simple approximate long-memory model of realized volatility. *Journal of Financial Econometrics* 7, 174–196.

[36] Cox, J.C., Ingersoll, J.E. and Ross, S.A. (1985). A theory of the term structure of interest rates. *Econometrica* 53, 385–407.

[37] Cox, J.C. and Ross, S.A. (1976). The valuation of options for alternative stochastic processes, *Journal of Financial Economics* 3, 145–166.

[38] de Boor, C. (2001). *A Practical Guide to Splines, revised edition*. Springer-Verlag, New York.

[39] De Rossi, G. (2010). Maximum likelihood estimation of the Cox-Ingersoll-Ross model using particle filters. *Computational Economics* 36, 1–16.

[40] Dokuchaev, N.G. and Haussmann, U. (2001). Optimal portfolio selection and compression in an incomplete market. *Quantitative Finance* 1, 336–345.

[41] Dokuchaev, N.G. (2005). Optimal solution of investment problems via linear parabolic equations generated by Kalman filter. *SIAM J. of Control and Optimization* 44, 1239–1258.

[42] Dokuchaev, N. (2006). Two unconditionally implied parameters and volatility smiles and skews. *Applied Financial Economics Letters* 2, 199-204.

[43] Dokuchaev, N. (2007). Mean-reverting market model: Speculative opportunities and non-arbitrage. *Applied Mathematical Finance* 14, 319–337.

[44] Dokuchaev, N. (2007). Bond pricing and two unconditionally implied parameters inferred from option prices. *Applied Financial Economics Letters* 3, 109–113.

[45] Dokuchaev, N. (2011). Option pricing via maximization over uncertainty and correction of volatility smile. *International Journal of Theoretical and Applied Finance* 14, 507–524.

[46] Dokuchaev, N. (2012). *Dynamic Portfolio Strategies: Quantitative Methods and Empirical Rules for Incomplete Information.* Kluwer Academic Publishers, Boston.

[47] Dokuchaev, N. (2014). Volatility estimation from short time series of stock prices. *Journal of Nonparametric Statistics* 26, 373–384.

[48] Dokuchaev, N. (2015). Modelling possibility of short-term forecasting of market parameters for portfolio selection. *Annals of Economics and Finance* 16, 143–161.

[49] Dokuchaev, N. (2017) A pathwise inference method for the parameters of diffusion terms. *Journal of Nonparametric Statistics* 29:4, 731–743.

[50] Dokuchaev, N. (2018). On the implied market price of risk under the stochastic numéraire. *Annals of Finance* 14:223–251.

[51] Duffee, G.R. (2002). Term premia and interest rate forecasts in affine models. *The Journal of Finance* 57, 405–443.

[52] Duffee, G.R. (2011). Information in (and not in) the term structure. *The Review of Financial Studies* 24, 2895–2934.

[53] Fama, E.F. (1976). Forward rates as predictors of future spot rates. *Journal of Financial Economics* 3, 361–377.

[54] Fan, J., Jiang, J., Zhang, Z. and Zhou, Z. (2003). Time-dependent diffusion models for term structure dynamics. *Statistica Sinica* 13, 965–992.

[55] Federico, M.B. and Russell, J.R. (2005). Microstructure noise, realised variance, and optimal sampling. *Review of Economic Studies* 75, 339–369.

[56] Fergusson, K. and Platen, E. (2015). Application of maximum likelihood estimation to stochastic short rate models. *Annals of Financial Economics* 10, 1550009.

[57] Föllmer, H. and D. Sondermann. (1986). Hedging of non-redundant contingent claims. In: W. Hildenbrand and A. Mas-Colell (eds.), *Contribution to Mathematical Economics.* North Holland, New York, 205âĂŞ-223.

[58] Geman, H., El Karoui, N. and Rochet, J.C. (1995). Changes of numéraire, changes of probability measure and option pricing. *Journal of Applied Probability* 32, 443–458.

[59] Gibbons, M.R. and Ramaswamy, K. (1993). A test of the Cox, Ingersoll, and Ross model of the term structure. *Review of Financial Studies* 6, 619–658.

[60] Gourieroux, C. and Monfort, A. (2013). Pitfalls in the estimation of continuous time interest rate models: The case of the CIR Model. *Annals of Economics and Statistics* No. 109/110, 25–61.

[61] Hagan, P.S., Kumar, D., Lesniewski, D. and Woodward, A. (2014). Arbitrage-Free SABR. *Wilmott* 69, 60–75.

[62] Heston, S. (1993). Closed-form solution for options with stochastic volatility, with application to bond and currency options. *Review of Financial Studies* 6, 327–343.

[63] Hin, L.Y. and Dokuchaev, N. (2014). On the implied volatility layers under the future risk-free rate uncertainty. *International Journal of Financial Markets and Derivatives* 3, 392–408.

[64] Hin, L.Y. and Dokuchaev, N. (2016a). Short rate forecasting based on the inference from the CIR model for multiple yield curve dynamics. *Annals of Financial Economics* 11, No. 1 1650004.

[65] Hin, L.Y. and Dokuchaev, N. (2016b). Computation of the implied discount rate and volatility for an overdefined system using stochastic optimization. *IMA Journal of Management Mathematics* 27, 505–527.

[66] Hu, Z., Xiong, S., Su, Q. and Zhang, X. (2013). Sufficient conditions for global convergence of differential evolution algorithm. *Journal of Applied Mathematics* Article ID 193196, 1–14.

[67] Hull, J. and White, A. (1987). The pricing of options on assets with stochastic volatilities. *Journal of Finance* 42, 281–300.

[68] Iacus, S.M. (2008). *Simulation and Inference for Stochastic Differential Equations with R Examples.* Springer-Verlag, New York.

[69] Issaka, A. and SenGupta, I. (2017). Analysis of variance based instruments for Ornstein-Uhlenbeck type models: swap and price index. *Annals of Finance* 13, 401–434.

[70] Ivanov, A.E., Kazmerchuk, Y.I. and Swishchuk, A.V. (2003). Theory, stochastic stability and applications of stochastic delay differential equations: A survey of results. *Differential Equations and Dynamical Systems* 11, 55âĂŞ-115.

[71] Jeanblanc, M., Yor, M. and Chesney, M. (2009). *Mathematical Methods for Financial Markets.* Springer, Heidelberg, London, New York.

[72] Jourdain B. and Kohatsu-Higa A. (2011). A review of recent results on approximation of solutions of stochastic differential equations. In: Kohatsu-Higa, A., Privault, N., Sheu, S.J. (eds). *Stochastic Analysis with Financial Applications. Progress in Probability.* vol 65. Springer, Basel. 121–144.

[73] Karatzas, I. and Kardaras, C. (2007). The numéraire portfolio in semimartingale financial models. *Finance and Stochastics* 11, 447–493.

[74] Kardaras, C. (2010). Numéraire-invariant preferences in financial modeling. *Annals of Applied Probability* 20, 1697–1728.

[75] Kessler, M. (1997). Estimation of an ergodic diffusion from discrete observations. *Scandnavian Journal of Statistics* 24, 211–229.

[76] Kim, D.H. (2007). The Bond Market Term Premium: What is it, and How can We Measure It? *BIS Quarterly Review*. Available at http://www.bis.org/publ/qtrpdf/r_qt0706e.pdf.

[77] Kim, Y.J. and Kunitomo, N. (1999). Pricing options under stochastic interest rates: A New Approach. *Asia-Pacific Financial Markets* 6, 49–70.

[78] Kloeden, P.E. and Platen, E. (1992). *Numerical Solution of Stochastic Differential Equations*. Springer-Verlag, Berlin, Heidelberg

[79] Krylov, N.V. (1980). *Controlled Diffusion Processes*. Springer-Verlag, New York.

[80] Lakner, P. (1998). Optimal trading strategy for an investor: the case of partial information. *Stochastic Processes and Applications* 76, 77–97.

[81] Lambertone, D. and Lapeyre, B. (1996). *Introduction to stochastic calculus applied to finance*. Chapman & Hall, London.

[82] Latané, H.A. and Rendleman, J. (1976). Author's correction: Standard deviations of stock price ratios implied in option prices. *Journal of Finance* 34, 1083.

[83] Latané, H.A. and Rendleman, J. (1976). Standard deviations of stock price ratios implied in option prices. *Journal of Finance* 31, 369–381.

[84] Lewis, A. L. (2000). *Option Valuation under Stochastic Volatility*. Finance Press, Newport Beach.

[85] Liptser, R.S. and A.N. Shiryaev. (2000). *Statistics of Random Processes. I. General Theory, 2nd ed.* Springer-Verlag, Berlin, Heidelberg, New York.

[86] Longstaff, F.A. (2000). The term structure of very short-term rates: New evidence for the expectations hypothesis. *Journal of Financial Economics* 58, 397–415.

[87] Longstaff, F.A., Mithal, S. and Neis, E. (2005). Corporate yield spreads: Default risk or liquidity? New evidence from the credit default swap market. *The Journal of Finance* 60, 2213–2253.

[88] Luong, C. and Dokuchaev, N. (2016). Modelling dependency of volatility on sampling frequency via delay equations. *Annals of Financial Economics* 11, 1650007.

[89] Luong, C., and Dokuchaev, N. (2014). Analysis of market volatility via a dynamically purified option price process. *Annals of Financial Economics* 9, 1450006.

[90] Luong, C., and Dokuchaev, N. (2014). On the implied volatility from a "purified" option price process. *Vietnam Journal of Mathematical Applications* 12, 71–82.

[91] Madan, D.B. and Schoutens, W. (2012). Tenor Specific Pricing. *International Journal of Theoretical and Applied Finance* 15, 1–21.

[92] Mao, X. and Shah, A. (1997). Exponential stability of stochastic differential delay equations. *Stochastics and Stochastics Reports* 60, 135–153.

[93] Mao, X., Koroleva, N. and Rodkina, A. (1998). Robust stability of uncertain stochastic differential delay equations. *Systems and Control Letters* 35, 325–336.

[94] McAleer, M. and Medeiros, M.C. (2008). A multiple regime smooth transition heterogeneous autoregressive model for long memory and asymmetries. *Journal of Econometrics* 147, 104–119.

[95] Mercurio, F. (2009). Interest rates and the credit crunch: New formulas and market models, Bloomberg Portfolio Research Paper No. 2010-01-FRONTIERS. Available at SSRN: http://ssrn.com/abstract=1332205.

[96] Moreni, N. and Pallavicini, A. (2014). Parsimonious HJM Modelling for multiple yield curve dynamics. *Quantitative Finance* 14, 199–210.

[97] Palandri, A. (2014). Risk-free rate effects on conditional variances and conditional correlations of stock returns. *Journal of Empirical Finance* 25, 95–111.

[98] Parentich, W. (2018). Modelling the dependence of volatility from sampling frequency. Available at SSRN: https://ssrn.com/abstract=3319212.

[99] Piazzesi, M. (2005). Bond yields and the Federal Reserve. *Journal of Political Economy* 113, 311–344.

[100] Renault, E. and Touzi, N. (1996). Option hedging and implied volatilities in a stochastic volatility model. *Mathematical Finance* 6, 279–302.

[101] Ross, S. (1976). Options and efficiency. *Quarterly Journal of Economics* 90, 75–89.

[102] R Core Team. (2012). *R: A Language and Environment for Statistical Computing*. R Foundation for Statistical Computing, Vienna, Austria.

[103] Randow, J. and Kennedy, S. (2016). Negative Interest Rates: Less Than Zero. Available at http://www.bloomberg.com/quicktake/negative-interest-rates.

[104] Rubinstein, M. (1994). Implied Binomial Trees. *Journal of Finance* 49, 771–818.

[105] Ruf, J. (2013). Negative call prices. *Annals of Finance* 9, 787–794.

[106] Schlögl, E. and Schlögl, L. (2000). A square root interest rate model fitting discrete initial term structure data. *Applied Mathematical Finance* 7, 183–209.

[107] Schroder, M. (1999). Changes of numeraire for pricing futures, forwards, and options. *Review of Financial Studies* 12, 1143–1163.

[108] Schweizer, M. (2001). A guided tour through quadratic hedging approaches, In: Jouini, E., Cvitanic, J. and Musiela, M. (eds.), *Option Pricing, Interest Rates and Risk Management*. Cambridge University Press, 538–574.

[109] Shiryaev, A.N., Kabanov, Yu. M., Kramkov, O.D. and Melnikov, A.V. (1994). Towards the theory of pricing of options of both European and American types. II. Continuous time. *Theory of Probability and Its Applications* 39, 61–102.

[110] SIRCA. (2013). *Thomson Reuters Tick History*. Retrieved from http://www.sirca.org.au/.

[111] Slepaczuk, R. and Zakrzewski, G. (2009). High-frequency and model-free volatility estimators. *Technical Report 2009-13*, Faculty of Economic Sciences, University of Warsaw, Poland.

[112] Sorensen H. (2000). *Inference for Diffusion Processes and Stochastic Volatility Models*. Ph.D. thesis. University of Copenhagen. Denmark.

[113] Stoica, G. (2004). A stochastic delay financial model. *Proceedings of the American Mathematical Society* 133, 1837–1841.

[114] Sugiyama, S. (1969). On the stability problems on difference equations, Bull. *Sci. Eng. Research Lab*. Waseda Univ. 45, 140–144.

[115] Turvey, C.G. and Komar, S. (2006). Martingale restrictions and the implied market price of risk. *Canadian Journal of Agricultural Economics* 54, 379–399.

[116] Uhlenbeck, G.E. and Ornstein, L.S. (1930). On the theory of Brownian motion. *Physical Review* 36, 82–841.

[117] Vasicek, O. (1977). An equilibrium characterisation of the term structure. *Journal of Financial Economics* 5, 177–188.

[118] Vecer, J. and Xu, M. (2004). Pricing Asian options in a semimartingale model. *Quantitative Finance* 4, 170–175.

[119] Vecer, J. (2011). *Stochastic Finance: A Numeraire Approach.* CRC Press, Boca Raton, London, New York.

[120] Wang, H., Zhou, H. and Zhou, Y. (2013). Credit Default Swap Spreads and Variance Risk Premia. *Journal of Banking & Finance* 37, 3733–3746.

[121] Weron, R. (2008). Market price of risk implied by Asian-style electricity options and futures. *Energy Economics* 30, 1098–1115.

[122] Wilson, G. (2018). Estimating volatility using the CIR Model. Available at SSRN: https://ssrn.com/abstract=3290379.

[123] Won, D., Hahn, G. and Yannelis, N.C. (2008). Capital market equilibrium without riskless assets: Heterogeneous expectations. *Annals of Finance* 4, 183–195.

[124] Yan, B. and Zivot, E. (2003). Analysis of high-frequency financial data with S-Plus. *UWEC-2005-03* University of Washington, Department of Economics, WA.

[125] Zhang, J. and Sanderson, A.C. (2009). JADE: Adaptive differential evolution with optional external archive. *IEEE Transactions on Evolutionary Computation* 13, 945–958.

[126] Zhou, H. (2001). Finite sample properties of EMM, GMM, QMLE, and MLE for a square-root interest rate diffusion model. *Journal of Computational Finance* 2, 89–122.

Index

P-augmentation, 52
σ-algebra, 1
σ-algebra, complete , 2

appreciation rate, 23
arbitrage, 30

backward Kolmogorov-Fokker-Planck
    equation, 19
Black–Scholes parabolic equation, 37
bond, 24
Brownian motion, 9
buy-and-hold strategy, 25

Clark Theorem, 19
Clark-Haussmann-Ocone Theorem, 19
completion of a measure, 2
conditional expectation, 5
continuous time process, 7
Cox–Ingersoll–Ross model / CIR model,
    96

Delta of the Strike, 145
Differential evolution, 169
diffusion processes, 15
diffusion type processes, 16
discounted stock price, 26
discounted bond prices, 116
discounted wealth, 26
dynamically purified option price, 144

ecvolutionary optimisation, 169
elementary event, 2
equivalent measures, 4
Existence and Uniqueness Theorem for
    Ito equations, 16
expectation, 4

filtration, 8

forward rate, 193

Girsanov's theorem, 21

Hilbert space, 54

implied volatility, 125
integrable random variable, 4
Ito differential, 13
Ito equation, 15
Ito formula, 14
Ito formula (Ito Lemma), 13
Ito integral, 11
Ito process, 13

Kalman–Bucy filter, 110
Kolmogorov axioms, 2

market price of risk, 27
market price of risk (vector case), 40
Markov process, 8
Markov property, 9
martingale, 8
martingale measure, 27
Martingale Representation Theorem, 20
mean, 4
model, Chan–Karolyi–Longstaff–
    Sanders – CKLS,
    96
model, CIR, 194
model, Cox-Ross-Ingersoll, 119
model, multi-bond, 122
model, multi-stock, 38
model, one-factor, 115
model, Vasicek, 124

Novikov condition, 20

Ornstein–Uhlenbek process, 124

over-defined, 175

pathwise continuous process, 7
portfolio strategy, 39
probability distribution, 7
probability measure, 2
probability space, 2
put-call parity, 143

Radon–Nikodim Theorem, 5
random event, 2
random variable, 3
random vector, 3
replicable claim, 29
replicating strategy, 29
risk neutral measure, 27

sampling frequency, 84
self-financing strategy, 25
sensitivity analysis, 155
short rate forecast, 198
stochastic differential, 13
stochastic differential equation, 15
stochastic integral, 11
strategy (portfolio strategy), 25

unconditionally implied risk free rate,
         140
unconditionally implied volatility, 140

variance, 4
volatility, 23

Wiener process, 9

zero coupon bonds, 113
Zhang–Sanderson algorithm, 170